EXPERIMENTAL METHODS IN TRIBOLOGY

TRIBOLOGY SERIES
Editor
D. Dowson (Gt. Britain)

Vol. 13	Tribology of Miniature Systems (Rymuza)
Vol. 14	Tribological Design of Machine Elements (Dowson et al., Editors)
Vol. 15	Encyclopedia of Tribology (Kajdas et al.)
Vol. 16	Tribology of Plastic Materials (Yamaguchi)
Vol. 17	Mechanics of Coatings (Dowson et al., Editors)
Vol. 18	Vehicle Tribology (Dowson et al., Editors)
Vol. 19	Rheology and Elastohydrodynamic Lubrication (Jacobson)
Vol. 20	Materials for Tribology (Glaeser)
Vol. 21	Wear Particles: From the Cradle to the Grave (Dowson et al., Editors)
Vol. 22	Hydrostatic Lubrication (Bassani and Piccigallo)
Vol. 23	Lubricants and Special Fluids (Stepina and Vesely)
Vol. 24	Engineering Tribology (Stachowiak and Batchelor)
Vol. 25	Thin Films in Tribology (Dowson et al., Editors)
Vol. 26	Engine Tribology (Taylor, Editor)
Vol. 27	Dissipative Processes in Tribology (Dowson et al., Editors)
Vol. 28	Coatings Tribology – Properties, Techniques and Applications in Surface Engineering (Holmberg and Matthews)
Vol. 29	Friction Surface Phenomena (Shpenkov)
Vol. 30	Lubricants and Lubrication (Dowson et al., Editors)
Vol. 31	The Third Body Concept: Interpretation of Tribological Phenomena (Dowson et al., Editors)
Vol. 32	Elastohydrodynamics – '96: Fundamentals and Applications in Lubrication and Traction (Dowson et al., Editors)
Vol. 33	Hydrodynamic Lubrication – Bearings and Thrust Bearings (Frêne et al.)
Vol. 34	Tribology for Energy Conservation (Dowson et al., Editors)
Vol. 35	Molybdenum Disulphide Lubrication (Lansdown)
Vol. 36	Lubrication at the Frontier – The Role of the Interface and Surface Layers in the Thin Film and Boundary Regime (Dowson et al., Editors)
Vol. 37	Multilevel Methods in Lubrication (Venner and Lubrecht)
Vol. 38	Thinning Films and Tribological Interfaces (Dowson et al., Editors)
Vol. 39	Tribological Research: From Model Experiment to Industrial Problem (Dalmaz et al., Editors)
Vol. 40	Boundary and Mixed Lubrication : Science and Applications (Dowson et al., Editors)
Vol. 41	Tribological Research and Design for Engineering Systems (Dowson et al., Editors)
Vol. 42	Lubricated Wear – Science and Technology (Sethuramiah)
Vol. 43	Transient Processes in Tribology (Dalmaz and Lubrecht, Editors)
Vol. 44	Experimental Methods in Tribology (Stachowiak et.al.)
Vol. 45	Tribochemistry of Lubricating Oils (Pawlak)

TRIBOLOGY SERIES, 44
EDITOR: D. DOWSON

EXPERIMENTAL METHODS IN TRIBOLOGY

Gwidon W. STACHOWIAK
School of Mechanical Engineering,
The University of Western Australia,
Perth, Australia

Andrew W. BATCHELOR
School of Engineering,
Monash University Malaysia,
Bandar Sunway, Selangor, Malaysia

Grazyna B. STACHOWIAK
School of Mechanical Engineering,
The University of Western Australia,
Perth, Australia

2004

ELSEVIER

Amsterdam – Boston – Heidelberg – London – New York – Oxford
Paris – San Diego – San Francisco – Singapore – Sydney – Tokyo

ELSEVIER B.V.
Sara Burgerhartstraat 25
P.O. Box 211, 1000 AE
Amsterdam, The Netherlands

ELSEVIER Inc.
525 B Street, Suite 1900
San Diego, CA 92101-4495
USA

ELSEVIER Ltd
The Boulevard, Langford Lane
Kidlington, Oxford OX5 1GB
UK

ELSEVIER Ltd
84 Theobalds Road
London WC1X 8RR
UK

First edition 2004

Library of Congress Cataloging in Publication Data
A catalog record is available from the Library of Congress.

British Library Cataloguing in Publication Data
A catalogue record is available from the British Library.

ISBN: 0-444-51589-5 (Vol. 44)
ISSN: 0167-8922 (Series)

∞ The paper used in this publication meets the requirements of ANSI/NISO Z39.48-1992 (Permanence of Paper).
Printed in The Netherlands.

To Professor, Sir Duncan Dowson

whose passion and dedication to
Tribology is an inspiration to us all

PREFACE

Upon completion of our first book, Engineering Tribology, it became evident that many important topics in tribology still remained to be presented and discussed. So a sequel to Engineering Tribology was immediately initiated to address this problem. Several different subject areas appeared suitable for a new book and the question was to choose the most important topic.

After being engaged over the years in several test programs dedicated to study various forms of wear and lubrication, the complexity of tribological experimentation became increasingly evident. When introducing students to tribological testing, it became obvious that hurried sketches and oral descriptions of experimental methods to students were inadequate. A book containing carefully prepared diagrams and exact descriptions of the experimental problems in tribology was required. There are many problems facing a student or researcher who has to perform tribological experiments for the first time. For example, what test rigs to use? How to prepare the samples for the experiments? How to assess the wear mechanism occurring? What type of experimental data need to be collected, and more importantly, what to do with it? How to check the validity of the data and whether any true information could be obtained? For instance, in one laboratory, mysterious lubricating effects from a highly purified mineral oil were claimed. The reasons for this strange result were never fully investigated but it was widely suspected that airborne sulphur from vehicle exhaust fumes contaminated the oil. This perhaps trivial example shows how essential is a thorough understanding of tribology and its experimental methods.

This book, Experimental Methods in Tribology, is intended to provide a basic guide for young or newly inducted researchers. The subjects covered range from the basic technology of experimental tribometers, to the theory of friction and wear testing. Much of the information on tribological experimentation is gained by experience and is not usually discussed fully in research papers or textbooks. Careful application of the principles outlined in this book should ensure valid test data and accelerate progress in tribology. If some doubts over the tribological experimentation still remain please let us know so that the book can be updated.

Gwidon W. Stachowiak, Andrew W. Batchelor and Grazyna B. Stachowiak

ACKNOWLEDGEMENTS

Any book depends on the efforts of many different people and this book is no exception. We would like to thank Professor Duncan Dowson for his great personal input, enthusiasm, encouragement and meticulous checking of the manuscript and very many constructive comments and remarks. Thanks are to Professor Irwin Singer for useful discussions on the effects of tilt angle on the apparent friction force, A/Professor Brett Kirk for Figure 8.2, Dr Simon Graindorge for Figure 8.9, Dr Ksenija Topolevec-Miklozic for Figure 8.21 and Dr Pawel Podsiadlo for help with dimensional analysis, images and finding useful references. Special thanks are to Dr. Nathan W. Scott for the preparation of the illustrations for the book. Without Nathan's illustrations the book would not be the same. We would also like to thank Professor Nic Spencer for useful comments and providing nice environment at the Swiss Federal Institute of Technology during the final checking of the manuscript. Finally we would like to thank the School of Mechanical Engineering, University of Western Australia and School of Engineering, Monash University Malaysia, for their help and unfailing support during the preparation of the manuscript.

We would also like to thank the following publishers for granting us permission to reproduce the figures listed below:

Figure 8.21: Austrian Society of Tribologists. From the Proceedings of the 2nd World Tribology Congress, 3-7 September 2001, Vienna, Austria, (editors: F. Frank, W.J. Bartz, A. Pauschitz), 2001, pp. 179-186.

Figures 10.22 and 10.23: Japanese Society of Tribologists. From the Proceedings of the International Tribology Conference in Nagasaki, 2000, Japanese Society of Tribologists, Vol. 1, 2001, pp. 251-256.

CONTENTS

1 INTRODUCTION

1.1 HISTORICAL ORIGINS OF EXPERIMENTATION IN TRIBOLOGY

Long before the initiation of historical records, palaeolithic people experimented with a basic frictional phenomenon, the temperature rise in sliding contact. Sticks, stones and any other available materials were rubbed together to create sparks, sound and debris [1]. The palaeolithic people then realized that the sparks generated might create that invaluable commodity, fire, which gave heat, light and protection from wild animals. Elementary methods of generating fire by banging two stones together were superseded by more advanced methods involving sliding surfaces which were probably developed by trial and error. The importance of high sliding speed was recognized by rotating the round end of a stick against another piece of wood, in order to create fire. To raise the sliding speed and frictional heat generated at the end of the stick still further, the string of a bow was threaded around the stick and the bow was pulled backwards and forwards to rotate the stick faster. Small splinters of wood or dried leaves were then placed around the rotating end of the stick. The leaves or wood fragments were known to catch fire when touching the hot surfaces of the rotating contact. When the wood fragments or leaves had started to burn, more wood or leaves were added to build up a fire of useful size. Fire ignition by a rotating stick is illustrated in Figure 1.1.

The process of deduction that led to the rotating stick method to start a fire is unknown. Perhaps some adults observed children playing with sticks or perhaps they noted that hands warm up when rubbed together and then made a few tests of their own.

The pre-Christian history contains many other examples of the applied study of friction and wear, for example, the development of a wheel or use of sledges and lubricants to move large masses for the construction of pyramids at Giza [1,2]. Although the development of a wheel is often heralded as the greatest invention in the human history it seems that the development of an axle with a set of bearings was a far greater and more practical invention. Contrary to the early progress of practical applications the scientific study of friction and wear in terms of abstract concepts and theories is far more recent. The earliest systematic studies of the laws of friction are ascribed to Leonardo da Vinci in the 16th century [2]. Starting with a notion of force represented by weight and a test apparatus consisting of a smooth surface, some weights, cord and a pulley, da Vinci observed and recorded the amount of weight hanging by a cord that was required to move a block placed on the smooth surface. This experiment led to the deduction of the proportionality between friction force and weight of the block and the independence of friction force on contact area. These experiments were refined later by Amonton and Coulomb. It is important to realize that these early observations are merely the approximations to the modern view of frictional mechanics and that many complex phenomena between two interacting surfaces would have occurred in this apparently simple experiment. Similar experiments performed more recently have revealed the presence of micro-sliding when a frictional force insufficient to cause gross sliding is applied [3]. The basic principles of tribological experimentation are, however, best illustrated by da Vinci's experiments where a usefully simplified model of complex phenomena was obtained.

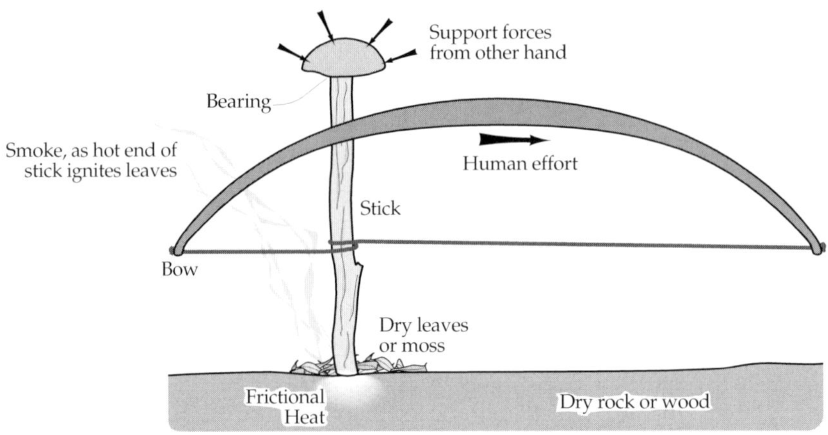

Figure 1.1 Palaeolithic method of igniting a fire by using a rotating stick as an early example of tribological experimentation.

Tribology, like any other field of science, provides the researcher with only a limited number of concepts or models that can be tested with available experimental techniques. In friction and wear experiments the observed friction and wear phenomena are usually interpreted in terms of these concepts even though an incomplete interpretation results. Until now, most of the experimental studies were conducted on a macro scale aiming at explaining the tribological phenomena related to everyday engineering problems. Although satisfactory solutions were often found, the theoretical explanation provided was often wrong or not complete. Technological improvements to the experimental techniques have subsequently led to the gradual modification or changes to the fundamental theories of friction and wear.

In recent years there was a great leap from macro to nano scale observation resulting in the development of a new field of nanotribology [4]. The new technologies developed allow the study of friction and wear processes at the fundamental level involving a single molecular contact interface. The techniques include the surface force apparatus [5], the atomic force microscope (AFM) [6,7], the scanning tunnelling microscope [8,9], the quartz-crystal microbalance [10] and nonlinear optical techniques [11]. Among these new techniques the AFM is the most widely used. The operating principle of the AFM involves the traversal of a microscopic stylus across the surface of a specimen. The tip of the stylus is so fine that it can contact individual atoms on the specimen surface where quantum effects between the stylus and the specimen enable information about the nature of the surface atoms to be obtained. By measuring the forces on the stylus it is also possible to measure frictional forces on an atomic scale. Localised force measurements of this nature have revealed that frictional forces vary considerably for even very small movements of the stylus and that most surfaces have a heterogeneous frictional characteristic. As a consequence of new discoveries in nanotribology classical tribological laws of friction are being modified or re-written as they are found not to be applicable on the atomic scale [10].

Recent developments in computer technology allow to confirm and explain many experimental observations and also to advance theoretical calculations of complex phenomena occurring within the tribological interface. One such an area where computer modelling is widely employed is molecular-dynamics (MD) simulation [12].

Significance of Tribological Experimentation

In the past, progress in tribology has usually been preceded by some experiment or experimental observation. One reason for the slower development of theoretical concepts in tribology is its close relationship to technology, which uses a more empirical methodology than science [2]. In this sense tribology is different from, for example, fundamental physics where theoretical predictions are often stated long before their experimental confirmation. For example, the electro-weak theory discovered in the 1970s is still used today to search for the Higg's boson [13,14].

A classical example of the experimentally based process of discovery in tribology is provided by the first observations of a hydrodynamic pressure field operating in a lubricated bearing. The original axle bearings for railway carriages were fitted with numerous oil holes in an effort to supply as much oil as possible to the bearing. During the operation, however, these bearings often became very hot due to excessive friction. As a result some of them caught fire or else ceased to rotate causing the wheel to develop a flat spot on its rolling surface. Both these phenomena were rather undesirable and much research effort was directed towards rectifying this problem. There was no shortage of ideas for an improved bearing but little progress was made in finding a real solution until the engineer, Beauchamp Tower, performed a detailed study of the friction in these bearings. A test bearing was constructed and a study of friction as a function of the lubrication condition was competently performed. The bearing was meticulously constructed with oil holes to feed the vital lubricant to the sliding surfaces. The problem of high friction in these bearings was, however, not solved even though all the experiments were carefully executed. As sometimes happens in research, chance intervened with a lucky discovery. One day the technician, frustrated with the mess the leaking bearings were making, first plugged the persistently leaking oil holes with rags and then with bungs but this failed to stop the leakage. Tower then realised that the oil in the bearing must be under a considerable pressure. When pressure gauges were fitted around the bearing, it was found that the oil pressure was capable to support the bearing load [15], i.e. the pressure generated was sufficient to separate the axle from its bearing as schematically illustrated in Figure 1.2 [16].

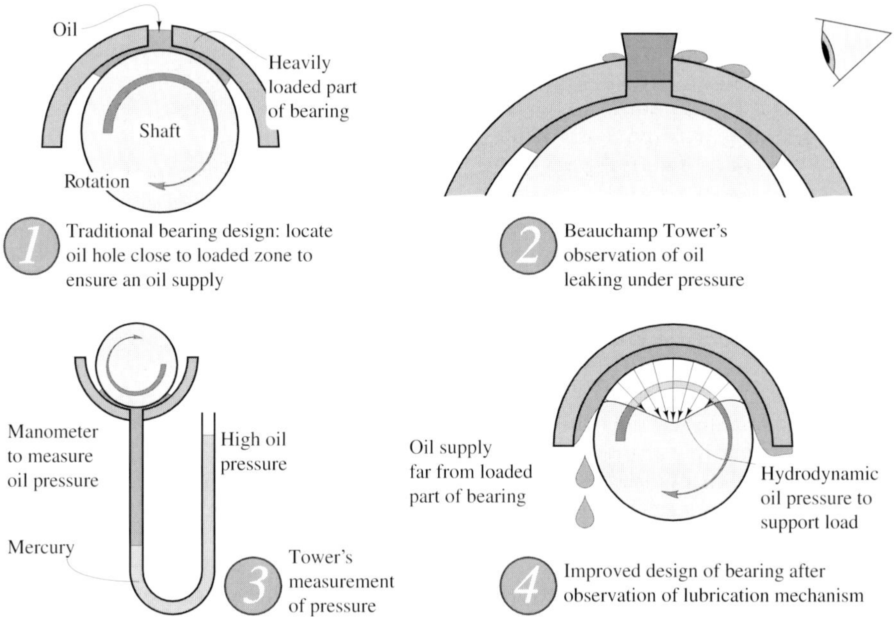

Figure 1.2 Accidental discovery of lubrication mechanism of axle bearings (adapted from [16]).

The publication of Tower's results led in consequence to the development of the hydrodynamic theory of lubrication by Osborne Reynolds [17]. Although the concept of a film of lubricant separating two sliding surfaces was not new (it had been earlier proposed by Leupold in 1735, Leslie in 1804, Rennie in 1829, Adams in 1853 and Hirn in 1854 [2]), it was virtually impossible to resolve the mechanism of lubrication until Tower's precise measurements were applied by

Reynolds to a detailed theoretical analysis. A combination of Tower's and Reynolds efforts replaced innumerable empirical ideas on railway axle bearing lubrication and effectively solved the problem. It was realized only later that Tower's measurements did not include other fundamental bearing parameters, such as operating temperatures and elastic deformation of the bearing under load [18]. Tower also performed an extensive series of, relatively unknown, experiments on thrust bearings with parallel surfaces. For some reason during the experiments conducted no pressures generated were measured [19]. Why were these pressures not measured? Nobody knows. Perhaps Tower, being a proud engineer, feared that his carefully prepared parallel surfaces might be deemed by his academic colleagues to be converging [19]. A precise model of the lubrication mechanism of ostensibly parallel surface bearings was only achieved in 1918, some thirteen years after Tower's death, by Lord Rayleigh [19,20].

A similar process of experimental observation forcing the development of basic tribological concepts is illustrated by the discovery of elastohydrodynamic lubrication. It was known for many years that a viable oil film existed between lubricated gear teeth. However, there appeared to be a contradiction between the oil film thickness predicted by classical hydrodynamic theory and the oil film thickness actually required for effective lubrication of gear teeth [21]. For speed and contact conditions operating in gears the predicted hydrodynamic films were so thin that it was inconceivable for the contacting surfaces to be separated by a viscous liquid film. In fact, the film thicknesses calculated suggested that the surfaces were lubricated by films only one molecule in thickness, clearly far too thin to be effective. The entire problem acquired an aura of mystery and remained unsolved for some time. There was an urgent need for a new theory explaining the lubrication mechanism operating under these conditions and many elaborate experiments and theories were developed as a result. The mysterious lubrication mechanism was solved by Ertel and Grubin [22], who provided a quantitative model of gear teeth lubrication in terms of a theory that is now known as elastohydrodynamic lubrication. Ertel and Grubin were however unable to fully describe the nature of the elastohydrodynamic film and the remarkable nature of the elastohydrodynamic film remained poorly understood until the observations of Cameron and Gohar [23,24]. The elastohydrodynamic film was rendered visible by an optical interference pattern where prismatic colours indicated the thickness of oil trapped between a rolling ball and glass disc. The experimental measurements were later analysed by Dowson and Higginson who were able to reproduce the elastohydrodynamic film as a numerical model [25].

Further historical review confirms the converse argument that not only the utility of tribological experiments rest on adequate modelling but also that scientific models of friction and wear depend on detailed experimental confirmation. In the 18th century, Desaguliers hypothesized that friction and wear between clean surfaces depended on the mutual adhesion of the contacting solids [26]. However, it was not until the middle of the 20th century that the experiments of Bowden and Tabor [27] into adhesion and friction between clean metals enabled this theory to advance understanding of friction and wear. Despite its initial universal acceptance, the theory linking friction to adhesive bonding between contacting surfaces and wear has since been modified in the light of new experimental evidence conducted at the atomic level [10].

In 1804, Leslie provided the first model of the friction of contaminated surfaces where waves of deformed material are pushed across the surface by asperities from the opposing surface [28,29]. This theory remained obscure until experimental confirmation of deformed material surface waves formation was provided by Challen, McLean and Oxley [30] in 1984. Slow-moving slabs of material that resembled the model waves generated by Challen and co-workers were recently observed by X-ray microscopy of a sliding contact [31]. It was found that the formulation of models of wear is greatly facilitated by observational studies, in a manner directly comparable to the experiments of Tower, Cameron and Gohar described above.

1.2 COMPLEXITY OF TRIBOLOGICAL PHENOMENA

The reason for interdependence between experiment and theory in tribology may lie in the fact that friction and wear are essentially chaotic processes [32,33]. While it might be possible to model the deformation of two contacting asperities from basic mechanical principles, the chaotic nature of tribological processes prevents extension of this model to predict wear and friction on a macroscale. Until the theory of complex systems is fully established [33] it is necessary to apply a phenomenological approach to the study of tribology. Phenomenological approach means that friction and wear processes are described in terms of a set of specific phenomena which are systematically analyzed to provide an engineering model of friction, wear and lubrication.

When reviewing the history of tribology it becomes apparent that the development process of an effective means of friction or wear control depends on the following steps. First, an unequivocal experimental observation of the underlying phenomena is required. Data from experimental observation then allows the formulation of an appropriate theoretical model. Thirdly and finally, refinement of the theory, which involves further experimentation, leads to the specification of effective methods of controlling friction or wear. This process is illustrated schematically in Figure 1.3.

Most tribological phenomena, e.g. friction, wear, frictional heating and triboemission of electrons, are not intrinsic material properties but depend on a complex balance between many competing factors. The simplest example of the complexity of tribological problems is provided by the hydrodynamic bearing, where it is necessary to consider viscous heating of the lubricant, cavitation and turbulent lubricant flow as well as elastic deformation of the bearing structure before an accurate value of load capacity can be calculated. This basic characteristic of all tribological phenomena imposes two restrictive conditions on most analyses or investigations:

- our limited capacity to analyse friction, wear and related phenomena, and
- a limited program of experimental tests during which the variability of friction and wear can easily be neglected.

The latter condition is less readily appreciated than the former. Even a brief survey of existing knowledge reveals the need for experimentation in tribology but it is often assumed, without justification, that experimental results will provide the required answers. The following example illustrates the problem associated with the validity of results obtained from the tribological tests. Lubrication engineers are frequently asked to evaluate effectiveness of a lubricant and in many cases this means finding the lowest wear rate from a range of available lubricants.

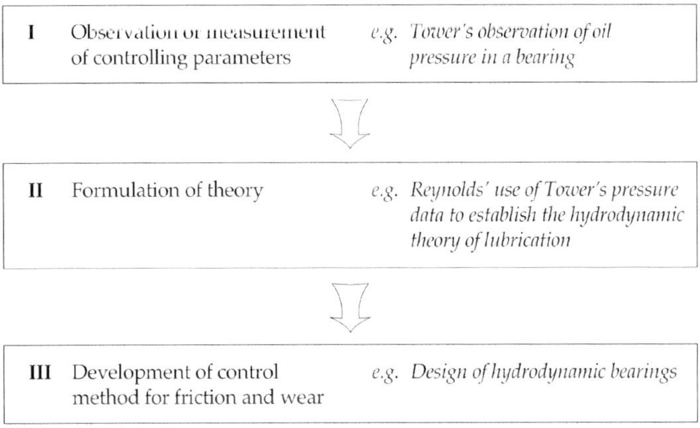

Figure 1.3 Experimental development of applied tribology.

In one investigation, the wear rate of steel discs lubricated by oil with and without lubricant additives was measured. The experiment involved two steel discs loaded against each other and rotated with one of the discs being constrained to produce combined sliding and rolling at the contact. This apparatus is often referred to as the 'twin disc' or 'two disc' test apparatus. An example of the measured wear rate on the discs versus sliding speed in the contact is shown in Figure 1.4 for an oil without additives and an oil containing dibenzyl disulphide (a mild E.P. additive) [34].

Without discussing the details of possible lubrication and wear mechanisms involved in this experiment, it can be seen from Figure 1.4 that at low sliding speeds less than 0.17 [m/s], the common additive dibenzyl disulphide (DBDS) is effective in reducing wear rate compared to plain oil. However, the same additive accelerates wear compared to plain oil above a sliding speed of 0.2 [m/s]. If the effect of sliding speed was not investigated and instead a sliding speed arbitrarily chosen during the lubricant testing, then it would be possible that the result obtained would be entirely misleading as to the merits of various additive-enriched lubricating oils. In most cases, there is no uniquely superior lubricant or wear-resistant material. The optimum combination of a lubricant and material will vary according to the characteristics of the tribosystem. Apart from material properties, parameters such as load, contact stress, temperature, sliding speed and environment (see the effects of replacing air with nitrogen in Figure 1.4) will all influence wear and friction.

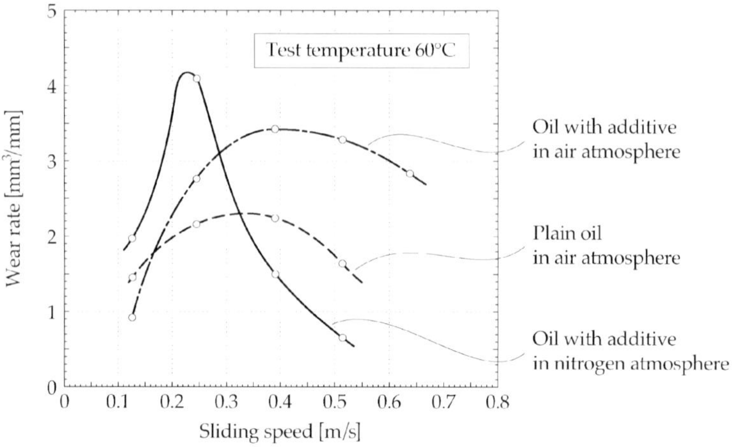

Figure 1.4 Example of effect of sliding speed on the wear rate of additive enriched oil compared to plain oil [34].

Purpose of Tribological Experimentation

Tribological experiments can be very expensive and time-consuming, e.g. for testing of the wear resistance of orthopaedic endoprostheses (artificial knee and hip joints), a series of identical test machines must be operated for several months at a time. A question could be asked - what is the precise purpose of this large effort? The purpose of tribological experimentation can range from pure research, e.g. investigation of how wear and friction mechanisms change as sliding speed is increased from several meters per second to about 1000 [m/s] or the prediction of wear rates, to practical industrial problems. An engine manufacturer may wish to know whether an improved microstructure of cylinder liner material will allow extended engine life to be obtained. A computer memory disk manufacturer may wish to know what is the minimum surface roughness required on a ceramic disc to avoid high friction at the commencement of sliding by a recording head. Very often a researcher in tribology is required to interpret the complex needs of users who usually know very little of the specialised experimental techniques required to obtain reliable tribological data. To summarise, the purpose of tribological investigation can be assigned to two categories which are:

- fundamental research into the basic mechanisms of friction and wear, and
- applied research to resolve specific industrial, scientific and medical friction and wear problems.

The purpose of this book on tribological experimentation is to help investigators solve the problems described above and obtain meaningful information from test data. The book begins with Chapter 1 where a brief history of tribological experimentation is described with some examples chosen to show the significance of experimental studies in tribology. The complexity of tribological processes, leading to the need of a phenomenological approach in their studies, is then emphasised. Chapter 2 discusses the importance of various parameters or variables that control friction, wear and lubrication. Operating parameters (load, speed, temperature, surface finish), material parameters (mechanical, physical and chemical properties) and lubricant and environmental parameters (humidity, surrounding gases, lubricant properties) are discussed and their significance for the tribological contact is described. The chapter ends with a condensed list of tribological contact conditions matched with the most appropriate experimental system (tribometer) for their studies. Chapter 3 contains detailed description of numerous tribometers for typical tribological contacts. The majority of the tribometers described are bench- or laboratory-type rigs and several of them are standardized. Some of the tribometers can accommodate a wide range of experimental conditions and can be used for different investigations. On the other hand, several tribometers specifically designed for investigation of a particular problem are also described. Since friction and wear are two main tribological phenomena being investigated, Chapter 4 is devoted to the methods used in their measurement. Apart from a short description of the principles of each method involved in these measurements, specific problems often encountered during the application of these methods are mentioned. Control of the testing environment is necessary to obtain meaningful results and Chapter 5 shows how to control and measure temperature, humidity and the surrounding atmosphere. Examples from practical experimentation are given that describe data errors resulting from poorly controlled environment. Chapter 6 focuses on the characterization of test specimens or samples. The emphasis is put on the measurement of surface roughness (finish) and measurement of hardness at various scales, i.e. bulk hardness, microhardness and nanohardness. Traditional (stylus profilometry) and newer noncontact methods such as optical interferometry, speckle analysis, confocal microscopy and SEM stereoscopy used to measure surface roughness are described. Other material properties, often included in the specimen characterization, are also briefly described and the examples of their importance for specific tribological investigations are given. The ways the specimens or samples should be prepared for tribological testing are described in Chapter 7. They include sample manufacture and final cleaning. Chapter 7 also contains a description of the execution of tribological tests. Again, various problems particular to tribological testing are indicated.

Surface micrography and analysis form an important part of tribological investigation and the majority of relevant techniques used in tribology are described in Chapter 8. Illustrative examples taken from tribological studies are given to show various possible applications of these techniques. Common micrography techniques such as optical microscopy, SEM or TEM, and atomic scale techniques such as AFM or STM are included. The aim is to emphasise the growing importance of nanotribology in both basic studies and advanced industrial applications. Among various surface analysis techniques described in Chapter 8 the most important are AES, XPS, IR spectroscopy and Raman spectroscopy. Most of the techniques included in this chapter are used to examine worn surfaces (wear scars) or reacted surface films. However, many of these techniques are not suitable for lubricant analysis. Lubricants, often present in the tribological contact, usually require different methods for their characterization. The techniques used to measure lubricant properties are described in Chapter 9. The emphasis is again put on the properties most important for the tribological contact. Techniques to measure lubricant viscosity (at various conditions of temperature, pressure and shear rate) and oxidation stability (both standard and non-standard methods) are described and tribological examples of their use are included. Wear particles can provide important information about the wear severity and mechanisms occurring at the tribological contact. The methods to measure the amount of wear debris in used lubricants and to examine the morphology of extracted wear particles are described in Chapter 10. Most of the methods described here are used routinely in machine condition monitoring. The same methods can be used to examine wear debris from tribological testing. The chapter finishes with the description of new computer-based methods that facilitate the characterization of wear and abrasive particles. The last chapter, Chapter 11, contains a mix of topics. It begins with a discussion of the importance of similarity between the experimental and real tribological contact. This similarity can be confirmed by comparing frictional levels, wear rates, wear surfaces or wear particles from experimental and real contacts. Again, computer-based methods that can help the investigator with surface classification are described. Next, a mathematical technique of dimensional analysis is briefly introduced. Example of using this technique in modelling of erosive wear is shown. The difficulties associated with mathematical modelling of wear problems are also briefly discussed here. Short section showing various ways of presenting tribological data follows next. The chapter ends with some examples of industrial tribological testing. As the detailed description of industrial testing would result in a significant increase in book size this area was deliberately omitted from the contents of this book. However, references to other books and publications devoted to industrial tribology are included for the interested reader.

The main purpose of this book is to describe experimental testing at a level sufficient for a newcomer to the area of tribology, i.e. introductory level. Numerous references included at the end of each chapter should provide additional, more advanced information if needed. Specialised monographs should be also consulted for in-depth information, especially about the surface examination techniques described in Chapter 8. Any book has its limitations as more information can always be included. It was the intention of the authors to keep the book to a reasonable size.

1.3 SUMMARY

Archaeological discoveries have revealed that the study and application of tribological phenomena have been performed since the earliest stages of human development. Formulation and systematic testing of concepts and theories relating to friction, wear and lubrication, however, have a far more recent history. Tribology is an experimental science where progress has usually been initiated by certain critical experiments or observations. Current tribological experimentation has travelled a long way from da Vinci's simple block-on-flat surface tests. Apart from quite sophisticated tribometers new techniques such as electron microscopy, AFM and computer simulation are now often used to further elucidate complex tribological phenomena. This complexity must always be kept in mind while designing and executing tribological tests otherwise it is very easy to obtain misleading results. The formulation of concepts or theoretical models relating to friction and wear remains the limiting factor in tribology. Experimental data that is not used to confirm or extend precise models of friction and wear phenomena (however simplified) has proved to be less valuable than otherwise expected. The subtlety of tribological phenomena presents much experimental difficulty and it is very easy to obtain incomplete data.

REFERENCES

1 J.E. McClellan and H. Dorn, Science and Technology, World History, John Hopkins University Press, 1999, United States of America.

2 D. Dowson, History of Tribology, 2nd Edition, Professional Engineering Publishing, 1998.

3 M. Eguchi and T. Yamamoto, Dynamic behaviour of a slider under various tangential loading conditions, Proc. JSLE. Int. Tribology Conference, 8-10 July 1985, Tokyo, Japan, Elsevier, 1986, pp. 1047-1052.

4 B. Bhushan, J.N. Israelachvili and U.V. Landman, Nanotribology: Friction, wear and lubrication at the atomic scale, *Nature*, Vol. 374, 1995, pp. 607-616.

5 H. Yoshizawa, Y-L. Chen and J. Israelachvili, Recent advances in molecular level understanding of adhesion, friction and lubrication, *Wear*, Vol. 168, 1993, pp. 161-166.

6 C.M. Mate, G.M. McClelland, R. Erlandsson and S. Chiang, Atomic-scale friction of a tungsten tip on a graphite surface, *Phys. Rev. Lett.*, Vol. 59, 1987, pp. 1942-1945.

7 R.J.A. van den Oetelaar and C.F.J. Flipse, Atomic-scale friction on diamond (111) studied by ultra-high vacuum atomic force microscopy, *Surf. Sci.*, Vol. 384, 1997, pp. L828-835.

8 G. Binnig and H. Rohrer, The Scanning Tunnelling Microscope, *Scientific American*, August 1985, pp. 40-46.

9 D.A. Bonnell (editor), Scanning Tunnelling Microscopy and Spectroscopy, VCH Weinheim, 1993.

10 J. Krim, Friction at the atomic scale, *Scientific American*, Vol. 275, 1996, pp. 74-82.

11 Q. Xu, X-D. Xiao, D. Charych, F. Wolf, P. Frantz, D.F. Ogletree, Y.R. Shen and M. Salmeron, Nonlinear optical studies of monomolecular films under pressure, *Phys. Rev. B*, Vol. 51, 1995, pp. 7456-7463.

12 J.A. Harrison, S.J. Stuart and D.W. Brenner, Atomic-Scale Simulation of Tribological and Related Phenomena, Handbook of Micro/Nanotribology, editor: B. Bhushan, CRC Press, Boca Raton, FL, 1998.

13 K. Ziemelis, Dutch theoreticians win physics prize, *Nature*, Vol. 401, No. 6754, 1999, p. 626.

14 P. Weiss, Physics prize takes another tour de force, *Science News*, Vol. 156, No. 16, October 16, 1999, p. 247.

15 Beauchamp Tower, First report on friction experiments, *Proc. Inst. Mech. Engrs.*, Nov. 1883, pp. 632-669.

16 A.W. Batchelor and G.W. Stachowiak, Revealing the hidden world of wear and friction, *Mechanical Engineering Transactions, Journal of the Inst. of Engineers, Australia*, Vol. ME 20, No. 1, 1995, pp. 5-13.

17 O. Reynolds, On the theory of lubrication and its application to Mr Beauchamp Tower's experiments including an experimental determination of the viscosity of olive oil, *Phil. Trans. Roy. Soc., London*, Vol. 177, Part I, 1886, pp. 157-234.

18 C.M.McC. Ettles, M. Akkok and A. Cameron, Inverse hydrodynamic methods applied to Mr Beauchamp Tower's experiments of 1885, *Transactions ASME, Journal of Lubrication Technology*, Vol. 102, 1980, pp. 172-181.

19 A. Cameron, Beauchamp Tower Centenary Lecture, *Proc. Inst. Mech. Engrs.*, Vol. 193, 1979, pp. 1-6.

20 Lord Rayleigh, Notes on the theory of lubrication, *Philosophical Mag.*, Vol. 35, 1918, pp. 1-12.

21 H.M. Martin, Lubrication of gear teeth, *Engineering, London*, Vol. 102, 1916, pp. 199-221.

22 A.N. Grubin, Fundamentals of the Hydrodynamic Theory of Lubrication of Heavily Loaded Cylindrical Surfaces, Investigation of the Contact of Machine Components, Kh.F. Ketova, ed. Translation of Russian Book No. 30, Central Scientific Institute for Technology and Mechanical Engineering, Moscow, 1949.

23 A. Cameron and R. Gohar, Theoretical and experimental studies of the oil film in lubricated point contacts, *Proc. Roy. Soc., London*, Series A, Vol. 291, 1966, pp. 520-536.

24 A. Cameron and R. Gohar, Optical measurement of oil film thickness under elasto-hydrodynamic lubrication, *Nature*, Vol. 200, 1963, pp. 458-459.

25 D. Dowson and C. Higginson, Elastohydrodynamic Lubrication, Pergamon Press, Oxford, 1977.

26 J.T. Desaguliers, A Course of Experimental Philosophy, John Senex, London, 1734.

27 F.P. Bowden and D. Tabor, The Friction and Lubrication of Solids, Oxford University Press, Oxford, 1964.

28 B. Avitzur, The hydrodynamic model of sliding inclined planes and its two limits: Coulomb/Amonton friction and fluid slug rigid body flow, *Tribology Transactions*, Vol. 36, 1993, pp. 249-257.

29 J. Leslie, An Experimental Enquiry Into the Nature and Propagation of Heat, T. Gillet Printer, Salisbury Square, United Kingdom, 1804.

30 J.M. Challen, L.J. McLean and P.L.B. Oxley, Plastic deformation in sliding contact with a hard wedge: Its relation to friction and wear, *Proc. Roy. Soc., London*, Series A, Vol. 394, 1984, pp. 161-181.

31 M. Chandrasekaran, A.W. Batchelor and N.L. Loh, Frictional seizure of lubricated sliding observed by X-ray imaging, Parts 1 & 2, *Journal of Materials Science*, Vol. 35, 2000, pp. 1589-1596 & 1597-1602.

32 B. Mandelbrot, The Fractal Geometry of Nature, Freeman, San Francisco, 1982.

33 P. Winiwarter and Cz. Cempel, Life symptoms: The behaviour of open systems with limited energy dissipation capacity and evolution, *Systems Research*, Vol. 9, 1992, pp. 9-34.

34 M. Masuko, Y. Ito, K. Akatsuka, K. Tagami and H. Okabe, Influence of sulphur-base extreme pressure additives on wear under combined sliding and rolling contact, Proc. Kyushu Conference of JSLE, October 1983, pp. 273-276 (in Japanese).

2 SIMULATION OF WEAR AND FRICTION

2.1 INTRODUCTION

The accurate simulation of friction and wear is a recurrent problem in tribological research. An industry may present a researcher with the task of reproducing a particular wear problem under controlled laboratory conditions to find the causes of wear and necessary remedies. For example, scuffing is a particular mode of adhesive wear commonly occurring in gears and if it is not accurately simulated in laboratory conditions little can be learned about how to control it. Friction and wear mechanisms, because of their inherent complexity, necessitate great care in experimental simulation. The investigator needs to know the appropriate levels of controlling parameters for the specific experiment to be performed. For example, what is a suitable level of load to be used in the experiments, 10 [N] or 1000 [N]? What is a suitable range of sliding speeds? What level of temperature should be selected? Which type of material or lubricant should be tested as both materials and lubricants are usually intended for a limited range of conditions? What type of motion is involved? These questions and many others of a similar nature confront the investigator when planning a laboratory experiment. Unless some selection criteria are applied to determine the appropriate experimental conditions, an arbitrary choice of experimental conditions is almost certain. The purpose of this chapter is to describe the basic principles relating to the planning of tribological experiments.

2.2 REAL TRIBOLOGICAL CONTACTS AND THE OBJECTIVE OF THEIR STUDY

In actual tribological contacts complex phenomena take place subject to innumerable influences. Common examples are an internal combustion engine and a transmission of a vehicle. Wear in the piston rings, cylinder liners and elsewhere in the engine and transmission is influenced by factors such as road conditions, driving habits, weather, characteristics of lubricants used and level of maintenance. Analogous patterns of behaviour with similar complexity can be found in any industrial machinery. Biotribology provides examples of even more complex forms of wear as those found, for example, between opposing surfaces of synovial joints and between eyeball and its socket. In those cases, pathological factors such as the release of destructive enzymes and oxidizing agents by cells close to the sliding interface can radically change the course of wear [e.g. 1,2].

In most cases, the purpose of an experimental investigation is not to comprehensively analyze friction and wear occurring but to isolate the effects of certain controlling factors that are of interest. Most forms of friction and wear are still too complex to be rigorously analyzed and some degree of empiricism is nearly always necessary. The art of tribological experimentation is to devise an experiment which is not too empirical so that its data is compromised by non-causal factors such as e.g. weather fluctuations, but at the same time maintains relevance to real problems. In terms of practical experimentation, the objective of any tribological investigation is to simulate the effects of all the complex factors that are associated with friction and wear by a limited number of controllable experimental parameters that hopefully are causally related to friction and wear. Referring to the example of the engine and transmission, the circumstantial factors such as driving habit and weather would be interpreted in terms of basic parameters such as load, speed, temperature and humidity.

It is not usually difficult to determine the basic controlling parameters, e.g. contact stress, amplitude of sliding, velocity, etc., for a wearing contact. Once the initial stage of deciding on basic parameters controlling the friction and wear has been completed, the next stage is to estimate suitable values of these parameters to be used in the experiments. In some situations, however, it is impossible to measure or set values of controlling parameters. This problem is usually found in experiments simulating wear and friction occurring in actual industrial equipment, e.g. open gears, combustion engines, etc. Careful selection and use of tribometers (friction and wear measuring devices) can assure the accuracy of the obtained data and prevent measurement and control difficulties. A general guide to determining appropriate values of controlling parameters is outlined in the next sections.

2.3 CHARACTERIZATION OF REAL TRIBOLOGICAL CONTACTS

The characteristics of any tribological contact vary according to the imposed conditions. If a close simulation of a practical wear situation is desired then it is necessary to fully characterize the tribological contact. Characterization means that all major controlling factors are identified and given a suitable value. There are a large number of controlling factors, for example, the load on a contact is also related to parameters such as contact stress and time variation of load (if any). Similar considerations apply to temperature, sliding speed, etc. In order to achieve close match between test and real tribological contact, the operational, material, environmental and lubricant parameters should be fully characterized. The summary of main parameters necessary in the characterization of tribological contacts is outlined in Table 2.1.

Table 2.1 Commonly used parameters in the characterization of tribological contacts.

	Basic parameter	**Related parameters**	
Operating parameter	Load [N]	Contact stress [Pa] Temporal variation in load [dimensionless] Impact force [N]	 Oscillating load [N]
	Sliding speed [m/s]	Rolling speed [m/s] Average speed [m/s] Sliding/rolling ratio	Spinning speed [rad/s] Impact speed [m/s]
	Sliding distance [m]	Amplitude of reciprocating sliding [m], Frequency of reciprocating sliding [Hz],	
	Temperature [K]	Transient temperature [K]	Bulk temperature [K]
	Surface finish	CLA [m]	RMS value
	Type of contact	Conformal	Non-conformal
Material parameter	Hardness [Pa or VHN]	Microhardness [Pa or VHN] Shear strength [Pa]	 Microstructure
	Toughness [Pa×m$^{0.5}$]	Limiting strain [dimensionless]	Average grain size [m]
	Melting point [K]	Glass transition temperature [K]	Softening point [K]
	Thermal conductivity [W/m×K]	Thermal diffusivity [m^2/s] Temperature variation of conductivity	Thermal shock resistance Specific heat [J/kgK]
	Electrochemical potential	Exchange current density [A/m^2] Potential [V]	Zeta potential [V]
Material parameters for Abrasive and Erosive Wear Tests	Grit hardness [Pa or VHN] and/or mineral type	Ratio grit/substrate hardness	
	Size spectrum of grits [m]	Shape characteristics of grits Fractal dimension Acidity of carrier fluid for slurries [pH]	Radance Spike parameter
Environmental parameter	Relative humidity [%]	Absolute humidity [kg/m^3]	
	Local air pressure [Pa]	Partial pressure of oxygen [Pa] and other active gases	
	Radiation level [Bequerels/m^2] (relevant to studies in nuclear environments)		
Lubrication parameter	Viscosity [Pa×s]	Pressure-viscosity coefficient [Pa^{-1}]	
	Flow rate [m^3/s]	Supply pressure [Pa]	Supply velocity [m/s]
	Thermal conductivity [W/m×K]	Thermal diffusivity [m^2/s] Temperature variation of conductivity	Specific heat
	Acidity [pH] Boiling point [K] Solidification point [K] Heat of oxidation [J/kg×mole]	Chemical reactivity [no fundamental parameter] Dipole moment [Debye]	
	Water/oxygen solubility		

The list of parameters presented in Table 2.1 is not exhaustive as there will always be some other specialised parameters for specific experiments. This large number of parameters involved has two following consequences:

- the probability of achieving good comparability without adequate characterization of materials, environment and operating conditions is negligible
- a comprehensive test program covering all variables would require an impossibly large research effort hence the data available from the experiments provides only fragmentary information.

It is within this structure (comprising the parameters listed in Table 2.1) containing a very large number of degrees of freedom that an investigation must be carefully planned in order to obtain the meaningful data.

Operating Parameters

Criteria for the selection of values of each of the major operating parameters are discussed in this section.

- *Load and Contact Stress*

Load (normal load) is a parameter that is fundamental to wear and friction. Its magnitude can range from 1 [μN] in microscopic contacts and 1 [mN] for computer head/disk interfaces to 10 [MN] for large bearings which is a 10^{13} variation in scale. It is essential to accurately select a value of load as friction and wear mechanisms are very sensitive to it. One or more transitional points in wear rates are observed when the normal load is increased [3,4]. In the simplest case where the experiment involves a wearing contact which is identical in size to the original wearing contact, the same value of load can be used. An example of this practice is in a study of the fretting wear in wire ropes where two wires are loaded with a force equal to the estimated contact force between flexing wire ropes [e.g. 5-7]. When the contact size is larger or smaller than the original contact then it is necessary to consider a parameter related to the load which is the contact stress (pressure). In many experiments conducted in the laboratory, the size of the contact is much smaller than is found in the practical situations so that the original value of load cannot be used. As a general rule the contact stress in the experimental contact should closely approximate to the contact stress found in practice. The values of typical contact stresses range from 100 [kPa] for aerostatic bearings through 10 [MPa] for sliding unlubricated wear commonly found in the industrial machinery to 1 ~ 5 [GPa] for rolling contact bearings or gears. It should be noted, however, that similarity in contact stress is a necessary but not sufficient condition for experimental comparability as effects associated with the gross value of load cannot entirely be eliminated. A fundamental question is whether the amount of frictional power dissipation is also comparable and this is related directly to load not contact stress.

Many tests involve observations of wear changes as a result of load variation over a period of time. For example, a vehicle on a hilly road will experience continuously varying engine load. Although it is often important that the load variation in an experiment closely follows the practical problem some degree of simplification in the pattern of load variation in the test is usually acceptable, especially in cases where the load variation is complex or chaotic.

Tests involving impact and oscillating forces conform to the same requirement of close approximation to the forces found in practice. Where miniaturization or enlargement of the test contact area compared to the original contact is required, comparability of contact stress needs to be applied. Impact forces are associated with an impact energy which also needs to be simulated during the experiment. The impact energy/unit contact area can usually be used to ensure comparability if miniaturization or enlargement is involved.

· *Sliding, Rolling and Spinning Speeds*

Sliding speeds can vary enormously from a range of 10^{-4} [m/s] for fretting tests to 10^{3} [m/s] for experiments on shells in gun barrels. A simple rule is to use the same value of sliding speed in the experiments as occurs in practice. Miniaturization or enlargement of the contact does not affect this value of sliding speed and if a different sliding speed or lubrication regime is used, wear will be affected by the resulting change in frictional temperature rise [3]. An exception to this rule appears to be in fretting studies under conditions of gross slip, where sliding speeds and subsequent temperature rises are so small that a higher sliding speed can be used to accelerate testing. The higher sliding speed in fretting experiments is achieved by increasing the frequency of fretting oscillations.

In some cases a lower speed, than that actually occurring in real system, is used in the experiments. For example, in research on critical temperatures of lubricant additives used to control scuffing and in the studies of high temperature wear resistance of materials, lower sliding speeds are used while specimen temperature is deliberately raised to check whether the effect of sliding speed is purely to raise the surface temperature or it has some other effects. A similar principle of lowered sliding speed and external heating is also used to test the high temperature wear resistance of materials.

Rolling speeds are important in contact fatigue tests of rolling element bearings and in studies of combined rolling and sliding. Pure rolling has a much weaker effect on wear and friction than sliding so that it is quite acceptable to select a rolling speed according to experimental convenience provided that basic criteria, such as the maintenance of elastohydrodynamic lubrication, are satisfied. When a combination of rolling and sliding is involved, the rolling speed should match the rolling speed found in practice closely as friction, wear and lubrication are dependent on the magnitudes of both rolling and sliding. The spinning speed of one contacting surface relative to another can also affect friction and wear [8] so that the value of this parameter in practice should be assessed first and simulated in the experiment.

· *Sliding Distance, Amplitude of Sliding and Out-of-Contact Time*

Sliding distances can vary from 1 [m] in fretting problems to 100 [km] for tests on wear resistant polymers and in theory there is no upper limit on the sliding distance. The choice of the correct sliding distance is critical to the success of any tribological experiment. In many cases it is assumed that once the wear and friction stabilize after an initial period of rapid change (running-in), then stable and unchanging wear and friction coefficients will result. The required sliding distance according to this reasoning is merely the distance sufficient to allow measurement of a stable wear rate. Long-term sliding experiments [9] reveal that an apparently steady wear rate can suddenly change after a long sliding distance of many km. Tests of silicon nitride ceramics at high temperature revealed that friction coefficients varied in a systematic manner over sliding distances of 500 to 1000 [m] [10]. These changes are believed to be associated with the gradual formation of wear debris layers [10]. To obtain good comparability between test and problem, the same sliding distance should be used unless it can be confirmed by control experiments that a shorter sliding distance is acceptable. The necessity of long sliding distances is the principal reason why detailed wear experiments are often very slow or require a large number of simultaneous wear tests.

The amplitude of sliding is highly significant to fretting wear. It is found that the severity and mechanisms of fretting change as the amplitude of movement is increased from 1 [μm] to 100 [μm] [11,12]. The fretting wear increase with increasing sliding amplitude is attributed to a change from partial slip to gross sliding regimes. Sliding amplitude is closely related to the Mutual Overlap Coefficient (MOC) [13,14] which is a parameter defining the changes in conditions of wear as the amplitude of sliding varies from much smaller than the size of the contact length to much larger than the contact length. Since MOC is defined as the ratio of the contact area of the smaller of the sliding members to that of the wear track [13] the differences between fretting wear and sliding wear can be interpreted in terms of this parameter. The value of MOC varies from just less than unity in small amplitude fretting to near zero values in large amplitude sliding.

The out-of-contact time is the quotient of the amplitude of sliding or the circumference of a circular wear track divided by the sliding speed. This parameter expresses the amount of time available for protective films to form on a worn surface between successive episodes of film destruction within the wearing contact. Out-of-contact times range from 1 [ms] to 10 [s] for most observed contacts, are particularly important to the functioning of lubricant additives and need to be carefully reproduced during the experiments. A minimum out-of-contact time is required for the lubrication of large open gears (5 [m] in diameter or more). The volatile carrier fluid requires time to evaporate and leave a layer of non-volatile solid or semi-solid lubricant layer.

· *Temperature*

Temperature exerts a profound influence on friction and wear so that it can never be neglected in any tribological experiment [3]. There are two types of temperature, i.e.: the bulk temperature and the frictional temperature rise. The bulk temperature is usually the temperature imposed on the specimen by external heating or refrigeration and this temperature should closely follow the temperatures found in the practical problem. The range of temperatures studied extends from cryogenic temperatures of a few degrees K to 1500K or more in high temperature tests on ceramic materials. Experimental temperature can be allowed to deviate from the practical case when the earlier mentioned effect of sliding speed is studied. Frictional temperature rises are significant for almost all kinds of tribological tests apart from small amplitude fretting. These temperature rises can range from ~ 1K to several hundred K where the temperature is limited by the melting point of the worn material. In general, frictional temperature rises during the simulation experiments should be maintained at the same level as those found in practical problems.

· *Surface Finish*

The surface finish has a strong influence on friction and wear and a simple rule is to use the same value of surface finish, i.e. R_a or RMS, in the experiments as occurs in practice.

· *Type of Contact*

The contacts can be conformal, e.g. between rolling element and outer ring in rolling contact bearing, or non-conformal, i.e. between rolling element and inner ring. The type of contact affects the contact area and hence the contact pressures. The parameter characterizing the contact geometry, i.e. reduced radius, can easily be calculated using standard formulae [e.g. 14].

Material Parameters

Criteria for the selection of values of each of the major material parameters are discussed in this section.

· *Hardness and Microhardness*

Hardness and microhardness are some of the most important parameters describing the tribological characteristics of a material. Friction and wear tests involve almost all known materials ranging from the hardest diamond to extremely soft materials such as human cartilage. For example, hardness is of critical significance to abrasive and erosive wear and the hardness of experimental specimens should be similar or identical to the hardness of practical components for reliable measurements of abrasive wear resistance. Other wear mechanisms are also strongly affected by hardness so that, in general, if the hardness of the test specimens diverges from either practical values or an intended level for study, then misleading results will most probably be obtained.

· *Toughness*

Toughness is particularly important for ceramics where microfracture is a dominating wear mechanism. Brittleness also affects metals by limiting ductile junction growth in adhesive wear and promoting crack extension during delamination wear. As a general rule, the comparability of material toughness between experiment and practical problem should be maintained.

· *Melting Point and Thermal Conductivity*

The melting point of a material controls wear by limiting the maximum temperature in a wearing contact. When the combined frictional temperature rise and bulk temperature of a specimen reach its melting point a rapid form of wear, known as melting wear, will ensue. Thermal conductivity determines the dissipation of frictional heat, e.g. a lower thermal conductivity ensures a higher frictional temperature rise for a given sliding speed and load. If more than a small difference in thermal conductivity or melting point is allowed between the material in the practical problem and the experimental material, then both the frictional temperature and the reaction of the material to frictional temperature rise will be subject to experimental artefact.

· *Electrochemical Potential*

The electrochemical potential of metals in particular determines the rate of corrosive interaction with atmospheric oxygen, lubricants or reactive contaminants such as e.g. chlorine. When studying mechanisms of wear which are dependent on chemical reactions such as corrosive wear, oxidative wear and corrosive-abrasive wear the electrochemical potential of the worn material and related parameters such as exchange current and zeta potential should be analyzed for similarity to the practical problem.

· *Material Parameters for Abrasive and Erosive Wear Experiments*

The material characteristics and size of abrasive grits should be determined first for meaningful investigation of abrasive and erosive wear. Critical parameters are grit hardness, e.g. the hardness of silica which constitutes sand is approximately 1100 HV, and the size range of the grits, e.g. whether the grit size is 1 [mm] or 10 [μm]. The ability of a grit to abrade depends on its shape which varies from very sharp for freshly fractured grits to rounded for weathered grits. This particle feature can be characterised by parameters such as: the spike parameter which describes the angularity of the grit's boundary [15-17]; the radance, which is obtained from a Fourier analysis of the shape of the grit [18]; the angularity parameter, which incorporates the probability that the particle protrusion will interact with the surface [19].

In most cases the difficulty in precisely matching materials parameters between the practical problem and experiment forces the choice of materials to be very similar, if not identical, to the materials used in the original components or specimens. The reason why materials should always be specified precisely in any tribological investigation is that no single parameter such as hardness or yield strength can ever provide sufficient material characterisation for future replication of experiments or application of the data obtained.

Lubricant and Environmental Parameters

Lubricants comprise a very wide variety of fluids ranging from the conventional mineral oils to synthetic silicone-based fluids. Test environments range from ambient conditions to extremes of temperature, pressure and high levels of radiation. Lubricant and environment need to be well characterized in order to achieve valid experimental results.

Basic parameters which characterize the lubricant properties are viscosity and coefficients describing the dependence of viscosity on temperature and pressure. These parameters determine the formation of hydrodynamic and elastohydrodynamic lubrication films. The thermal conductivity of the lubricant affects the dissipation of frictional heat within the lubricant. The solubility and diffusivities of oxygen or other reactive gases influence scuffing, corrosive wear and other phenomena involving film formation by chemical reaction. The acidity of a lubricant and water solubility and concentration are also critical to corrosive wear. The presence of reactive species such as oxygen, sulphur and halogen compounds in the lubricant needs to be determined as these can either induce corrosive wear or else provide some degree of 'Extreme Pressure' lubrication and obscure the results obtained. Polar organic species, e.g. fatty acids and alcohols present in the lubricant may cause adsorbed films to form on the worn surface which can either reduce friction or impede the functioning of solid lubricants. Where water is present in significant quantities, e.g. 10% by volume, the emulsifiability of the lubricant becomes very relevant.

Operating parameters relating to the lubricant such as flow rate, supply pressure and supply velocity may be of critical significance particularly when there is some degree of lubricant starvation. It is therefore very important to know the values of these parameters in order to conduct the accurate simulation experiments.

In most studies, environmental parameters relate to the pressure of the local atmosphere or to partial pressure of oxygen and humidity as these quantities exert the strongest effect on friction and wear. The ambient temperature should also be monitored as this can vary significantly with geographical location and time of year. In specialized experiments involving lubrication in the presence of special gases, e.g. halons or chlorine gas, or nuclear radiation, the composition and pressure of the gas or the type and intensity of nuclear radiation should be determined. Nuclear radiation accelerates chemical degradation and oxidation of the lubricant which in turn affects basic wear and lubrication mechanisms.

2.4 SELECTION OF A TRIBOMETER FOR WEAR AND FRICTION SIMULATION

Once the purpose of an experiment is defined and the necessary information about controlling parameters is compiled, the next stage is to decide on the type of tribometer. As is discussed in Chapter 3, there is a large number of different tribometers available to the researcher and it is essential to accurately select the most appropriate type of tribometer for a particular type of experiment. Selection of a tribometer depends on the type of wearing contact and operating conditions needed to be studied. A summary of typical wear studies and suitable types of tribometer is shown in Table 2.2.

Almost all experiments of practical significance can be classified according to the categories described in Table 2.2. If the wrong category of tribometer is selected for the experiments then it is unlikely that the prevailing wear mechanisms will be simulated. It is important to realise that a generalized materials property of 'wear resistance' does not actually exist and that specific experiments are required to measure the response of materials to different types of wear. For example, an accurate indication of resistance to abrasive wear of a metal cannot be obtained from tests of dry sliding wear. The converse of this also applies. Once the basic type of tribometer is selected then the appropriate type of apparatus can be chosen. The detailed information on commonly used tribometers can be found in Chapter 3.

Table 2.2 Prevailing wear conditions and type of tribometer.

Mechanism of Wear or Lubrication	Tribometer
Abrasive	Apparatus involving abrasive paper or bed of sand
Erosive wear	Specimens mounted in stream of air or liquid jet mixed with abrasive particles
Cavitational wear	Specimens mounted in stream of fluid or else mounted on vibrating platform immersed in fluid
Dry or lubricated sliding wear in ambient conditions	Specimens slid against a moving counterface (which may also be a specimen) either by rotation or reciprocating movement. Lubricant supply system fitted for lubricated tests
Dry or lubricated wear under non-ambient conditions	Specimens and moving counterface enclosed in chamber fitted with heating, refrigeration and/or vacuum pumping system to maintain a specialised environment
Fretting/fretting fatigue	Specimens slid against the counterface at very small amplitude. Apparatus can be fitted with chamber for specialized environments
Wear under combined rolling and sliding or pure rolling	Specimens in form of rollers and spheres and constrained to move at specified speeds. Apparatus can be fitted with chamber for specialized environments
Impact wear	Apparatus containing hammer to impact wearing specimen. Can be fitted with enclosing chamber for non-ambient tests
Diffusion or solubilization wear	· Cutting tool as specimen. Cutting performed at high surface speed · Specimens immersed in molten test material and rotated to accelerate wear
Hydrodynamic and elastohydrodynamic lubrication	Idealized conformal and non-conformal contacts devised with purpose of displaying mechanism involved in the formation of lubricating films
Boundary and solid lubrication	Apparatus involving sliding wear at low velocities allowing the friction coefficient during the tests to be monitored

2.5 SUMMARY

Among the first decisions an investigator is required to make when planning a tribological study is to decide which parameters are likely to control the friction and wear process. Physical and chemical factors should also be considered either in isolation or in synergistic combination. The next stage is to select the appropriate type of experiment as the various friction and wear mechanisms differ in many fundamental aspects. Once the basic type of experiment and controlling parameters are determined, planning of the investigation can then proceed to a more detailed stage.

REFERENCES

1 A.W. Batchelor and G.W. Stachowiak, Arthritis and the interacting mechanisms of synovial joint lubrication, Part I - Operating conditions and the environment, *Journal of Orthopaedic Rheumatology*, Vol. 9, 1996, pp. 3-10.

2 A.W. Batchelor and G.W. Stachowiak, Arthritis and the interacting mechanisms of synovial joint lubrication, Part II - Joint lubrication and its relation to arthritis, *Journal of Orthopaedic Rheumatology*, Vol. 9, 1996, pp. 11-21.

3 M.B. Peterson, Design considerations for effective wear control, Wear Control Handbook, ASME, 1980, pp. 413-473.

4 N.C. Welsh, The dry wear of steels, *Phil. Trans. Roy. Soc. London*, Vol. 257A, 1965, pp. 31-50.

5 B.R. Pearson, P.A. Brook and R.B. Waterhouse, Fretting in aqueous media, particularly of roping steels in seawater, *Wear*, Vol. 106, 1985, pp. 225-260.

6 A.W. Batchelor, G.W. Stachowiak, G.B. Stachowiak, P.W. Leech and O. Reinhold, Control of fretting friction and wear of roping wire by laser surface alloying and PVD coatings, *Wear*, Vol. 152, 1992, pp. 127-150.

7 G.W. Stachowiak, G.B. Stachowiak and A.W. Batchelor, Suppression of fretting wear between roping wires by coatings and laser alloyed layers of molybdenum, *Wear*, Vol. 178, 1994, pp. 69-77.

8 B. Briscoe and T. Stolarski - Informal presentation at the Limits of Lubrication Conference, Imperial College, London, 1979.

9 K.J. Brown, J.R. Atkinson, D. Dowson and V. Wright, The wear of ultra high molecular weight polyethylene and a preliminary study of its relation to the in vivo behavior of replacement thigh joints, *Wear*, Vol. 40, 1976, pp. 255-264.

10 N. Roels, G. Kapelski, F. Platon and J. Desmaison, Effect of carbon-containing CVD Si_3N_4 coatings on the wear and friction of ceramic couples, Proc. High Performance Ceramic Films and Coatings, (7th CIMTEC-World Ceramics Congress), Montecatini Terme, Italy, 27-30 June 1990, editor: P. Vincenzini, Elsevier, 1991, pp. 603-612.

11 R.B. Waterhouse, Fretting Fatigue, Applied Science Publishers Ltd., London 1981.

12 R.C. Bill, Fretting wear and fretting fatigue - How are they related?, *Transactions ASME, Journal of Lubrication Technology*, Vol. 105, 1983, pp. 230-238.

13 D. Play, Mutual overlap coefficient and wear debris motion in dry oscillating friction and wear tests, *ASLE Transactions*, Vol. 28, 1985, pp. 527-535.

14 G.W. Stachowiak and A.W. Batchelor, Engineering Tribology, Butterworth-Heinemann, 2001.

15 M.G. Hamblin and G.W. Stachowiak, A multi-scale measure of particle abrasivity, *Wear*, Vol. 185, 1995, pp. 225-233.

16 M.G. Hamblin and G.W. Stachowiak, A multi-scale measure of particle abrasivity, and its relation to two body abrasive wear, *Wear*, Vol. 190, 1995, pp. 190-196.

17 M.G. Hamblin and G.W. Stachowiak, Description of abrasive particle shape and its relation to two-body abrasive wear, *Tribology Transactions*, Vol. 39, No. 4, 1996, pp. 803-810.

18 P.A. Swanson and A.F. Vetter, The measurement of abrasive particle shape and its effect on wear, *ASLE Transactions*, Vol. 28, 1984, pp. 225-230.

19 S.L. Palasamudram and S. Bahadur, Particle characterization for angularity and the effects of particle size and angularity on erosion in fluidized bed environment, *Wear*, Vol. 203-204, 1997, pp. 455-463.

3 TRIBOMETERS

3.1 INTRODUCTION

Tribometers, or devices to measure friction and wear, are the basic technology used in most of the tribological investigations. A carefully selected tribometer can simulate all the critical characteristics of a wear or friction problem without the difficulties associated with the experimentation on actual equipment. Conversely, a poorly designed or selected tribometer can provide entirely false results so that a thorough knowledge of the characteristics of tribometers is essential to any tribological test program. There is a large variety of tribometers available and the basic characteristics of the major groups of tribometers are described in this chapter. An engineer often faces the practical questions such as: should the tribometer provide rotating or reciprocating sliding? Does that tribometer need a chemically controlled chamber? Does the tribometer need a variable loading and velocity systems?, etc. All these questions and more have to be addressed and answered before a choice of a reliable tribometer is made.

The purpose of a tribometer is to provide simulation of friction and wear under controlled conditions. Friction and wear are very sensitive to factors such as variations in temperature, load or humidity, therefore it is essential to provide an apparatus where all these factors can carefully be controlled and monitored. There is also the practical benefit of facilitating experimentation without the difficulties accompanying studies of wear and friction occurring on the original equipment. For instance, wear between cylinders and pistons is of considerable economic importance. Full-scale tests on internal combustion engines are limited by time and resources so that much preliminary lubricant testing is performed on simplified test machines. A further reason for the use of tribometers is that they can be designed to allow precise measurement of friction, wear and related parameters (e.g. temperature) in a way that is not possible with typical industrial devices. This aspect of tribometers is becoming increasingly important as the new technologies allow for the application of more sophisticated means of observation and measurement.

Not all tribological studies require the application of typical tribometers with a dynamic contact. In some cases, particularly those relating to fundamental aspects of tribology, these traditional tribometers are not used. Instead, basic characteristics of friction, wear or lubrication are simulated by physical processes, which are believed to provide an experimental model of tribological phenomena. An example of this approach is the study of material deformation aimed to investigate the mechanical aspects of wear, i.e. mechanisms of asperity deformation. The purpose of this type of experiment is to gain information on mechanisms of wear and friction, which are normally concealed from direct observation or measurement.

3.2 BASIC FEATURES OF A TRIBOMETER

Although the majority of tribometers are bench-top devices that allow experimentation inside a laboratory, in certain cases industrial equipment may be modified to allow tribological measurements, for example, wear tests on real wire rope and sheave assemblies [1] or full-scale engine tests [2]. Apart from the general implication that tribometers are designed for a laboratory, there is a wide variation in tribometer design and technology. According to the Society of Tribologists and Lubrication Engineers, there are 243 recognized designs of tribometer [3]. Unfortunately, most of these designs follow the dictates of practical convenience rather than the principles of tribological experimentation. Despite the number of available designs the actual range of options for useful experimentation is more limited. The number of tribometers that have been regularly used by research groups is much smaller.

All tribometers contain a wearing contact, which is the focus of conducted tests and this forms the core of the apparatus. The average tribometer also contains various ancillary equipment providing motion, imposing load and supplying lubricant or special gas atmosphere. More advanced tribometers are equipped with instrumentation to measure friction coefficient, wear rates and often also a temperature. Apart from these basic features, tribometers differ considerably in design and function. The prime factor in necessitating a wide range of tribometers is the variety of wearing contacts occurring in the real situations that need to be simulated. Wear occurs under conditions of pure sliding, pure rolling, combined rolling and sliding, impact, abrasion, erosion and cavitation. Each form of wearing movement requires a specially designed tribometer, as shown in Table 2.2.

For instance, a popular tribometer designed to test lubricants under extreme contact stresses is the 'four-ball' apparatus which consists of a pyramidal combination of steel balls pressed together with three balls stationary and the fourth rotating. In contrast, a tribometer designed to test the coefficient of friction between animal footpads and a hard surface consists of a treadmill which can be pivoted at a range of angles in order to determine the coefficient of friction from the slope which is required for the animal to slip [4]. Variations in scale of basic operating parameters, such as sliding speed and amplitude also have a fundamental influence on the design of a tribometer. Most tribometers operate at sliding speeds in the range between **0.001 - 1** [m/s] and are designed to provide data on abrasive wear, adhesive wear, delamination wear, fatigue wear or any other related wear mechanisms. However, some tribometers are required to operate at sliding speeds outside of this range. For example, studies of melting wear in metals require very high sliding speeds of about **100** [m/s] or more and a different design of tribometer is required. In fretting studies, although fretting is a form of reciprocating sliding, the amplitude involved is very small in the range between **5 - 100** [μm] so that the construction of a fretting wear apparatus is significantly different from a tribometer designed to study macroscopic sliding.

Thus, although there is a certain number of tribometers which are described as having a comprehensive range of test capabilities, e.g. the 'universal wear test machine', almost all known test machines have a limited range of applications. Tribometers are usually designed to cover a specific range of wear mechanisms or operating conditions and are usually unsuitable or inefficient for experimental conditions which are outside of their intended operating range. In general terms, tribometers can conveniently be classified in groups which are based on wear mechanisms or operating conditions.

3.3 TRIBOMETERS FOR DRY OR PARTIALLY LUBRICATED SLIDING CONTACTS

This class of tribometers is probably the largest and most extensively developed as much commercial interest is vested in an apparatus which can conveniently evaluate lubricants. The most common sample configurations used in dry or partially lubricated sliding wear apparatus are schematically illustrated in Figure 3.1.

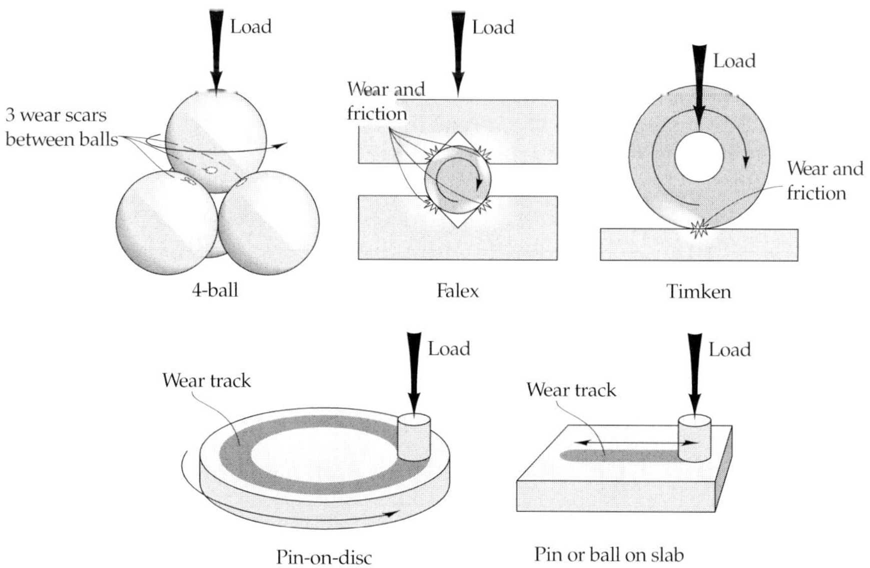

Figure 3.1 Schematic illustration of basic sample configurations used in simulations of dry or partially lubricated sliding contacts.

Four-Ball Tester

In this apparatus four 12.7 [mm] diameter balls are used, one of which is held in the chuck rotating at 1770 ± 60 [rpm] while three others are held rigidly in a pot (ASTM D 2783-88 [5], IP 239/97 [6]). During the tests the pot is filled with the lubricant and the fixed balls are then pressed against the rotating ball over a ten second interval. Testing involves an application of a series of ten seconds runs at pre-selected and successively higher loads until welding of the four balls occurs. The measurements of the wear scars are made on each of the three lower balls. From the measurements the load - wear scar curve is obtained and the weld point.

Falex Tester

The Falex tester, since it allows large contact stresses to be developed without the need for a strong mechanical support, is widely used in industrial research laboratories and elsewhere (ASTM D 3233-93 [7], IP 241/84 [8]). In this apparatus 4 contact forces balance each other so that the only force required is a V-block closure force and torque to drive the pin. The torque provides a direct measure of friction force without an error of friction generated in bearings. High contact pressures achieved during the tests render this apparatus suitable for seizure and scuffing studies. This instrument provides the data on the value of coefficients of friction for a particular material/lubricant combination, maximum load before seizure and wear scar diameter after testing.

Timken Apparatus

In this apparatus a ring ('test cup') rotating at 800 [rpm] is pressed against a stationary slab ('test block') in the presence of the lubricant tested (ASTM D 2782-94 [9], IP 240/84 [10]). The test is conducted for ten minutes at successively higher loads until 'OK' and failure loads are determined. The 'OK' load is the maximum load which can be applied on the lever arm without producing scoring marks on the slab. The load on the lever arm is used as the indication of the strength of the lubricating film.

The four-ball, Falex and Timken apparatus are used for standard lubricant tests that are specified by either scientific institutions or by industrial organizations. These tribometers, while frequently used for commercial testing and some research work, have the disadvantages of varying contact stress during tests and poorly controlled frictional temperature. The size of the contact areas between balls in a four-ball test can vary from the small area of Hertzian contact to a much greater size depending on the scale of the wear scars formed on the balls. The Falex machine is also subject to changes in contact force if severe wear occurs during a test [11].

Pin-on-Disc Apparatus

This apparatus is perhaps the most widely used. A pin is pressed against a rotating disc either on its flat surface, as shown in Figure 3.1, or on the circumference of the disc. The latter configuration is sometimes referred to as pin-on-drum. The dimensions of the pin and disc depend on the type of tests conducted. The pin-on-disc apparatus offers far better control of experimental conditions and for this reason is becoming increasingly used in preference to other tribometers. The apparatus allows experiments to be conducted under relatively steady conditions without systematic variations, e.g. in contact area or sliding speed, as they often occur in other tribometers. One of the problems encountered in these tribometers is the control of normal load on the pin. Initially, at the beginning of the test, the surface of the disc is smooth but later the transfer films may accumulate on the surface or wear roughening and wear anisotropy may take place causing the pin to vibrate or bounce over the surface of the disc. Wear anisotropy is often found in ceramics where one part of a ceramic disc wears more than other parts because of the change in direction of sliding relative to grain orientation. For consistent experimental data hanging weights should not be used as a load system and instead a spring or hydraulic system (with a pressure stabilizer) should rather be applied [12].

Pin-on-Slab Apparatus

This apparatus provides a convenient means of studying reciprocating sliding as opposed to unidirectional sliding which is studied on the pin-on-disc apparatus. The Bowden-Leben machine is probably the most famous application of the pin-on-slab principle and has been used in a large number of research projects [13]. A more recent development is the Cameron-Mills machine which is the short amplitude high frequency version of the pin-on-slab device [14]. The distinction between the two types of pin-on-slab device is schematically illustrated in Figure 3.2.

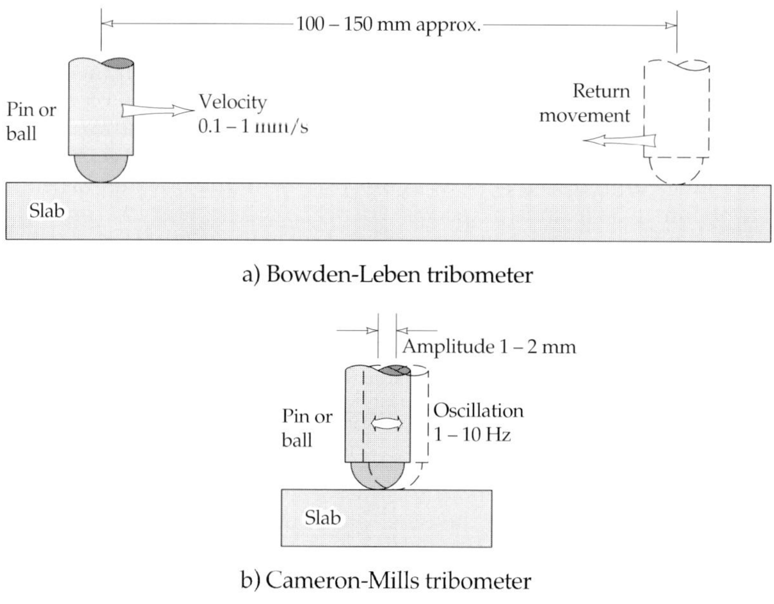

Figure 3.2 Distinction between Bowden-Leben long amplitude low frequency pin-on-slab tribometer and Cameron-Mills high frequency short amplitude pin-on-slab tribometer.

The Cameron-Mills arrangement has the advantage of a high repetition rate of sliding contact cycles which causes friction and wear processes to rapidly reach steady state conditions as opposed to the initial stages of wear and friction. On the other hand, the Bowden-Leben tribometer tends to produce friction and wear data relating only to the initial stages of wear because the pin makes fewer traverses over the slab in a specific period of time. Surface temperature rises are accurately controlled (and suppressed) by the limited sliding speeds involved in both devices. Also in both pin-on-disc and pin-on-slab tribometers the specimens, i.e. pins and discs or slabs, can be manufactured from a variety of materials, as opposed to standard materials, e.g. the steel balls, required in the four-ball apparatus.

Design of new tribometers for dry and partially lubricated sliding contacts

The existing designs of tribometers for dry and partially lubricated sliding contacts are largely based on arbitrary choices or convenience. In some situations it might be desirable to design a new type of tribometer dedicated to specific tests. The fundamental question is, what constitutes a good tribometer or what are the basic requirements for a good tribometer? The design of any tribometer should ensure sufficient mechanical rigidity to prevent vibration when there is a rapidly varying friction force and to allow accurate measurement of wear. It is often found that wear measurements based on movement of a specimen are corrupted by deflection of the structure surrounding a specimen. Contact load can be applied either hydraulically, pneumatically, by a spring or by a hanging weight. As mentioned already when a hanging weight is used, some form of weight suspension should be applied. This practice is most important when sliding on a rough surface is involved. In many cases it is usually sufficient to suspend the weight by a spring. If there is no suspension for the weight then the contact force will vary by an unknown amount whenever surface roughness or transfer films impose acceleration of the sliding surfaces in a direction normal to sliding. When low frictional temperature rise is required then it is often necessary to apply an extremely low sliding speed such as is practised with the Bowden-Leben machine [13]. To obtain reliable data a specialized drive system that can provide steady sliding at very low speeds, e.g. 1 [mm/s] or less with no backlash or vibration, is required.

3.4 TRIBOMETERS FOR ABRASIVE, EROSIVE AND CAVITATION WEAR

This class of tribometers is radically different in design from the tribometers used to study dry or partially lubricated sliding contacts. There may, however, be some exceptions. For example, a pin-on-disc or pin-on-slab tribometer can be used in the study of 2-body abrasive wear by attaching abrasive paper to the disc or slab or preparing a special disc covered with abrasive grits. In this case the pin becomes the sole wearing specimen. On the other hand, in studies of 3-body abrasive and erosive wear it is necessary to handle a stream of abrasive particles which either flow past or impinge against a test specimen. In erosive wear tests abrasive particles are usually propelled by an air, gas or liquid jet against the stationary sample while in abrasive wear tests the specimen can be moved in order to achieve flow of the abrasive particles over its surface.

In erosive wear tribometers, selection of nozzle parameters is important since abrasive particle speeds and trajectories can affect the data obtained. For example, it has been found that in tribometers where the erodent particles are delivered by a gas stream from a nozzle, the roughness of the nozzle interior surface needs to be carefully controlled [15]. An excessively rough nozzle causes the stream of particles emanating from the nozzle to fan out at a wider angle and a greater spread in the distribution of particle velocities [15]. In other words, particles are delivered to the test specimen with a large range of impingement velocities and impingement angles which complicates the interpretation of experimental results. To prevent progressive roughening of the nozzle interior surface, a hard erosion resistant material must be used for the nozzle interior.

Studies of erosive wear by liquids sometimes require a pumping system capable of propelling a liquid as a jet at speeds greater than 100 [m/s]. Also studies of cavitation wear require a specialised apparatus. Cavitation can be induced by rapid motion of a test specimen in fluid or more conveniently by ultrasonic vibration of an immersed specimen. The basic types of apparatus for the study of abrasive, erosive and cavitation wear are schematically illustrated in Figure 3.3.

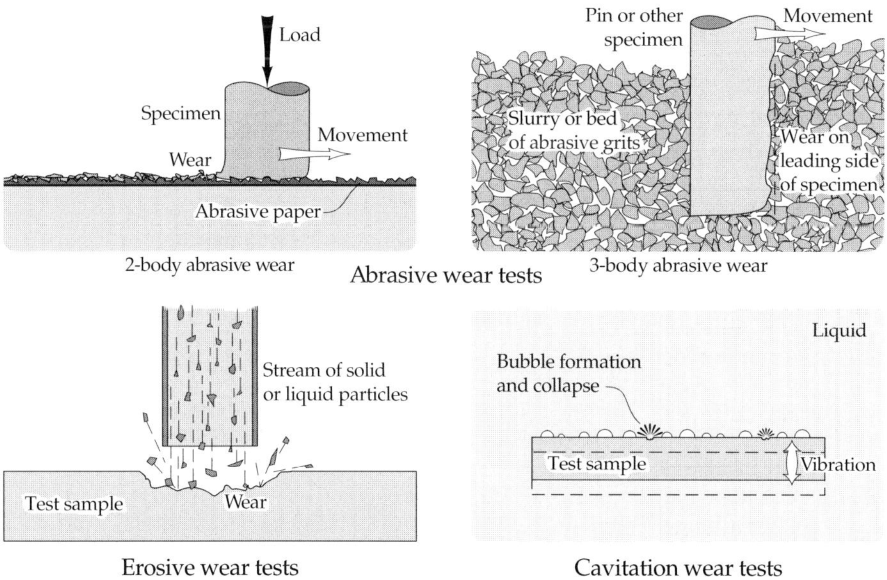

Figure 3.3 Schematic diagrams of the tribometers used in the studies of abrasive, erosive and cavitation wear.

The design of tribometers used for the studies of abrasive, erosive and cavitation wear depends on requirements of the specific research program. For example, to study a mechanism of 3-body abrasive wear a special apparatus which enables observation of the movement of one abrasive grit between two constraining plates has been built [16]. A schematic diagram of this apparatus is shown in Figure 3.4.

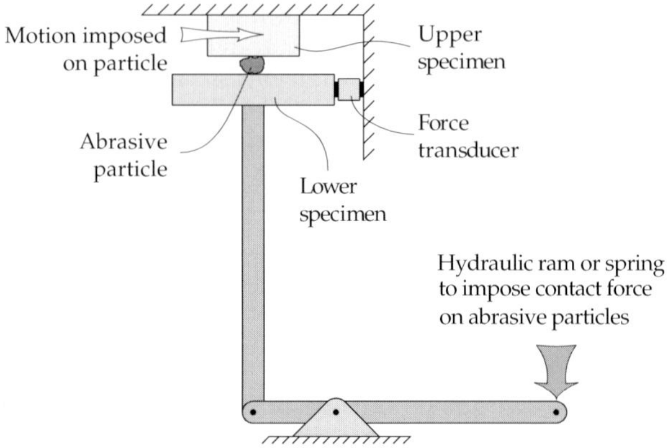

Figure 3.4. Schematic diagram of an apparatus to study mechanism of 3-body abrasive wear (adapted from [16]).

In this apparatus the pattern of rolling and sliding of the grit can be observed and later matched with the wear damage of the constraining plates. Apparatus of this kind allows the characteristics of 3-body abrasive wear, which are now known to be significantly different from 2-body abrasive wear, to be investigated.

In another apparatus a steel ball is rotated against a metallic wear plate with slurry passing between the contact point, as schematically illustrated in Figure 3.5. Slurry is made up of abrasive particles suspended in water solution. Wear scar generated on the test sample surface is in the form of a crater. The wear rates are assessed by measuring the crater diameter and determining the volume of material removed from the crater. Information about the wear mechanisms acting can be obtained from microscopic examination of worn surface morphology. This test rig has originally been designed and is mainly used for measurements of coating thickness and coatings abrasive wear resistance [17]. For these applications the ball diameters used are about 25 [mm] and the abrasive particles are less than 10 [μm] [17,18]. The design was subsequently modified to accommodate a bigger diameter ball, e.g. 41 [mm], which in turn allows studies of 3-body abrasion of wear resistant materials, or effects of particle parameters, e.g. size and shape, on wear [19]. The main advantage of a ball cratering apparatus is that it is simple to use, gives repeatable data and the tests are inexpensive.

In abrasive, erosive and cavitation wear tests, synergistic interaction with corrosion can be additionally studied by adding the appropriate corrosive fluid to the particle slurry or to the jet stream of particles. Also the effects of physical properties of liquid medium, e.g. viscosity, can be investigated by testing fluids with different viscosity.

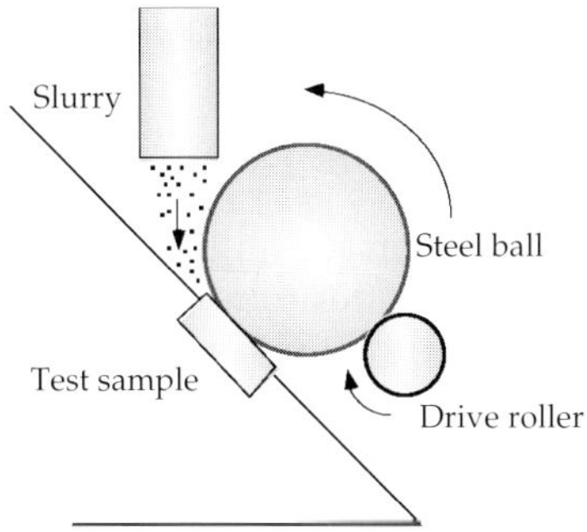

Figure 3.5 Schematic illustration of the ball cratering apparatus (adapted from [20]).

3.5 TRIBOMETERS FOR NON-AMBIENT CONDITIONS

The tribology of dynamic contacts situated in non-ambient conditions such as the extremes of temperature or a vacuum is of great practical and scientific interest. For example, liquid propellant rocket motors require rolling bearings to operate in a flow of liquid oxygen and it is essential to make sure that these bearings will not fail by wear before all the fuel in the rocket is burnt [21]. The performance of the adiabatic internal combustion engine or advanced hypersonic engine depends on the development of low friction coatings for ceramics operating at high temperatures [22-24] and clear understanding of ceramic tribology at elevated temperatures is needed for any progress to be made. Modern technology increasingly depends on reliable operation of mechanical components in severe environments, e.g. nuclear reactors, aircraft, space vehicles, etc. The dependence of component reliability on controlled wear and friction necessitates tribological studies of dynamic contacts in a wide variety of non-ambient environments.

A tribometer for non-ambient conditions allows the standard tribological measurements to be conducted in extreme environments, e.g. at high or low temperature or in a vacuum, etc. The basic mechanics of the dynamic contact, e.g. contact stress and sliding speed, remain unchanged by the presence of non-ambient conditions. Design considerations of the tribometer imposed by non-ambient conditions relate to isolation of the observer, instrumentation, load and power systems from the harmful effects of an extreme environment. A common example of this requirement is adequate thermal insulation of the driving shaft and electric motor from a tribometer operating in high temperatures [25,26] or in vacuum [27]. The rig components subjected to high temperatures must be fabricated from temperature resistant materials such as ceramics or nickel super alloys with thermal barrier coatings (TBC) [26,28]. Careful design is often required to ensure that friction and wear transducers can record accurate data while being protected from the harmful effects of temperature, corrosive chemicals, radiation and magnetic fields. The design of the experimental chamber must allow the execution of the experiments at precisely maintained conditions, e.g. at high temperatures, in high vacuum, while the drive and measurement systems are located outside the chamber. It is also important to ensure that the data recorded is not subject to any bias from e.g. magnetic fields or vibration emanating from the test environment.

High Temperature Tribometers

Ceramics are very attractive materials from the engineering view point due to their high strength, heat and wear resistance, chemical inertness, low thermal conductivity and maintenance of good mechanical properties at elevated temperatures. Thus there is an increasing interest in the use of advanced ceramics in internal combustion engines and gas turbines where temperatures up to 1000°C and 1400°C are expected respectively [28]. As conventional liquid lubricants break down at much lower temperatures, ceramic lubrication based on lubricious solid coatings is usually investigated at temperatures above 800°C [23,24]. To attain the temperatures range where advanced ceramics are intended to replace metals in many engineering applications and conduct the necessary tribological tests a specially designed and built tribometer is required [e.g. 25]. Some of the tribometers originally designed for ambient temperature applications, such as pin-on-disc machine or pin-on-slab apparatus, can be adapted to operate at moderate temperature by fitting the heating elements and insulating the test chamber. These modifications can be introduced without much effort, are relatively inexpensive and allow testing at temperatures up to 500°C [e.g. 24] to be conducted. For higher temperature ranges purpose built tribometers become necessary [e.g. 22,23,25,28]. These tribometers can be used for wear and friction measurements at temperatures as high as 1200-1500°C [26-28].

To account for various environments high temperature tribometers are designed to operate in vacuum [28,29], air [24,26,28] or other controlled gas atmospheres [23,26,28]. Modern tribometers often include computer data acquisition systems [26]. Most tribometers use a furnace or an insulated chamber with various heating elements (Kanthal, SiC) to obtain the desired high temperature in the specimen chamber [23,24,26,28]. In one work an electron beam heater was used in an ultrahigh vacuum apparatus to obtain temperatures as high as 1200°C [27]. The temperature is usually monitored by thermocouples placed in close proximity to the test specimens [23,25,26,28,29]. However, some tribometers are equipped with special windows allowing temperature measurements using infrared pyrometers to be made [24-26]. An example of the apparatus capable of reaching the temperatures of about 1000°C during friction and wear tests is schematically illustrated in Figure 3.6. A water cooled shaft provides isolation of both the bearing power system and instrumentation from the hot test chamber. The test specimen, a pin, sits at one end of a water cooled shaft which acts as a cantilever.

This apparatus has been used to measure friction and wear characteristics of ceramics over the temperature range from ambient to 1000°C. It was found that the coefficient of friction for many ceramics reaches a maximum around 500°C with moderate friction and wear around 100°C. This was an important finding since many ceramics are used in high temperature applications. It would not have been possible to detect this ceramic behaviour if only low temperature wear and friction experiments were performed.

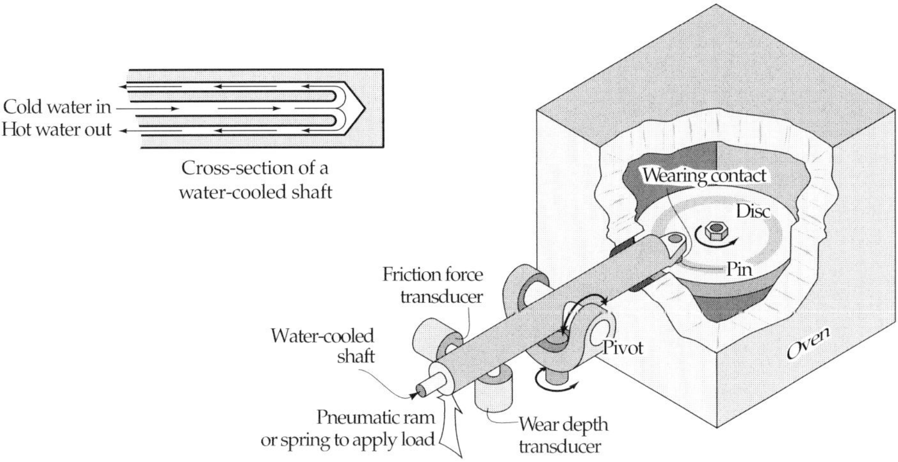

Figure 3.6 Schematic diagram of the high temperature tribometer (adapted from [25]).

Low Temperature Tribometers

The tribological properties of materials and lubricants are also tested at low temperatures, e.g. for polar conditions and in some cases at extremely low temperatures, e.g. for space applications [e.g. 30]. Schematic diagram of a tribometer to study the tribological behaviour of various materials under reciprocating sliding or fretting conditions at extremely low temperatures, about 4K, is shown in Figure 3.7.

In this tribometer liquid nitrogen and liquid helium were used to cool the contact between the materials tested. One of the problems facing the designer was to isolate the liquid nitrogen or helium from all moving parts apart from the dynamic contact. In the tribometer shown in Figure 3.7, long shafts were used to drive the mobile wear specimen and restrain the stationary specimen. Sufficient distance was maintained between the insulated flask containing liquid nitrogen or helium and the instrumentation allowing a stable temperature difference between the end of the shaft in contact with the cryogenic fluid and the other end connected to the drive system and friction transducer to be maintained. In some other designs of tribometers for extremely low temperatures, the dynamic contact is not immersed in a cryogenic fluid, instead gas (e.g. oxygen) is chilled by passage through liquid nitrogen before reaching the contact [31]. Since the tribometers for extremes of low and high temperatures conform to the same basic design principles, i.e. insulation of the heating or cooling source from the instrumentation, bearings and power system, they can be built, in principle, for the tests at both high and low temperature extremes [e.g. 32].

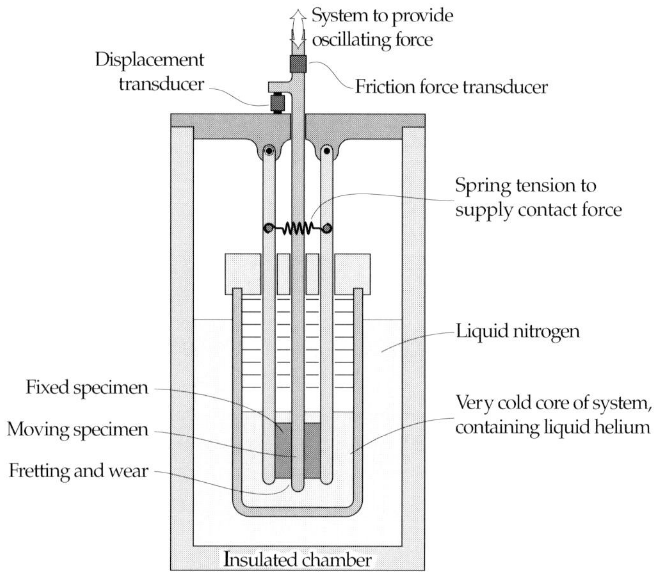

Figure 3.7 Example of tribometer for extremely low temperatures (adapted from [30]).

Tribometers for Operation in Vacuum

Studies of friction, wear and lubrication under vacuum can be classified into the following groups:

- fundamental studies of wear mechanisms, friction and lubrication, e.g. surface film formation, often involving nascent surfaces,
- wear, friction and lubrication of materials for space applications.

The latter studies are of great importance to space vehicle technology [33]. A tribometer for tests under vacuum usually consists of a vacuum chamber which contains the dynamic contact and surrounding ancillary equipment such as vacuum pumps and a drive system for the dynamic contact [e.g. 27,34-36]. The shaft driving the dynamic contact passes through the wall of the vacuum chamber via a seal to the electric motor. Friction and wear transducers can be fitted either inside the vacuum system as the lack of air does not usually interfere with the transducers [28,29,34] or outside the vacuum system. A schematic diagram of tribometer for vacuum operation is shown in Figure 3.8.

Some of these tribometers are fitted with heating devices allowing the tests to be conducted at elevated temperatures [28,29]. Often these tribometers are made very small to fit into the Scanning Electron Microscope (SEM) [36], Auger Spectrometer (AS) [35] or Electron Spectrometer for Chemical Analysis (ESCA) [29] as these devices allow direct observation of wear mechanisms and the formation of surface films taking place during the wear processes. A schematic illustration of tribometer for vacuum operation is shown in Figure 3.9.

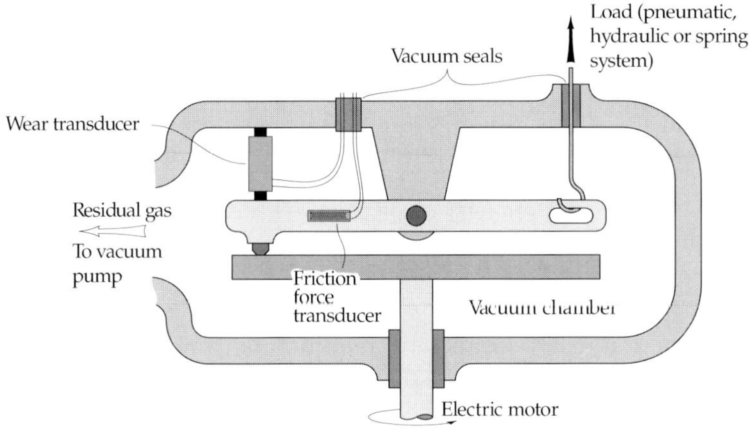

Figure 3.8 Schematic diagram of tribometer for operation in vacuum (adapted from [34]).

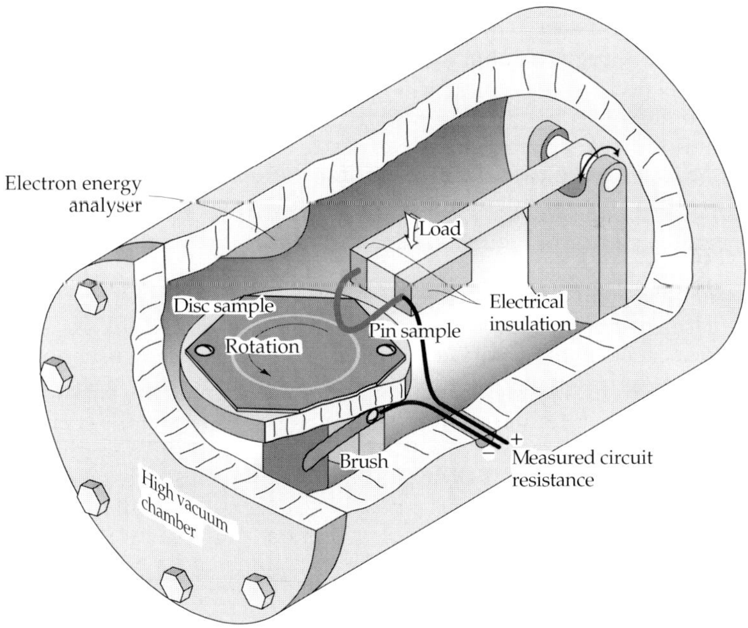

Figure 3.9 Schematic illustration of tribometer for operation in vacuum (adapted from [27 and 35]).

Friction coefficients tend to be more unstable in a vacuum than in open air since as soon as a protective oxide or lubricant film is removed by wear, the friction coefficient of most metals rises rapidly and becomes very large. For this reason a support structure to the dynamic contact, which is usually fitted with strain gauges and functions as a frictional transducer, should be sufficiently strong to withstand sudden rises in friction force. In these tribometers the lack of air prevents any cooling of the dynamic contact by convection so that frictional heat can only be dissipated by conduction and radiation. Also the lubricants in particular tend to release vapour when heated by friction and this affects the residual pressure in the chamber. Thus, in the design of these tribometers a provision has to be made for the adequate dissipation of heat, i.e. the application of a cooling system, and removal of oil vapours, i.e. the installation of a high capacity vacuum pump.

Tribometers for Operation in Corrosive Liquids

As the mechanical components are required to function in increasingly aggressive environments there is a gradually increasing demand for tribometers operating in corrosive fluids. For example, demands for pollution control lead to a need for extensive testing of mechanical seals in the presence of corrosive fluids [37]. Wear of seals is a direct cause of leakage of pollutants and cannot be tolerated and wear under these conditions needs to be investigated and controlled. Many important industrial wear problems involve corrosive slurries, as for example wear encountered in petroleum extraction pumps located at the bottom of petroleum wells [38], wear of pumps circulating hot nickel sulphate solution during nickel refining, wear of pumps circulating alkaline bauxite slurries in alumina production, etc. A typical tribometer used to simulate wear under these conditions contains a tub with a corrosive fluid which isolates the fluid from the instrumentation and driving and loading systems. The example of such a tribometer, based on a pin-on-disc design, for the operation in aqueous acid and alkali solutions at room temperature is schematically shown in Figure 3.10 [39].

In the apparatus shown in Figure 3.10 load is applied by forcing the tub assembly upwards against a rigidly mounted shaft which drives the disc. Wear is assessed from the measurements of the tub displacement and friction force is determined from torque measurements between motor and the driving shaft. As many wear problems associated with corrosive slurries involve non-ambient temperatures, a heating system is often provided in order to conduct the tests at elevated temperatures.

3.6 SPECIAL PURPOSE TRIBOMETERS

Tribometers for Fretting Studies

Although the simulation of fretting contact is relatively simple, e.g. a rotating shaft subjected to bending can provide a fretted contact surface, a precise control of fretting movement is more difficult to achieve and very small sliding amplitude fretting studies are usually performed on specialized apparatus. There are two basic types of tribometer developed to simulate fretting contact. The most common type consists of a clamped contact where microsliding is induced by an oscillating load on one of the specimens which cyclically stretches and contracts in response to the load. This type of tribometer is suitable for the simulation of microsliding contacts at very short amplitudes of about 1 [μm]. Larger amplitudes of about 60 [μm] can also be obtained by lengthening the stretched specimen [e.g. 40]. A schematic diagram of this tribometer is shown in Figure 3.11a.

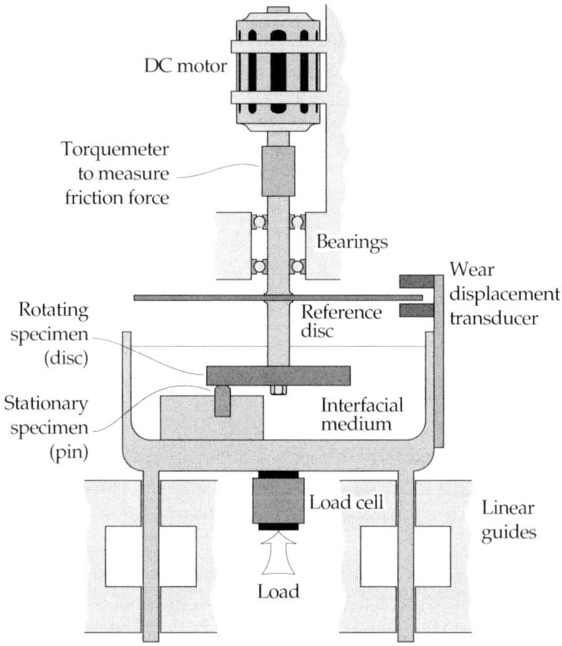

Figure 3.10 Schematic diagram of a tribometer for the operation in corrosive fluids at ambient conditions (adapted from [39]).

In the second type of tribometer a microsliding contact is driven by an electrodynamic exciter. Electrodynamic exciters provide precisely controlled high frequency reciprocating movement at small amplitudes, ranging from 10 [μm] up to several millimetres, and have been used for fretting studies [e.g. 41]. In some other designs crank mechanisms are used to drive the fretting contact. This type of tribometer used for simulation of microsliding contacts is illustrated schematically in Figure 3.11b.

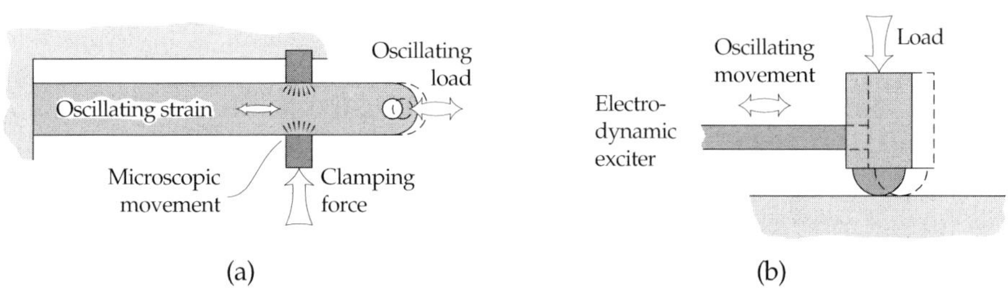

Figure 3.11 Schematic diagrams of basic types of tribometer for simulation of fretting contacts; a) clamped surface type, b) reciprocating slider type.

The two types of fretting tribometers described serve different experimental objectives. The clamped surface-oscillating load type tribometer is mostly used in fretting fatigue studies, i.e. studies of the effect of fretting on material fracture, while a small amplitude reciprocating sliding tribometer is often used in the studies of friction and wear in the gross slip fretting regime. The distinction between the two categories of these tribometers is illustrated schematically in Figure 3.12.

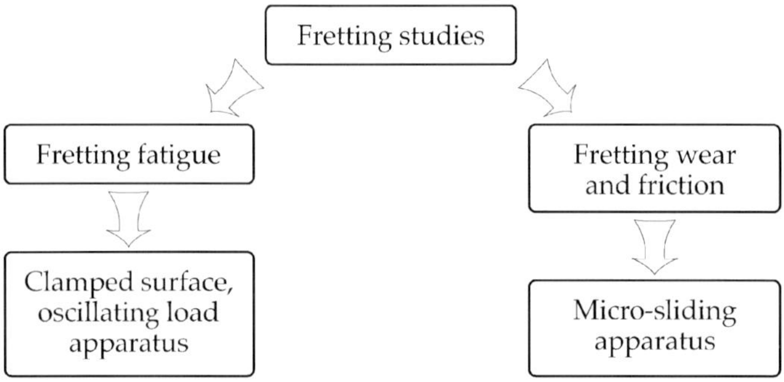

Figure 3.12 Classification of fretting tribometers according to experimental objective.

The temperature increase in fretting contacts is usually very small and therefore the tests can be conducted at a high frequency of oscillation in order to save time without the risk of frictional heat influencing the wear mechanism [42]. For example, it was found during fretting experiments at ambient temperature using a 12.5 [mm] steel ball oscillated against a sapphire disc, with frequency of 150 [Hz] and an amplitude ranging from 20 to 100 [μm], at a load varying between 4 [N] and 20 [N] that the frictionally induced temperature did not exceed 40°C [42].

In most cases, in fretting contacts the level of temperature rise does not cause fundamental changes in the plastic deformation and metal oxidation which control fretting wear and friction. There may, however, be exceptions particularly during tests close to the transition temperatures of metal oxidation. For example, the oxidation rates of steel change considerably around 200°C so that tests at temperatures in the range of 150-200°C should include checks for any experimental artefact caused by frictional temperature rise. A similar problem may occur when the effect of fatty acids and other adsorbents on fretting is investigated. For tests at room temperature, desorption of adsorbents caused by frictional temperature rise would be unlikely although it is still possible for some adsorbents. During tests at around 100°C desorption induced by frictional heat may also occur.

Although most fretting studies have been performed in ambient conditions, many tests have also been performed at high and cryogenic temperatures, or in the presence corrosive fluids. The design of the tribometer must be adjusted to suit the particular environment according to the already outlined design principles of tribometers for extreme environments.

Tribometers for Extremely High Sliding Speeds - Melting Wear Studies

Most of the tribometers already discussed are designed with a limiting speed in the range of 10 to possibly 30 [m/s]. In theory with careful design to control imbalance forces, it might be possible to design a tribometer with a sliding speed limit up to 100 [m/s] but some sliding phenomena occur at yet higher sliding speeds. With the new generation of high speed trains developed in France and elsewhere, sliding speeds between the power cable and pantograph currently reach 140 [m/s] and may attain even higher levels in future. When a gun is fired, the projectile inside the muzzle reaches speeds exceeding 500 [m/s]. In both cases, friction and wear at high sliding speeds need to be investigated experimentally to ensure reliable performance of the mechanical system.

The first tribometer allowing sliding speeds of about 1000 [m/s] to be reached was designed by Bowden and Freitag and is schematically illustrated in Figure 3.13 [43]. It can be noticed that the design of this tribometer differs considerably from the commonly used, low sliding speed, tribometers discussed so far [43,44]. The moving specimen is a steel ball which is suspended magnetically above the stationary wear specimens and rotated by an oscillating magnetic field.

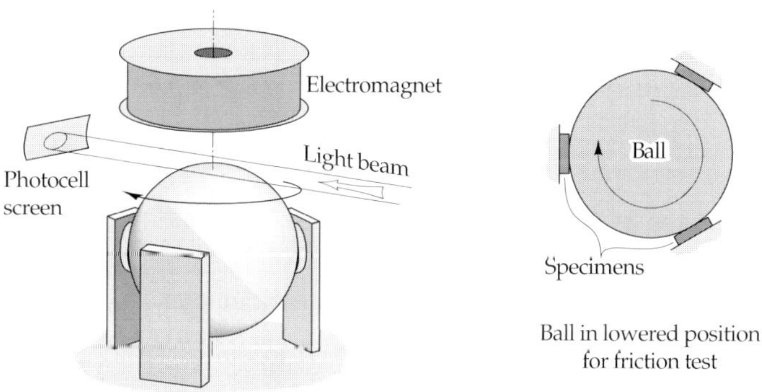

Figure 3.13 Schematic illustration of the operating principle of a high sliding speed tribometer of about 1000 [m/s] (adapted from [43]).

Friction measurements are obtained by momentarily lowering the spinning ball and suddenly clamping the three copper pads on the side of the ball as illustrated in Figure 3.13. A graph of friction changes vs velocity is obtained. The ball velocity is measured by applying a light beam and a photocell while from the angular deceleration and inertia of the ball the friction force is calculated. Contact forces are found from the difference between the weight of the ball and the reduced magnetic suspension force. Wear is determined from the size of the wear scars on the stationary specimens at completion of the test. The test is by necessity of short duration but the extreme sliding speed ensures that sufficient sliding distance for valid results is obtained. The contaminant films and debris are removed during the early stages of contact between ball and stationary pads and thereafter, friction and wear coefficients typical of long term trends are established. Due to the extremely high velocity frictional energy input into the surface is also very high, causing melting of the surface layers of the copper pads resulting in a reduced coefficient of friction. Hence this apparatus was used, for the first time, to study the phenomena of surface melting in high speed metallic contacts [43,44].

Tribometers for Impact Wear Studies

The design of tribometers for impact wear studies usually involves a mobile sample impacting against a flat or hemispherical counter-sample [45]. A typical tribometer for these studies allows the impact energy to be varied and contains a force transducer to measure impact force. Some of the tribometers are also able to accept specimens of varying geometry. Although almost all impact wear studies have been confined to ambient conditions there appears to be no reason why an impact wear tribometer should not be designed for tests in extreme environments. A schematic diagram of impact wear tribometer is shown in Figure 3.14.

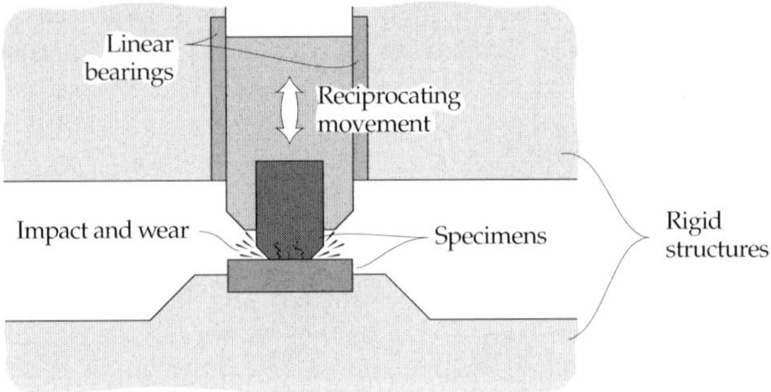

Figure 3.14 Schematic diagram of impact wear tribometer.

The design of an impact wear tribometer usually does not pose any severe problems providing that certain conditions are observed in order to obtain valid experimental data. Sufficient rigidity of the tribometer is necessary to prevent microsliding between impacting specimens. Some deflection of the support structure to the specimens during impact is unavoidable but it should be limited to the movement in the direction of impact. If a lateral movement caused by flexure of the support structure is allowed to occur, then microsliding between the specimens may result and fretting wear may then become a significant mode of wear during testing. Microsliding between specimens induced by lack of structural rigidity is illustrated schematically in Figure 3.15.

Impact tribometers must be made sufficiently robust to withstand the repetitive impact forces which render the specimen support structures prone to mechanical fatigue.

Tribometers for Combined Rolling and Sliding

Most tribometers are designed to simulate either pure sliding or pure rolling as these conditions represent easily identifiable limits of severity in wear and friction. However, many mechanical components contain dynamic contacts

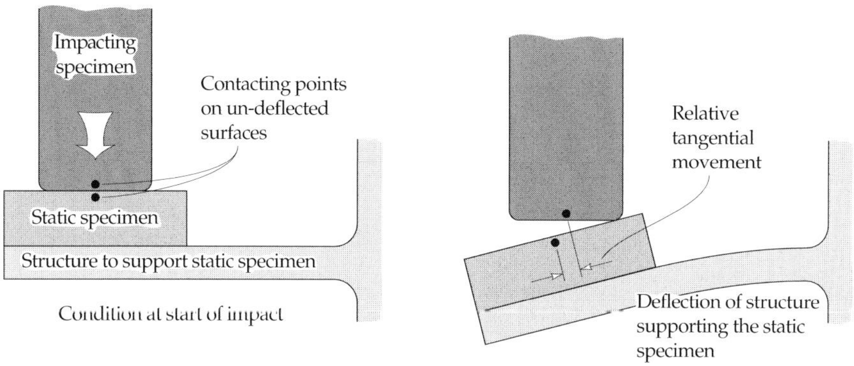

Figure 3.15 Microsliding between specimens occurring during impact.

which function under a combination of rolling and sliding and experiments conducted under conditions of either pure sliding or pure rolling do not always accurately simulate the friction and wear phenomena occurring. The typical example of combined rolling and sliding can be found between railway wheels and rails [46,47] or at the interface between meshing gears. The most commonly used design of the tribometer to simulate the combined rolling and sliding is the two-disc apparatus. In this apparatus two discs in contact are driven at different angular velocities which superimposes some sliding on rolling at the dynamic contact [48]. The amount of sliding is controlled by manipulating the velocity of one of the discs, from almost pure rolling, i.e. the discs rotate at the same circumferential speed, to pure sliding, i.e. one of the discs is stationary. The apparatus allows experimental data on relative wear resistance and behaviour of materials under combined sliding and rolling to be obtained. The apparatus is also often used to measure the traction coefficients of lubricating oils under elastohydrodynamic conditions and also to study scuffing in lubricated contacts. A schematic diagram of the two-disc apparatus is shown in Figure 3.16.

One of the discs is mounted in a rigid structure while the other disc is free to pivot and it can be loaded against the fixed disc. The sliding component of velocity can be obtained by using gears to impose a ratio of angular velocities on the discs or by varying the ratio of disc diameters [e.g. 48]. In other designs, the discs are driven independently so that there is no fixed ratio of disc rotational speeds [e.g. 49]. The latter method does not require a range of disc sizes and hence gives more flexibility in experimentation. Contact load is usually applied to the pivoted disc by, for example, hydraulic and pneumatic loading or by a hanging weight. Most two-disc apparatus operate in open air at ambient temperature but some apparatus are fitted with a controlled atmosphere chamber [50]. For example, studies of various lubricants in air and nitrogen atmosphere revealed a complex interaction between atmospheric oxygen concentration, additive chemistry and proportion of rolling and sliding in the contact [50].

Although virtually all experimental projects using two-disc apparatus involve metal to metal contacts, some other material combinations are also used. For example, a metal disc sliding against discs of sandstone was used to study the effect of rolling and sliding on abrasive wear [51]. Ripple formation on abraded metal surfaces was observed at critical ratios of rolling to sliding [51].

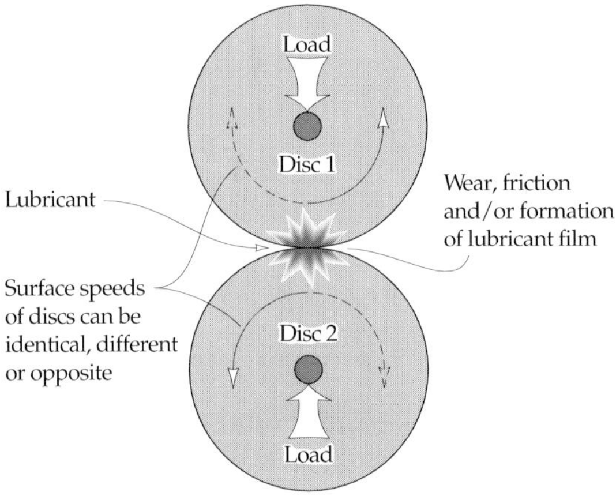

Figure 3.16 Schematic diagram of the two-disc apparatus.

Tribometers for In-Situ Studies of Friction and Wear

Recent advances in measuring technology are gradually making accessible the hitherto obscure contact processes between wearing surfaces. In particular, the development of high-resolution X-ray microscopes with the capacity to resolve hidden detail to a dimension of approximately 1 [μm] has created much opportunity to observe friction and wear processes in-situ [52-54]. A planar map or view of the contact interface can be obtained by allowing the X-rays to pass through both contacting surfaces at normal incidence angle to an X-ray detector located below the wearing contact. The characteristics of the wearing contact are revealed by the variation of X-ray absorption within the contact. Asperity contacts which generate a continuous material path for the X-ray are revealed as dark spots due to higher absorption of the X-rays while the gaps between surfaces are revealed as light areas due to greater transmission of X-rays through the air or vacuum in the gap. The apparatus allowing in-situ studies of wear and friction is schematically illustrated in Figure 3.17 [52].

X-rays are nearly perfect medium for these types of experiment as they efficiently penetrate the wearing bodies without side-effects such as heating, vibration or initiation of tribochemical reactions. Virtually all materials can be penetrated and imaged by X-rays. The main limitation for a tribometer is size, as the entire apparatus must be contained within the confined chamber of an X-ray microscope. Another consideration is shielding the heat emission from a motor and ensuring that ancillary systems and components such as bearings, shafts, motors do not block the path of the X-rays from their source through the contact and finally to the detector. These requirements dictate that a custom-designed and built tribometer has to be used. A critical parameter of the X-ray microscope is the level of current supplied to the Roentgen X-ray source. It is essential to provide current sufficient to enable complete penetration of the sliding components. A high current or strength of X-rays improves the signal to noise ratio of the X-rays passing through the contact allowing higher resolution and better image quality to be obtained. Apart from these limitations, most tribometer configurations can be used in-situ wear experiments with an X-ray microscope.

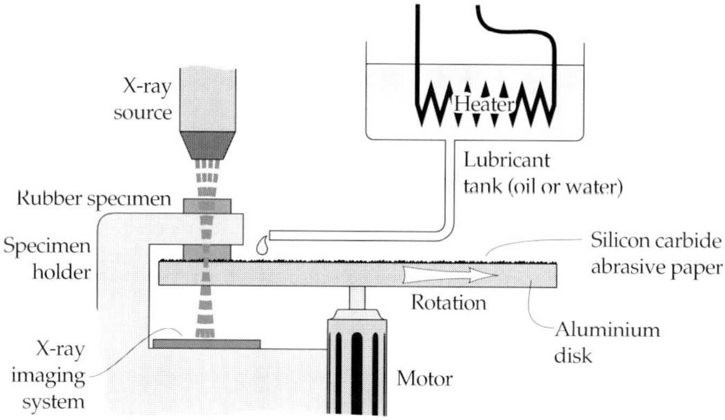

Figure 3.17 Schematic diagram of the apparatus allowing in-situ studies of wear and friction (adapted from [52]).

Another development is the tribometer for in-situ chemical analysis of a wearing contact. The lubricant films controlling friction inside a wearing contact are often so thin that they can only be identified by their chemical content. A micro-Raman spectrometer has been used to observe the persistence and progressive removal of boric acid and molybdenum disulphide films during reciprocating sliding [55,56]. A sapphire hemisphere was used as the transparent sliding member to allow passage of infra-red radiation between the contact and microscope. As there is virtually no restriction on the material for the sliding counterface this tribometer enables a range of realistic bearing materials to be tested. In this system a 514 [nm] Argon laser acts as the source of radiation and an Olympus microscope with long working distance optics is used for observations of the changes occurring in the contact area [57]. The specimen tested is mounted on the reciprocating stage fitted underneath the optics of the microscope. The friction force is monitored by a piezo sensor [56]. The system provides a real-time image of the distribution of specific chemical species forming inside the contact during sliding. It also allows visual monitoring of dynamics of liquid or solid lubricants in the contact [57]. The system is illustrated schematically in Figure 3.18.

The identity of a chemical compound, e.g. boric acid, is determined from an intensity peak at a characteristic wave number. This method of determining chemical compounds is quite similar to the related technique of Infra-Red Spectroscopy. Data from the micro-Raman spectrometer, friction records and electrical resistance measurements were simultaneously collected to show that changes in electrical contact resistance coincided with localised removal of the lubricant film [55,56]. It was also shown that the localised removal of the lubricant film preceded a rise in the friction coefficient by several cycles of sliding. This finding confirmed the model of gradual film failure in solid lubrication as opposed to a sudden fracture of the film. Some other tribometers designed for tribological measurements or observations in-situ have already been described in the section on tribometers for operation in vacuum [e.g. 29,35,36].

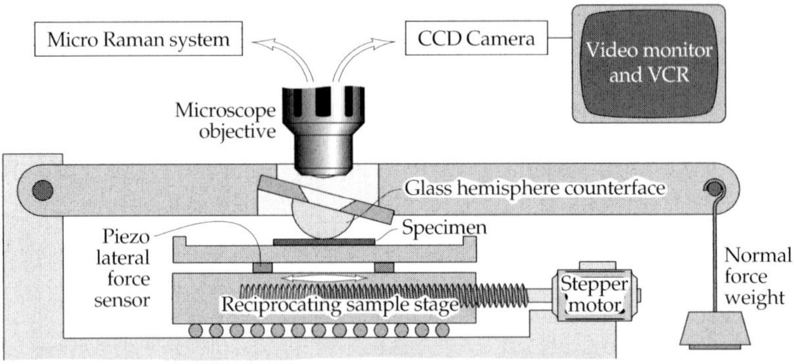

Figure 3.18 Schematic diagram of micro-Raman tribometer for in-situ analysis of chemical content of surface films during sliding (adapted from [57]).

Other Special Purpose Tribometers

A general purpose tribometer is often unable to provide much specialised information about friction and wear. The wear of thin coatings provides a good example of the need for specialised test apparatus. Non-stick coatings on kitchenware are usually based on PTFE and are extremely soft, i.e. with a Vickers Hardness Number less than 10. Wear of the non-stick coating is a common mode of failure of the kitchenware and it is of commercial importance to be able to enhance the wear resistance of the coatings. The relevant wear parameter is the volume of coating that is removed by scratching. This volume is calculated from profilometric measurements of the worn surface, e.g. of a frying pan [58]. In order to accurately determine the wear of a coated frying pan a special purpose tribometer needed to be designed and built. There are two different designs of this tribometer to provide a standard abrasion test for a coated frying pan [59]. In one design, called the 'mechanical tiger-paw test' apparatus, three ball-point pens fixed in a rotating tripod which is mounted on a rotating arm are rubbed against the test pan subjected to a reciprocating motion [59]. In the other design, called the 'Nord test' apparatus, steel-wire whisks are rotated against the test pan, which is irrigated by a slurry of margarine and salt crystals. The schematic illustration of these tribometers is shown in Figure 3.19.

It is found, surprisingly perhaps, that the 'mechanical tiger-paw test' provides better discrimination between different coating materials than the 'more realistic' Nord test. It has been found that the Nord test is sensitive to not only the coating material but also to the coating thickness [59]. Both tests were run continuously for up to 1000 cycles of scratching and with a test temperature in the range of 180 to 200°C. This means that neither test approximated to hand-washing of non-stick pans and it would be interesting to see whether a team of cleaners with the task of scrubbing the non-stick pans would produce comparable results.

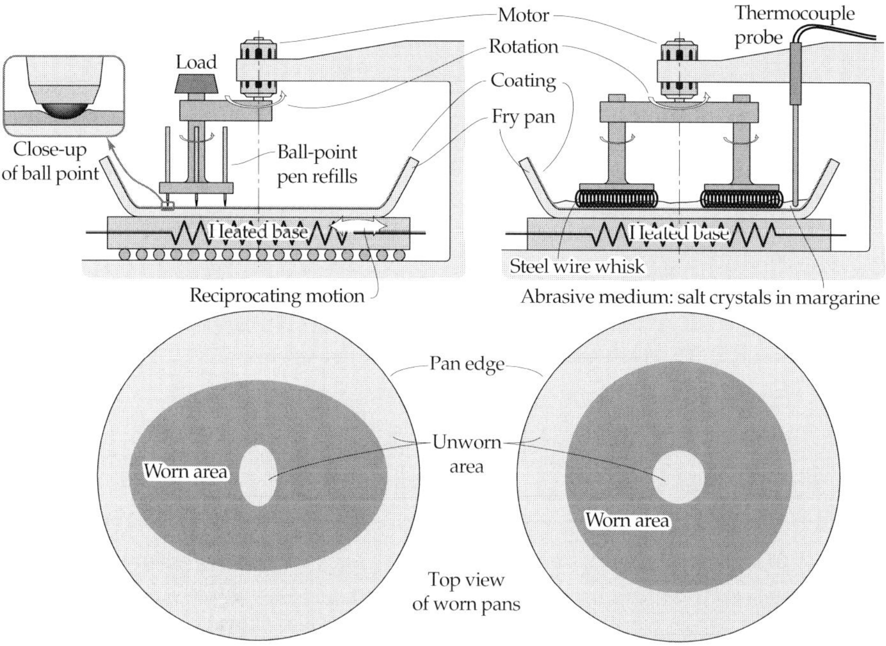

(a) Mechanical tiger-paw test apparatus (b) Nord test apparatus

Figure 3.19 Schematic illustration of the special purpose tribometers for testing wear resistance of non-stick coatings in frying pans (adapted from [59]).

3.7 TRIBOMETERS FOR STUDIES OF LUBRICATION MECHANISMS

Tribometers for the studies of lubrication mechanisms can be described as the test rigs for dynamic contacts where friction and wear is controlled by lubrication. In the apparatus described so far, friction and wear proceed largely uncontrolled and an observational type of experiment is performed. In tribometers to study hydrodynamic lubrication, EHL and boundary lubrication conditions are deliberately arranged to ensure that one of these specific lubrication mechanisms occurs. In almost all studies the experimental objective is to measure the film thickness of lubricant and also the coefficient of friction since these are the critical parameters that dictate the effectiveness of lubrication.

Tribometers for Studies of Hydrodynamic Lubrication

Experimentation in hydrodynamic lubrication has a long history. The early type of experiment involved measurements of either pressure or friction in hydrodynamic bearings [60,61] followed later by studies of vibrations in bearings [e.g. 62] and hydrodynamic film formation [63].

- *Apparatus to Study Wear and Friction Measurements in Hydrodynamic Bearings*

In most cases the bearings used in these types of experiments were the same as used in real engineering equipment. The apparatus usually consisted of a journal bearing with a rotating shaft and bush loaded by a hanging weight. Friction forces were measured from a side arm connected to the bearing that pulled on a spring or balancing weight. This type of test apparatus has periodically been upgraded by successive research groups [e.g. 64,65] and a recent design is schematically illustrated in Figure 3.20.

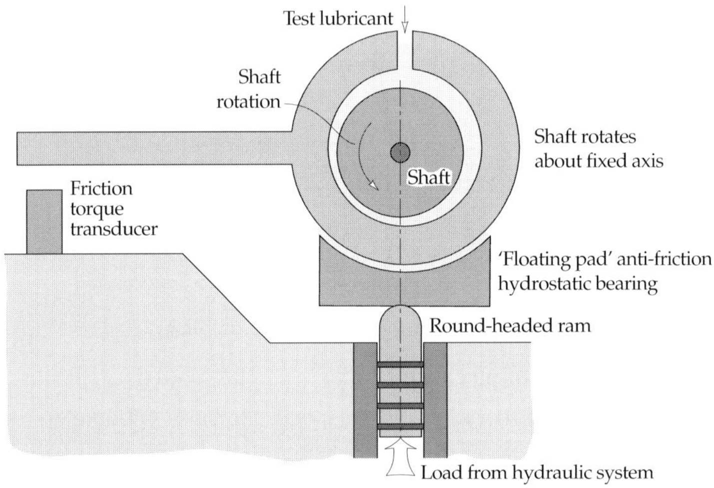

Figure 3.20 Schematic diagram of the test apparatus for studies of pressures and friction coefficient in journal bearings.

The apparatus is fitted with a hydraulic loading system instead of hanging weights allowing tests under variable load to be conducted with computer controlled loading cycles. Friction force is measured by a force transducer. The electrical contact resistance between shaft and journal can also be monitored in order to determine when hydrodynamic film failure occurs. Misalignment between journal and shaft is suppressed by fitting a hydrostatic bearing between the bush and the loading system as shown schematically in Figure 3.20. The curved surface of a hydrostatic bearing allows the bush to pivot until alignment with the shaft is attained.

- *Apparatus to Study Vibration in Hydrodynamic Bearings*

The vibrational characteristics of hydrodynamic bearings have also been investigated experimentally. The experimental apparatus consists of a journal bearing similar to that discussed in the previous section specially modified to induce vibration. The bearing is made to operate at a high rotational speed with low bearing eccentricity while a large rotating mass is connected to the shaft. Under these conditions, a troublesome feature of journal bearings 'oil whirl' or unsteady hydrodynamic film formation, can easily be initiated. Shaft movement is recorded by displacement transducers in contact with the shaft together with the additional data such as friction force which usually increases during vibrations [e.g. 62]. The apparatus is quite costly since a heavy construction is required to withstand the vibrational forces. The schematic diagram of such apparatus is shown in Figure 3.21.

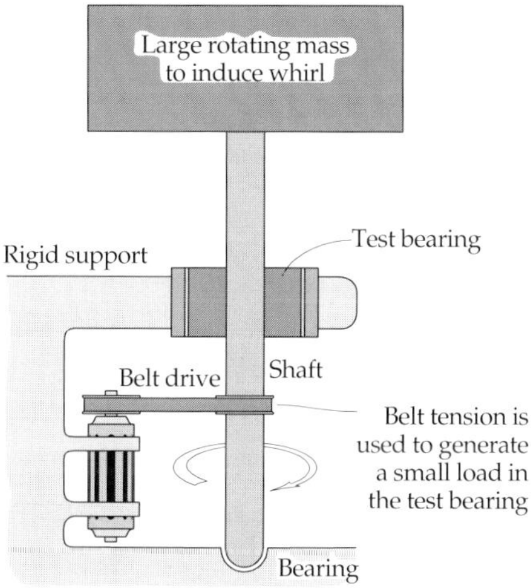

Figure 3.21 Schematic illustration of an apparatus to study oil whirl.

· *Apparatus to Study Hydrodynamic Films in Bearings*

In studies of hydrodynamic lubrication conducted in the early 1950s, film thickness measurement had become of interest as the efforts to extend the operating range of hydrodynamic film formation gained priority. As a result a convenient method of measuring the film thickness in a hydrodynamic contact was developed. The method is based on the application of a transparent material for one of the contacting surfaces and the application of a coloured lubricant. One of the first of these pioneering experiments involved the use of a highly polished metal shaft and a glass bush to observe the hydrodynamic film formation in a journal bearing [63]. The formation of a hydrodynamic film, under the variety of operating conditions, can be studied by observing the coloured oil layer present between the bush and the shaft. The intensity of the colour observed depends on the thickness of lubricant film and hence the latter can easily be assessed. Adding an intensely coloured dye to the lubricant in sufficient quantities allows the investigator to detect and measure even very thin films of lubricant.

The more accurate measurements of the hydrodynamic film thickness were achieved by irradiating the lubricant with a laser emitting visible light. Certain dyes fluoresce under laser irradiation which ensures some emission of colour by even a very small quantity of dye and lubricant. The wavelength of the fluorescence (light emitted by the dye after irradiation by laser) has a characteristic value which facilitates quantitative assessment of the film thickness as it is possible to detect the fluorescence without interference by stray radiation. The technique gives very accurate results and has, for example, been applied to measure the film thickness changes between a piston ring and cylinder liner in an operating internal combustion engine [e.g. 66] and also in many other hydrodynamic contacts [e.g. 67,68]. The studies of hydrodynamic films forming between a piston ring and cylinder liner are quite difficult because the film thickness varies in a complex manner due to changes in sliding speed and gas load on the ring. The principle involved in film thickness measurements by laser fluorescence is schematically illustrated in Figure 3.22.

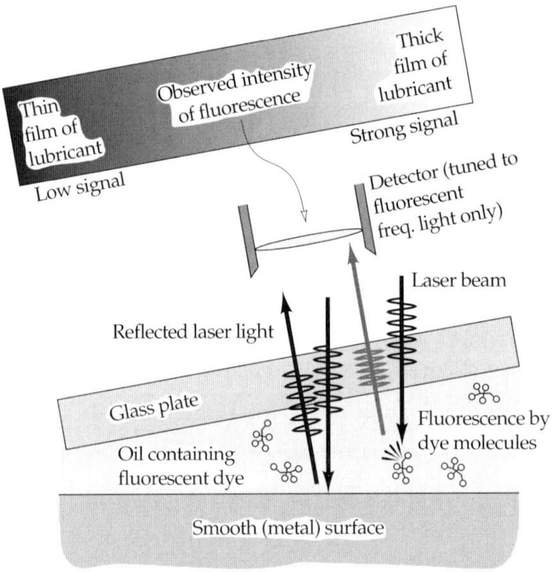

Figure 3.22 Schematic illustration of the principle involved in film thickness measurement by laser fluorescence.

The measurable film thickness is in the range between 5 to 20 [μm] and the higher limit of film thickness can be simply adjusted by varying the concentration of fluorescent dye in the lubricant [69]. More detailed information about laser fluorescence can be found in specialized literature on hydrodynamic and elastohydrodynamic lubrication [e.g. 69].

Tribometers for Studies of Elastohydrodynamic Lubrication

Film thicknesses in elastohydrodynamic lubrication (EHL) are usually so low that the fluorescence technique discussed above is unsuitable. There are several methods available to determine either an average value of film thickness or local values of film thickness within the EHL contact area. Each method has its own particular advantages and disadvantages compared to the other methods.

· *Capacitance Method of EHL Film Thickness Measurement*

Probably the most versatile method used to assess the thickness of the thin films is the capacitance technique. The capacitance of the oil film is measured and film thickness determined by external calibration against a standard oil filled gap between two plates. This technique requires relatively simple equipment to accurately measure film thickness in real bearings [e.g. 48,70,71]. The capacitance method is illustrated schematically in Figure 3.23a. A major limitation of the capacitance method is that the dielectric constant of any lubricating oil, which must be determined before the film thickness measurement, is sensitive to temperature and pressure. This variation in dielectric constant necessitates simultaneous measurements of temperature and pressure in the EHL contact to obtain meaningful results. Capacitance measurements of a lubricating oil film or direct resistance measurements can also be used to determine film thickness in hydrodynamic lubrication [e.g. 64].

· *X-Ray Method of EHL Film Thickness Measurement*

In this technique a beam of X-rays is passed through the interacting surfaces and lubricant film as illustrated schematically in Figure 3.23b [e.g. 72-74]. A relatively low energy X-ray beam is used so that the most of the rays are blocked by the metallic structure of the bearing. However, some X-rays will pass through the lubricated contact and the amount of energy of X-rays transmitted provides a measure of film thickness. With increasing film thickness, a larger quantity of X-rays are able to pass through the film and to reach the X-ray detector. This technique exploits the fact that most lubricants are made of hydrocarbons while most bearings are made of metals which have much higher atomic numbers than carbon and hydrogen. Absorption of X-ray energy by any irradiated material is proportional to its atomic weight, so that high atomic weight elements, i.e. metals, will absorb more X-ray energy than a carbon-based material.

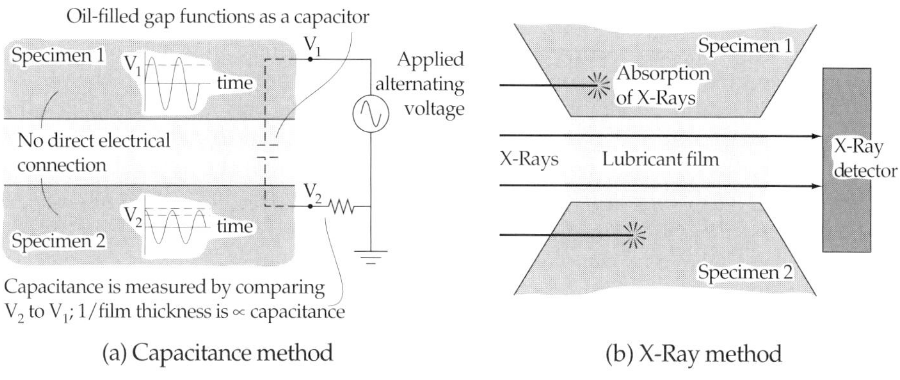

Figure 3.23 Measurement of film thickness in an elastohydrodynamic contact by electrical capacitance and X-ray methods.

In EHL film thickness measurements, X-ray absorption is equivalent to non-transmission or opaqueness to X-rays. A similar principle applies to medical imaging of the human body by X-rays. Bones contain calcium which has a higher atomic weight than the carbon and hydrogen of the surrounding tissues. This difference in elemental composition makes the bones visible by X-ray imaging. Therefore, if a bearing is made of a polymer, e.g. ultra-high molecular weight polyethylene, then the X-ray technique is unlikely to be successful because polymers are also made of hydrocarbons. A major technical difficulty associated with this method is to ensure that the X-rays are precisely aligned tangentially with the plane of the elastohydrodynamic film to prevent misleading results [75].

· *Optical Interferometry Method of EHL Film Thickness Measurement*

The most exact measurements of EHL film thickness are obtained using a specialized apparatus utilizing the principle of optical interferometry [76-78]. The method exploits the fortunate coincidence that the wavelength of visible light is close to the range of EHL film thicknesses occurring in most bearings or gears. In this method a highly polished bearing ball or cylinder rolls on a glass disc coated on one side with approximately 10 [nm] thick semi-reflecting layer of chromium.

To ensure a stationary EHL contact, the ball or cylinder is supported by either an air bearing or set of rollers while the glass disc is rotated. A beam of white light is shone through the glass and a layer of lubricant. The semi-transparent coating allows one fraction of the light to be reflected off the surface of the rolling ball (or cylinder) and a similar fraction to be reflected by the coating itself. As a consequence of this division of incident light by the coating, the reflected light consists of two beams of similar intensity with differing path lengths and this causes an optical interference. When white light is used optical interference generates graduation of colours depending on the film thickness. The thickness of the lubricating film is found from the variation of colour in the interference pattern. However, the film thickness found is the optical film thickness and the real thickness of the lubricating film is obtained by dividing the optical film thickness by the refractive index of the lubricant. The principle of EHL film thickness measurement by means of optical interferometry is schematically illustrated in Figure 3.24.

Since the colour observed corresponds to a particular film thickness the method must be calibrated before its application. Calibration is performed with the aid of monochromatic light. With the monochromatic light, as the speed and the film thickness increase, there is phase change from dark to light and vice versa every quarter of the wavelength. Running the test, by increasing the speed from zero, first with the monochromatic light and then repeating the tests with the white light allows a calibration curve to be obtained, i.e. to correlate the observed fringe colours with the thickness of the film.

The resolution of this method depends on the ability of the human eye to distinguish colours and is then quite poor. Only a few, quite widely spaced film thicknesses can easily be measured and the technique cannot be used to measure films that are less than approximately one quarter the wavelength of the visible light, i.e. of about 80 [nm] [79]. To overcome these limitations, further advances to this technique have recently been introduced. A new method for the evaluation of thickness of very thin films, called an ultra-thin film technique, has been developed [80]. In this technique the surface of the glass disc is coated with a 20 [nm] thick layer of chromium and an additional 500 [nm] thick transparent spacer layer of silica (SiO_2). The presence of this layer allows interference fringes to be obtained even in the absence of an oil film, since the layer behaves as a permanent solid coating of oil [80]. The principle of this technique is illustrated schematically in Figure 3.25. To improve the resolution even further, instead of relying on the human eye, a spectrometer is employed, so that it is possible to determine with much greater accuracy which wavelength have been interfered [80]. The application of colorimetric image analysis to optical interference images enabled to obtain an accurate 3-D map of lubricating film distribution [81]. The combination of spacer layer method and colorimetric image analysis allows for mapping of very thin films in boundary or mixed lubrication regimes [81].

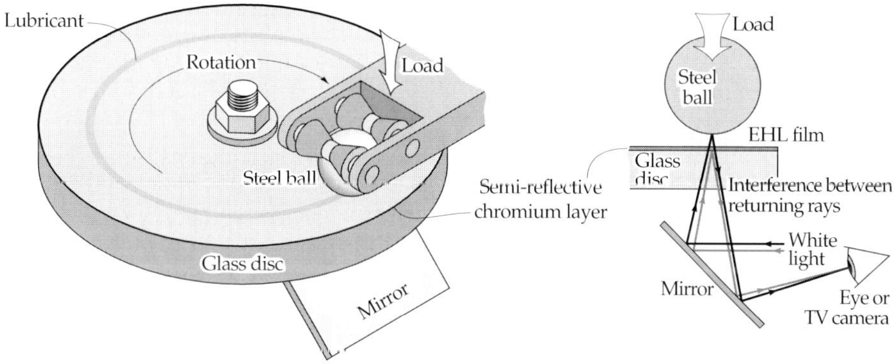

Figure 3.24 Principle of film thickness measurement by the application of optical interferometry to an elastohydrodynamic contact.

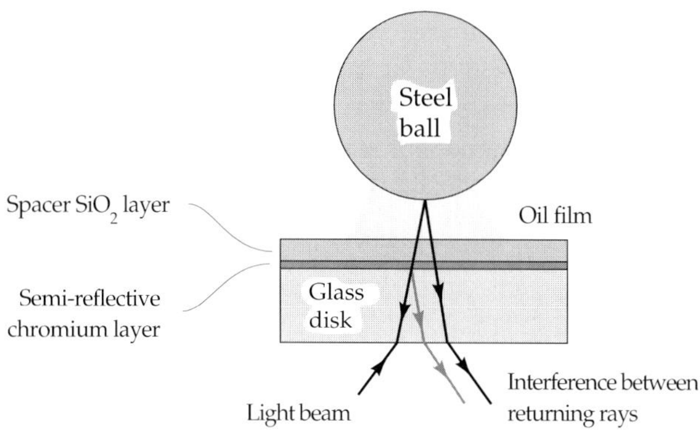

Figure 3.25 Principle of ultra-thin film technique in elastohydrodynamic film thickness measurement (adapted from [80]).

Unfortunately the glass discs are only suitable for moderate contact stresses. For the studies under heavy load a sapphire disc is employed which permits higher contact stresses reaching 7.5 [GPa] [69]. The semi-transparent coating is prone to wear and in the design of apparatus a provision is usually made to move the ball or roller to another 'fresh' track on disc.

The main advantage of optical interferometry is that an accurate and detailed map of film thickness distribution within the EHL contact can be obtained with a relatively simple and cheap apparatus. The variations of film thickness within an EHL contact, which are considerable, are revealed in full clarity by an experimental system which closely simulates conditions operating in a real bearing [77]. Optical interferometry measurements of EHL film thickness can also be used to determine the pressure-viscosity coefficient of lubricants at conditions similar to those operating in the real EHL contacts, i.e. at high shear rates and high contact pressures. Pressure-viscosity coefficient can be determined from the EHL film thickness measurements by working backwards from an EHL film thickness equation. These formulae, for the central and minimum EHL film thicknesses, numerically derived by Dowson and Hamrock [75] are in the following form:

$$\frac{h_c}{R'} = 2.69\left(\frac{U\eta_0}{E'R'}\right)^{0.67}\left(\alpha E'\right)^{0.53}\left(\frac{W}{E'R'^2}\right)^{-0.067}\left(1 - 0.61\,e^{-0.73k}\right) \tag{3.1}$$

$$\frac{h_0}{R'} = 3.63\left(\frac{U\eta_0}{E'R'}\right)^{0.68}\left(\alpha E'\right)^{0.49}\left(\frac{W}{E'R'^2}\right)^{-0.073}\left(1 - e^{-0.68k}\right) \tag{3.2}$$

where:

h_c is the central film thickness [m];

h_0 is the minimum film thickness [m];

U is the entraining surface velocity [m/s], i.e. $U = (U_A + U_B)/2$, where the subscripts '**A**' and '**B**' refer to the velocities of bodies '**A**' and '**B**' respectively;

η_0 is the viscosity at atmospheric pressure of the lubricant [Pas];

E' is the reduced Young's modulus [Pa];

R' is the reduced radius of curvature [m];

α is the pressure-viscosity coefficient [m^2/N];

W is the contact load [N];

k is the ellipticity parameter defined as: $k = a/b$, where '**a**' is the semiaxis of the contact ellipse in the transverse direction [m] and '**b**' is the semiaxis in the direction of motion [m]. The approximate value of this parameter can be calculated with sufficient accuracy from:

$$\bar{k} = 1.0339\left(\frac{R_y}{R_x}\right)^{0.636}$$

where:

$\mathbf{R}_x$, $\mathbf{R}_y$ are the reduced radii of curvature in the 'x' and 'y' directions respectively.

Film thickness generated at a specific velocity is measured and the application of the EHL film thickness formula yields the pressure-viscosity coefficient. It can be noticed that all the variables in the above formulae are known or can be measured.

The principle of film thickness measurement utilizing an optical interferometry can be extended beyond simple rolling through the apparatus which allows transient effects in EHL to be studied by using an impacting ball. A basic characteristic of high speed EHL contacts is that the lubricant, when entrained in the contact, is subjected to extremely rapid changes in pressure. Precipitate changes in pressure can conveniently be studied by allowing a ball to impact on a glass surface and measuring the change in film thickness as a function of time [82,83]. The ball can also be made to spin about an axis inclined to the glass hemisphere (and passing through the centre of gravity of the ball) so that on impact, the effect of shearing in a transient contact can be investigated [84]. The principle of the optical inteferometry method involved in studies of transient EHL films is schematically illustrated in Figure 3.26.

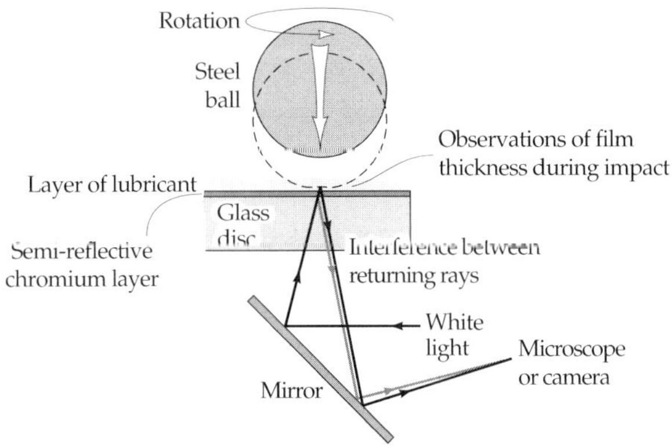

Figure 3.26 Schematic illustration of optical interferometry method involved in studies of transient effects in elastohydrodynamic lubrication at the impact between ball and glass disc.

When the ball impacts against the disc, the elastic deformation of both the ball and disc cause entrapment of the lubricant to occur. A small quantity of lubricant is momentarily retained in a lens shaped cavity that forms between the ball and disc. The dimensions of the cavity can be determined from interferometric film thickness measurements as the depth of the cavity is identical to the lubricant film thickness. The subsequent expulsion of lubricant from the cavity is slower than the formation of the cavity and the change in film thickness at successive intervals can be recorded by high speed photography [82-85]. The mechanism of lubricant entrapment during ball impact is illustrated in Figure 3.27.

During the cavity collapse, the lubricant is subjected to extremes of shear stress between the glass disc or hemisphere and ball. Recorded data of film thickness versus time can reveal much about the rheology and load carrying capacity of the lubricants under these conditions [e.g. 86-88].

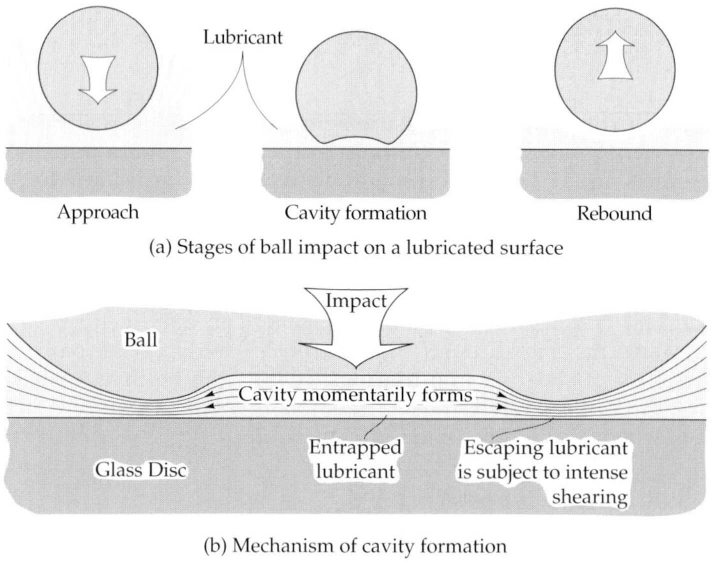

Figure 3.27 Schematic illustration of lubricant entrapment mechanism during ball impact against a flat surface.

Based on these principles an impact microviscometer has been developed [82]. A small amount of the test fluid is placed on a glass hemisphere. A glass ball is dropped onto the fluid tested from a small height. The contact between glass ball and hemisphere is illuminated by a laser light at two different incidence angles. Part of the light is reflected from the semi-reflecting surface of the hemisphere/fluid interface while the rest of the light after passing through the fluid is reflected from the fluid/ball interface. The two reflected beams interfere generating a pattern of Newton's fringes. The impact microviscometer uses two beams of light at two different angles of incidence. As a result two sets of Newton's fringes are generated each illuminating a different traverse across the fluid film. If only one beam of light is used then only the optical film thickness can be found at the point of illumination. Using two beams of light allows to evaluate the refractive index and the film thickness across the contact independently. Once the film thickness across the contact is known it is then possible to calculate pressure distribution and to solve the Reynolds equation for viscosity.

However, in this viscometer, since the glass ball is dropped onto the fluid tested from a small height, the shear rates result purely from squeezing the fluid out of the contact region and hence they are extremely low, i.e. generally less than 1 [s^{-1}]. In a modified design the ball could be rotated after an entrapment was formed allowing to achieve the shear rates of about 100 [s^{-1}] [86]. Further modifications involved placing the shaft with the glass ball in an air bearing and spinning it with high angular velocity [84]. This allowed to achieve the shear rates of 10^6 [s^{-1}], which are typical of the elastohydrodynamic lubrication. In this viscometer a tiny droplet of fluid is subjected to intense shearing characteristic of high speed-high pressure contacts at a controlled contact temperature since the frictional energy dissipation is extremely limited. The operating principles of this viscometer are schematically illustrated in Figure 3.28 while the details of contact illumination are shown in Figure 3.29.

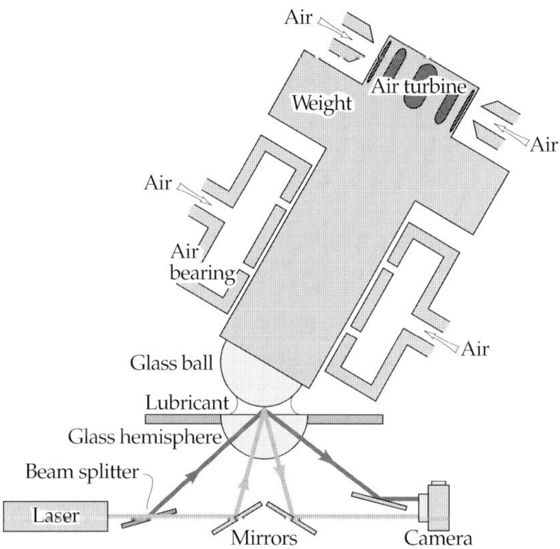

Figure 3.28 Schematic illustration of the impact viscometer (adapted from [84]).

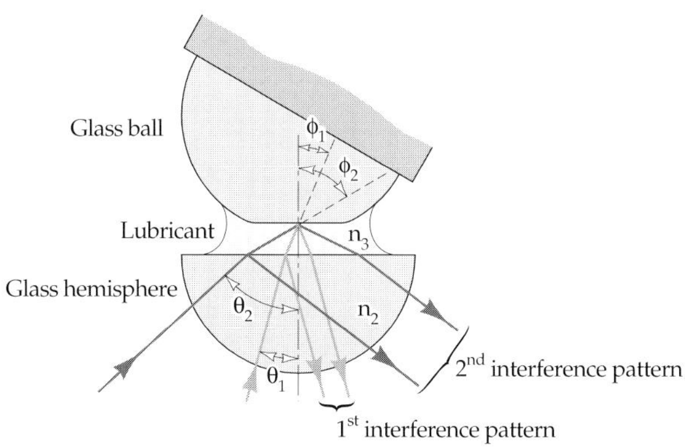

Figure 3.29 Schematic illustration of contact illumination details; where θ_1, θ_2 are the angles of illumination of the incident laser light, ϕ_1, ϕ_2 are the refraction angles of the two beams and $\mathbf{n}_2$, $\mathbf{n}_3$ are the refractive indices of the glass hemisphere and oil film respectively (adapted from [84]).

The example of the contact interference pattern obtained together with the image of the deformed ball under impact is shown in Figure 3.30.

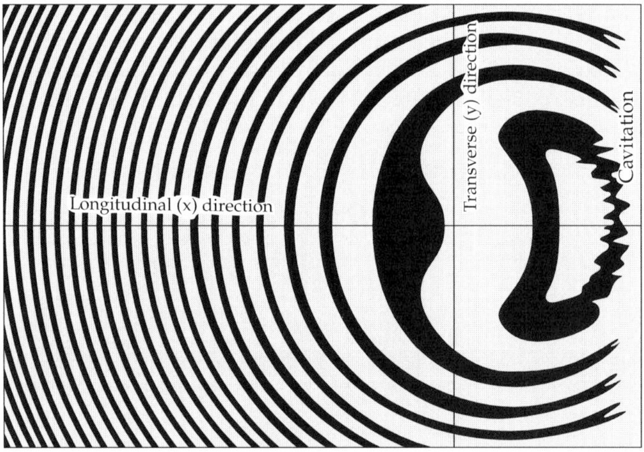

(a) Image from high shear rate, high pressure microviscometer

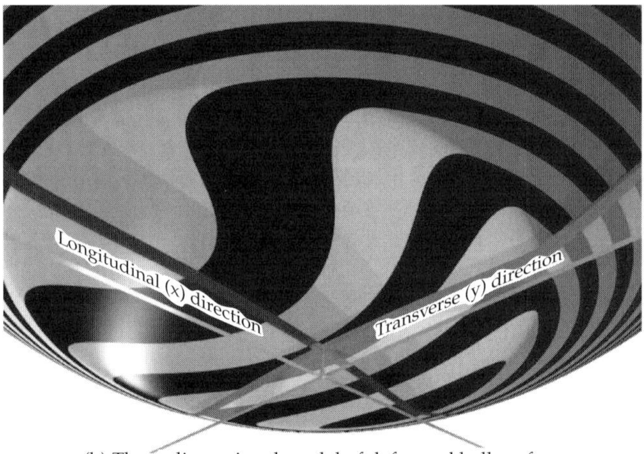

(b) Three-dimensional model of deformed ball surface

Figure 3.30 Example of the data obtained from the impact viscometer; a) example of the contact interference pattern, b) schematic illustration of the deformed ball under impact.

From the interference patterns pressure and film profiles in 'x' and 'y' directions can be obtained as illustrated in Figure 3.31.

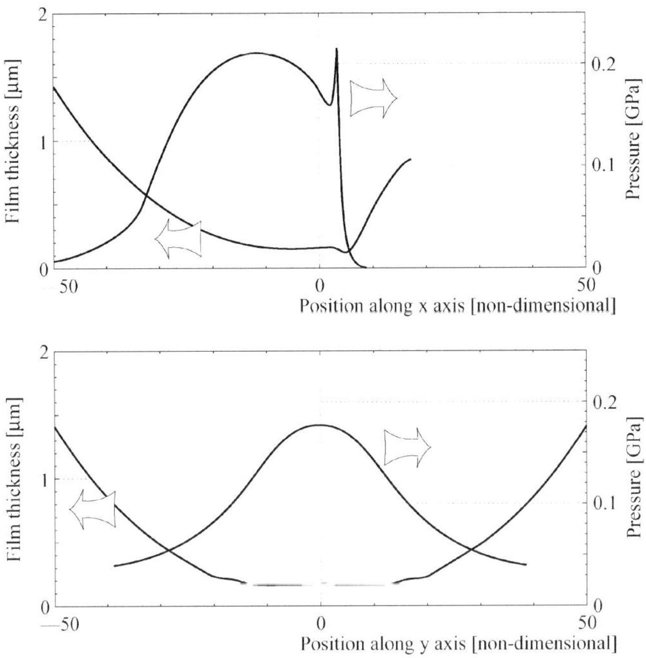

Figure 3.31 Example of the pressure and film thickness profiles in 'x' and 'y' directions obtained from the contact interference pattern (adapted from [84]).

There are also many other modifications or variations of the EHL interferometry technique that are possible. A comprehensive review of interferometry techniques applied to studies of EHL can be found in [69].

Tribometers for Studies of Mechanical Properties of Adsorbed Films in Boundary Lubrication - Surface Force Apparatus

Interpretation of many friction and lubrication phenomena depends on accurate knowledge of interfacial adhesive forces and values of the mechanical strength and modulus of adsorbed films. It is known that the interfacial adhesive forces play an important role in modifying the contact area beyond the value expected from pure elastic deformation [89]. These forces can resist a negative load (pull-off force) which needs to be applied to separate the surfaces. An accurate value of contact area between elastic solids is essential for the determination of their adhesive properties and for a better understanding of the mechanisms of friction [89-92]. The mechanical strength and modulus of adsorbed films are also important. For example, if the normal stiffness of an adsorbed film is known, then the significance of stiffness measurements in studies of extremely thin EHL films can be more easily interpreted. In other words, if a high stiffness is measured at very small EHL film thicknesses, is this stiffness comparable to the stiffness of an adsorbed film?

To measure the interfacial adhesive forces a special apparatus called a 'surface force apparatus' has been developed. The apparatus is based on the pioneering work of Tabor [89,93,94] and determines the adhesive forces by monitoring the growth of contact area with applied load [89,95-99]. This apparatus has been further modified to accurately measure the forces and displacements between metallic surfaces in liquid medium (an apparatus called a 'molecular tribometer') [100], the stiffness of adsorbed films [101,102], nanoviscosity of adsorbed layers and the effects of surface roughness in lubricated contacts [102]. It comprises a piezo-electric actuator system to move the sample, a capacitative position detector to record the displacement of the sample and a stiffly sprung capacitative force transducer to measure the normal and tangential forces. The schematic diagram of the apparatus is shown in Figure 3.32.

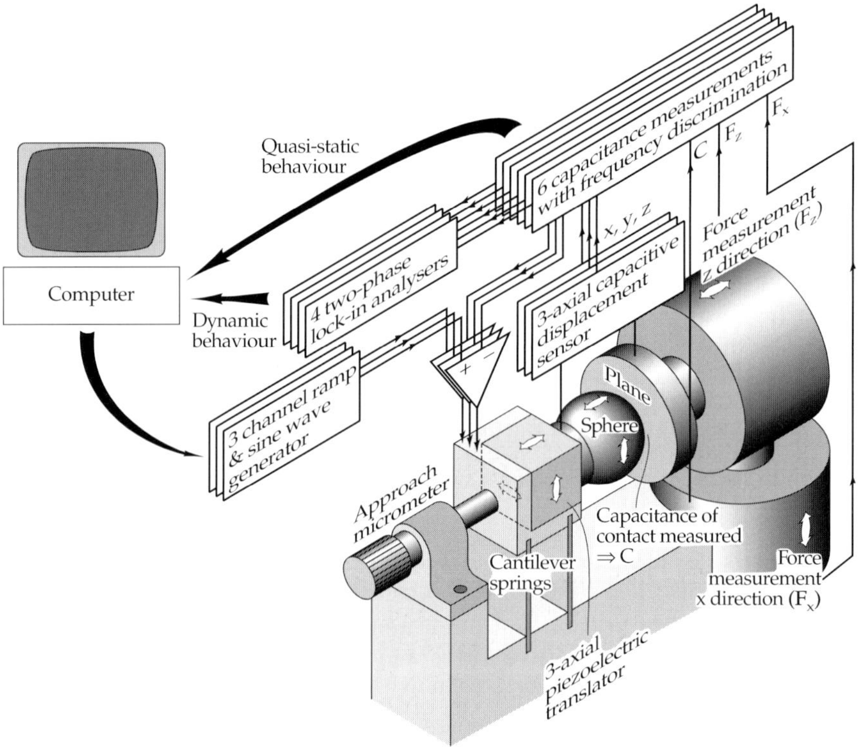

Figure 3.32 Schematic diagram of the apparatus to measure contact forces and contact stiffness over very small displacements (adapted from [102]).

The apparatus also allows measurement of microhardness in a variety of specimens with detailed record of the contact force versus depth of indentation. The variation of hardness versus depth, which is available for depths as small as a few nanometres, enables detection of a soft film on top of a hard substrate. The indenter can be substituted by a hard sphere which enables measurements of the stiffness in compressive contact of a bare surface and of a surface covered by an adsorbed film. It has been demonstrated that an adsorbed film has a non-linear stiffness, which rises sharply with depth of compression [102]. A lateral force can also be applied to the sphere allowing the tangential stiffness of the film to be found as well as its limiting tangential shear stress.

The application of an oscillating voltage to the piezo-electric transducer allows the viscous damping properties of the adsorbed films to be determined [102] in both the normal and tangential directions. It was shown that the damping characteristics of an adsorbed film can be predicted by a model of the adsorbed film as a dense 'brush' of adhering chains [103]. It was also shown that the Reynolds theory of squeeze film hydrodynamics is valid even for extremely thin film of a few nanometres thick [102]. This device, although extremely sensitive and requiring careful isolation from vibration and dust, provides much useful data which cannot otherwise be obtained by any other known means.

3.8 TRIBOMETERS FOR STUDIES NOT INVOLVING A SLIDING CONTACT

A basic problem associated with experimentation in tribology is that several processes occur simultaneously in any dynamic contact which impedes study of any single process. A common example of this is the study of a sliding contact lubricated by an oil containing additives. In these contacts mechanical processes of plastic deformation producing wear particles are combined with chemical processes involved in protective film formation resulting in the overall friction and wear characteristics. The chemical reaction between lubricant additives and the worn surface is affected in an unknown manner by frictional heat and mechanical activation caused by plastic deformation. If fundamental information about the mechanism of chemical interaction between lubricant and worn surface is required, it is often necessary to devise special experimental models where the complicating effects of friction and wear are removed. Similar considerations apply to many other processes known to contribute to friction and wear such as adhesion mechanisms and tribometers used in such studies are described in this section.

Tribometers Used to Study Adhesion Between Clean Surfaces

Adhesion studies are required to demonstrate the basic cause of adhesive wear, i.e. strong adhesion between clean surfaces that are generated in the dynamic contact. The experimental principle used involves first a generation of clean surfaces on two opposing surfaces and then the measurement of the attractive force when these two surfaces are made to contact and then to separate [104]. The experiment is performed by placing two specimens in a vacuum chamber and subjecting them initially to a cleaning process. One specimen is usually fixed while the other specimen is allowed to move. The specimens are then first brought into close proximity followed by contact and finally by retraction. Forces of attraction and repulsion between the specimens are monitored continuously during the test in order to determine whether contact has occurred which is revealed by a sharp rise in repulsion force after an initial attraction force. Adhesion is monitored by a tensile or attractive force transducer during the retraction of a specimen. The mode of dynamic contact used in these experiments is usually normal approach without sliding, i.e. typically a pin or stylus is lowered onto a flat surface. Control of the mobile specimen is usually performed by electrostatic forces generated by the electronic control system allowing for an accurate measurement of the contact forces [104]. The schematic diagram of the apparatus is shown in Figure 3.33.

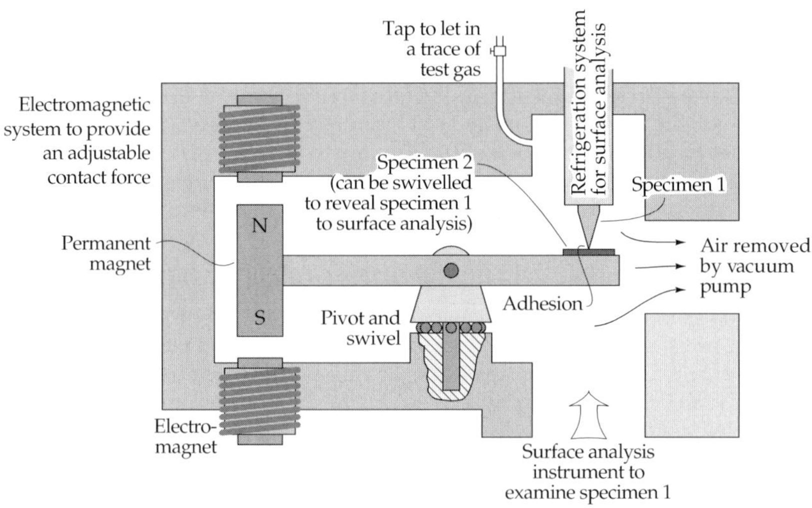

Figure 3.33 Schematic diagram of an apparatus for measurement of adhesion forces between clean surfaces in a vacuum.

The Atomic Force Microscope and Scanning Tunnelling Microscope (discussed in Chapter 8) are, in part, developments of the original adhesive contact apparatus based on a stylus contacting a flat surface.

Adhesive contact is usually accompanied by transfer of material between contacting specimens. In many cases these experiments are conducted in the vacuum chamber of a surface analysis apparatus to enable detection of changes occurring in the specimen surfaces after the adhesive contact occurred. Vapours of lubricant or contaminant gases such as oxygen can be admitted to the vacuum chamber in small quantities to test their effect on adhesion and material transfer [104].

Adhesive contact in a tribological context refers to quasi-instantaneous adhesion between contacting surfaces as a result of van der Waals forces or electron transfer. Adhesion is necessarily a rapid process because periods of contact between asperities in sliding are extremely short. There is also an accompanying phenomenon of the gradual strengthening of adhesion forces between solids in static contact. This form of adhesion often involves exchange of atoms by diffusion between contacting solids. Experimental tests for this type of adhesion usually involve a twist and compress test between two cylindrical specimens [105]. However, these experimental results cannot be validly applied to tribological problems as the phenomena involved are different from those commonly occurring during friction and wear.

Tribometers Used to Study Asperity and Surface Deformation

Asperity deformation plays a fundamental role in wear and friction yet it is largely hidden from view during dynamic contact. The obscurity of asperity deformation impedes understanding the underlying mechanism involved and is a major experimental problem in tribology. In order to provide some experimental simulation of asperity deformation in a manner amenable to observation, a special apparatus to model asperity deformation was devised where a hard wedge was drawn over a soft slab [106]. Deformation of surface material in the worn slab was viewed in cross-section while tangential and normal contact forces can simultaneously be measured. A schematic diagram of this apparatus is shown in Figure 3.34.

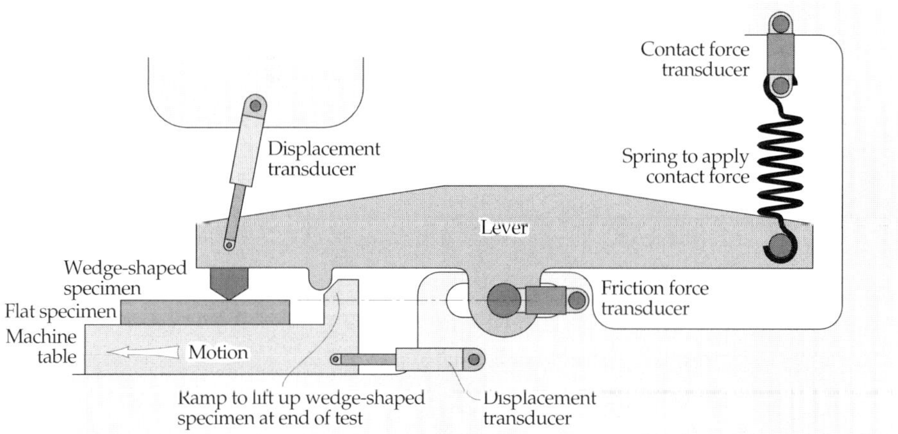

Figure 3.34 Schematic diagram of an apparatus providing a cross-sectional model of asperity deformation in a sliding contact.

An earlier design of apparatus to model asperity contact involved two wedges made of similar material which were forced to move past each other while held in a rigid frame as illustrated schematically in Figure 3.35. This apparatus allowed the differences in asperity deformation in adhesive and non-adhesive contacts to be measured [13].

A miniaturized version of the contacting wedges was also developed to allow deformation at an asperity scale of size to be observed in a Scanning Electron Microscope and is described in Chapter 8.

Although most of the experiments involving asperities modelled by wedges have already been completed a new application for this type of experiment could be found in investigations of the influence of surface coatings on asperity deformation. A microscopic apparatus containing either a wedge on slab or pair of wedges could be used to test wedges coated with protective films such as titanium nitride. The deformation of the coated wedges could then be examined using microscopy techniques.

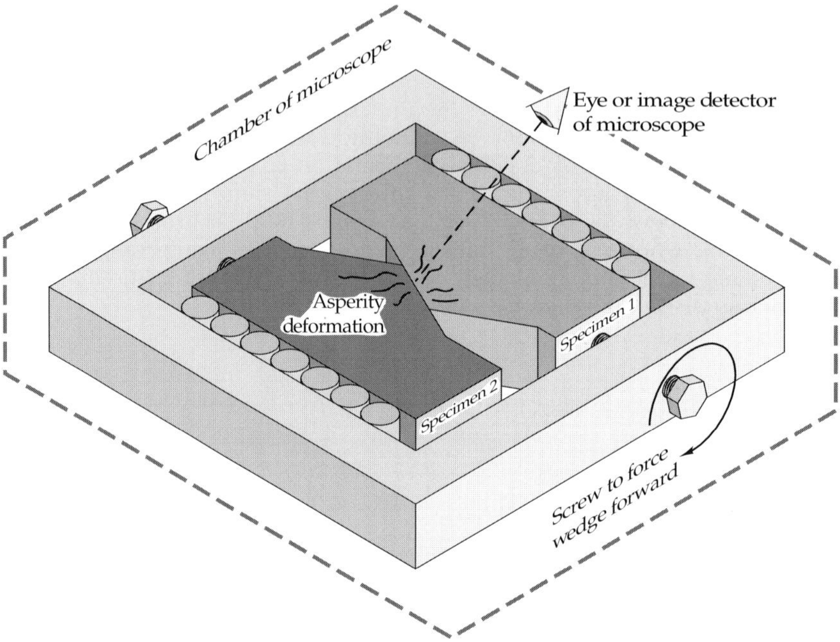

Figure 3.35 Schematic diagram of an apparatus involving two wedges in contact to simulate asperity deformation.

Tribometers Used to Study the Chemistry of Clean or Freshly Exposed Surfaces

Chemical aspects of friction and wear are very often best studied in an apparatus which does not involve a dynamic contact. Conditions in a dynamic contact are complex and usually poorly controlled which often prevents conclusive data from being obtained. When viewed from the perspective of chemistry, the basic features of wear and friction are extremes of temperature and exposure of clean mechanically activated metal surfaces.

The 'hot-wire' apparatus has been developed and used to study the chemistry of reaction between metallic surfaces and lubricants to temperatures exceeding 500°C [107]. Electrical current is used to heat a thin wire immersed in a lubricating oil sample. Variation of electrical resistance of the wire is used to determine the reduction in wire cross section caused by corrosive reaction with oxygen or additives present in the oil. However, the problem with this device is that unless the lubricant sample and wire are enclosed in a sealed vessel maintained at high pressure, the wire becomes surrounded by a lubricant vapour film during tests which obscures the obtained data. Since the experimental conditions in this apparatus are not typical of lubricated wear, the validity of these tests can be questioned. Another difficulty with this experimental set up is that corrosion of the wire tends to be non-uniform and is often concentrated at one point on the wire. This non-uniform corrosion prevents the accurate assessment of corrosive reduction in wire cross section by electrical resistance measurements. The schematic diagram of 'hot-wire' apparatus is shown in Figure 3.36.

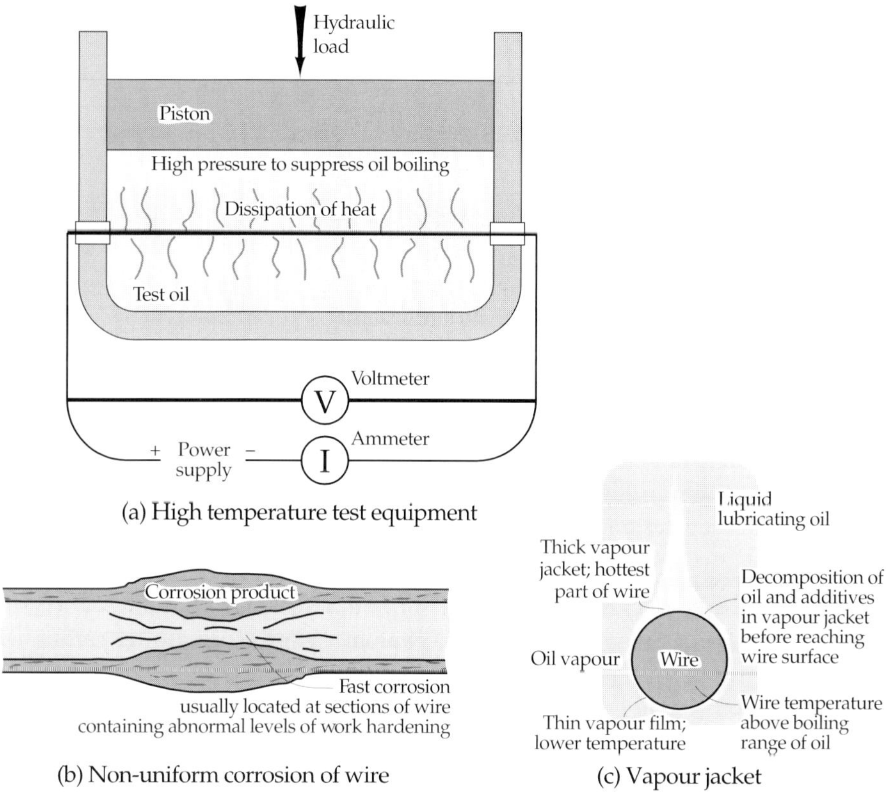

Figure 3.36 Schematic diagram of the 'hot-wire' apparatus.

The chemistry of freshly exposed clean surfaces (known as nascent surfaces) can be studied by drawing a cutting tool over a metallic surface while it is immersed in the lubricant tested [108]. The cutting tool exposes the nascent surface by removing layers of oxides and contaminants. Oxygen has only limited solubility in most lubricants so the lubricant or its additives can react with the clean surface for a considerable period of time before oxidation restores the oxide film removed by cutting. It was found that by the application of this method it is possible to study chemical reactions with nascent surfaces at very short times, even of a few milliseconds, which are typical of sliding contact [109]. Temperature rises that are inevitable during the cutting process can be controlled by reducing the cut depth to a few micrometres. This minute cut depth ensures that localised high temperatures occurring around the cutting tool are unable to affect the lubricant tested [109]. A schematic diagram of such apparatus suitable for studies of nascent surface chemistry at short contact periods is shown in Figure 3.37.

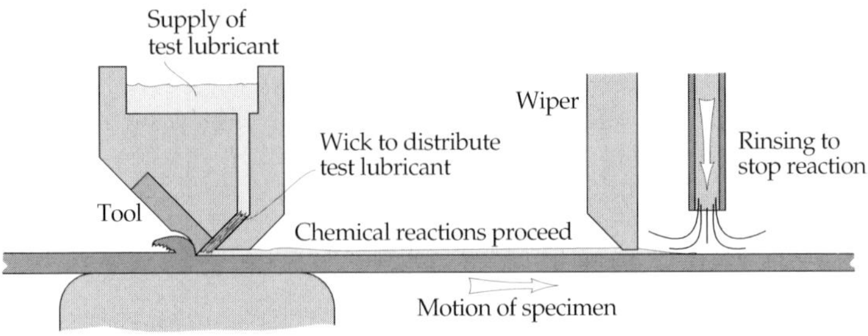

Figure 3.37 Schematic diagram of the apparatus to study chemistry between nascent surface and lubricant.

Reaction between nascent surface and lubricant is terminated by washing the metallic surface with a jet of solvent after a set period of time. Samples containing reacted nascent surface can then be extracted for surface analysis to determine the extent of chemical reaction which took place. Comparison of surface analysis data between samples reacted for different periods of time enables an information about surface film formation to be obtained.

A simpler way of measuring nascent surface reactions without the complexity of the lubricant deposition and rinsing systems is to release the lubricant as a vapour inside a vacuum chamber [110] as schematically illustrated in Figure 3.38. The quantity of vapour released is determined from the pressure rise in the chamber and the progress of reaction with the surface is monitored by the pressure decline of gas with time. The decline in gas pressure during cutting can be used as a measure of nascent surface chemical activity.

If the limitations of vapours used instead of realistic liquid lubricants are accepted much useful information about reactions between organic compounds and nascent metal surfaces can be obtained. Nascent surfaces of, for example, nickel [111] and iron [112] were found to display strong chemisorption activity for benzene, 1-hexene and other organic compounds at room temperature while uncut oxidized surfaces are inactive. It was also shown that stainless steel exhibits stronger chemisorption activity for many organic compounds than plain steel despite its accepted perception as virtually inert [113]. Even gold, which is considered to be a noble metal, displays in a nascent state strong chemisorption of organic compounds and causes their decomposition to release hydrogen [114].

Using a similar experimental set-up, e.g. a pin-on-disc or ball-on-disc apparatus placed inside the vacuum chamber, allows to study triboemission of the charged particles (negative or positive ions and photons) from the freshly worn surface as there is no intervening air to block the ion or photon paths [115]. A charge detector or a photomultiplier can be placed inside the vacuum chamber at a suitable location to collect charged particles or photons from the worn surface.

If the extreme surface deformation of cutting is considered to be a source of experimental artefact, nascent surfaces can also be studied by depositing a film of clean metal on the interior surface of a prepared vacuum chamber. The chamber can be made of glass for experimental convenience [113]. A gas or vaporised lubricant is then introduced in small quantities into the chamber. The decline in pressure after the introduction of test gas provides a measure of the extent of reaction taking place [113]. It is often found that a clean metallic surface is capable of decomposing chemicals and lubricant vapours to simpler organic compounds [e.g. 116]. Samples of vapour can be extracted during the tests and analyzed to further investigate this phenomenon. Placing the test apparatus inside a surface analysis device allows surface analysis to be conducted during controlled reaction tests on nascent surfaces [e.g. 104]. Experiments of this type enabled resolution of the very early stages of sulphidation and oxidation of nascent metal where there was less than one monolayer of either sulphur or oxygen present on the nascent surface. It was found that even such small amounts of sulphur and oxygen can significantly influence friction and wear [104].

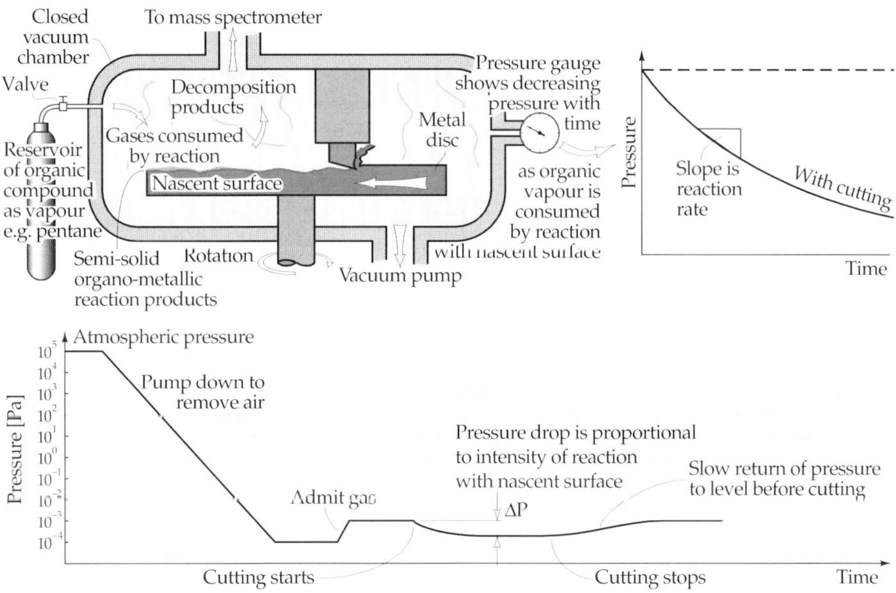

Figure 3.38 Schematic illustration of a vacuum based system to study the chemical activity of nascent surfaces.

Differential scanning calorimetry has also been used to study the chemistry between metals and lubricant additives [117]. The principle of differential calorimetry is described later in Chapter 9. In these studies, metal powders were used as an experimental model of worn surfaces. The surfaces of the powders were then reduced by passing hot hydrogen through the sample in order to generate a nascent surface. Basic tests involve metal powders but compound powders can also be used to simulate more subtle aspects of lubrication. For example, metal oxide and metal sulphide powders were used to simulate the interaction between oxide films or films formed by E.P. additives and adsorbates such as fatty acids on surfaces where the contact load is too low to expose nascent surface [118,119].

An impure form of nascent surface may be created when a sliding wear apparatus is fitted inside the vacuum chamber of a surface analysis instrument [35,120]. The high quality of vacuum ensures that oxide and contaminant films are progressively removed by wear, without significant re-oxidation or hydration. The surface chemical characteristics of the worn surface can be examined by, for example, Auger electron spectroscopy, without contamination from the outside atmosphere.

Apparatus Used in the Studies of Adsorbed Films

Basic properties of adsorbed films found in lubricated contacts can conveniently be studied on specially prepared surfaces under controlled conditions. For example, the energy of adsorption was found to correlate well with the friction coefficient of contacts lubricated with fatty acids under prescribed friction test conditions [121]. The energy of adsorption can conveniently be measured by immersing a capacitor in a solution of the fatty acid [121]. The capacitor consists of two metal plates one of which is made of a test material, e.g. steel, and the other is made of a chemically inert material such as brass coated with a thin layer of polytetrafluoroethylene [121]. The difference in adsorption energy between the active metal electrode and the inert electrode creates a differential voltage of adsorption. However, this voltage is normally suppressed by polarization across the capacitor so that it is not possible to measure it directly. An indirect measurement of voltage is obtained by exploiting the characteristics of capacitors under imposed oscillating voltage. The schematic diagram of the apparatus is shown in Figure 3.39.

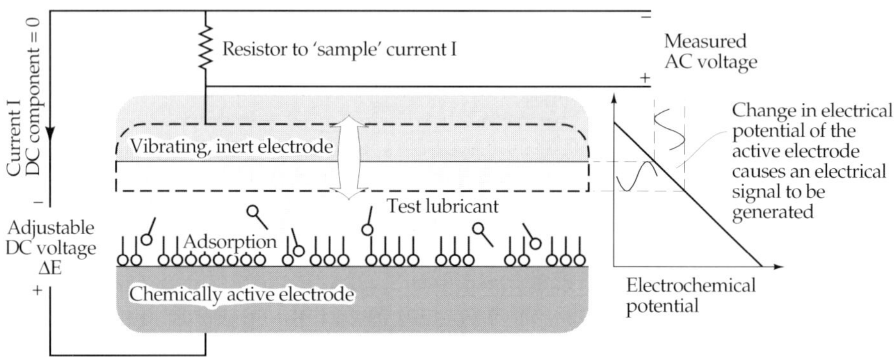

Figure 3.39 Schematic diagram of the apparatus used to evaluate the adsorption energy by the measurement of contact potential difference in a capacitor.

Balancing the circuit (by adjusting ΔE until I is zero, with the vibration stopped) ensures that the measuring circuit draws almost no current from the test electrode 'battery', and so does not affect the measurement. The measurement itself is actually of the slope of the curve shown in Figure 3.39. This dynamic measurement technique is needed because any attempt to measure the potential of the surfaces in a static way will cause immediate double-layer ionic effects which will distort the measurement. The whole procedure can be seen as a way of characterising a 'battery' (the electrode/lubricant system) which has a very limited current capacity.

When the capacitor is subjected to an oscillating voltage an oscillating current across the capacitor is generated by variations in electrical charge storage between the capacitor plates. This current is proportional to the product of the differential voltage of adsorption and the capacitance of the system. By imposing a reverse voltage across the capacitor just sufficient to reduce the oscillating current to zero, the differential voltage of adsorption can then determined since it is equal to this reverse voltage [121,122].

Parallel tests using the capacitor and a standard tribometer can be performed with a variety of metals and fatty acids in order to determine whether the lubricating effect of fatty acids is determined by the energy of adsorption. Data of this kind can be used to determine the mechanism involved in lubrication of fatty acids, i.e. to confirm whether fatty acids lubricate by the formation of thin adsorbed films on metallic surfaces. It should be noted, however, that the unworn metallic surface in the test system would be different from a worn surface. For example, the unworn surface is most likely to be smoother that a worn surface. Its surface area would be smaller that the corresponding true surface area of a worn surface. The high levels of plastic deformation found on worn surfaces would create a higher dislocation density in the worn surface. This in turn would cause the adsorption characteristics of the unworn surface and the worn surface to differ.

3.9 APPARATUS TO MEASURE THE INTEGRITY OF WEAR RESISTANT COATINGS - SCRATCH TEST

Tests for the integrity or mechanical strength of wear resistant coatings are gaining more importance as the use of these coatings becomes widespread. Such tests are of importance to coatings research and production quality control. The most commonly used test is the scratch test. The concept of scratching the surface with a sharp instrument, e.g. knife, to determine the adhesion of paints or other surface coatings is not new. The actual tests to measure adhesion strength were introduced by Heavens [123,124]. Adhesion strength is the stress required to remove the coating from a substrate [123]. Indentation and scratch tests on micro and nanoscales are commonly used to determine adhesion of thin hard films and coatings [e.g. 125-129]. The mechanical strength of coatings depends on three factors: the brittleness of the coating, coating adhesion to the stylus and coating adhesion to the substrate.

In the scratch test a diamond stylus, e.g. Vickers indenter for macro and micro scratching [e.g. 123,130,131] and a conical diamond indenter for nanoscratching [e.g. 123,132,133], is drawn over the surface coating as illustrated schematically in Figure 3.40. The load on the stylus is progressively increased during scratching until coating film failure is detected by acoustic emission (noise) [134,135]. Load can be either progressively applied up to a specified indentation depth or up to some pre-set maximum value. Both load and scratch depth are monitored. The minimum load at which the coating is removed or detached is used as a measure of adhesion [e.g. 123,130,131,136-144]. This technique is commonly used to measure adhesion of hard coatings with strong interfacial adhesion (>70 [MPa]) [123]. Microscopic observations are often used to determine the moment of coating removal. Sometimes the friction force is measured during scratching and a sharp increase in its value during the test has been used as an indicator that the diamond tip has broken through the coating [145-147].

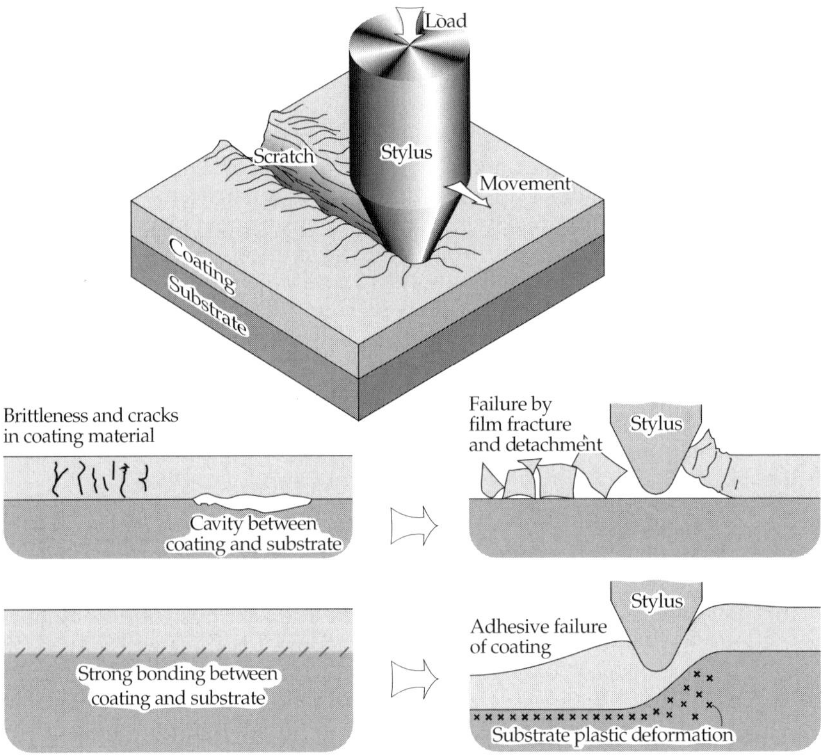

Figure 3.40 Schematic illustration of the principle involved in a scratch test of wear resistant coatings.

The shape of the diamond stylus is typically a cone with a rounded end. A tip radius of 0.2 [mm] and a cone angle of 120° was found to be effective in assessing the integrity of thin coatings [148]. For very thin coatings below 1 [μm] thickness, coating failure can only be determined by microscopic examination as the acoustic emission is too weak to be detected [148]. Mechanical resistance of coatings to damage by scratching was found to correspond to a wear resistance by mechanical deformation [134,135]. The mechanism of coating failure under scratching can later be investigated by microscopic examination. Failure due to cavities present between the substrate and coating or excessive brittleness of the coating is characterized by distinctive types of scratching damage and is clearly visible under the microscope [134,135]. Adhesion of the coating to the stylus, indicating potentially high friction in sliding contact, can also be easily detected by examination of both the scratched coating and the stylus itself.

Information about the adhesion of thin hard coatings to the substrate can also be obtained by subjecting the sample to a rolling contact in a relative motion. In this way the performance of various coatings and coating techniques can be rapidly assessed [149]. More information on nanoscratch testers can be found in [e.g. 123].

3.10 SUMMARY

An enormous variety of tribometers have been developed, partly for scientific reasons and also because researchers often wished to devise their own improved test apparatus suiting their particular research objectives. As a result tests are conducted under different conditions and contact geometries hence there are frequent problems in comparing research data reported in the literature. Despite the large number of designs available, most tribometers are variations of a few basic geometrical configurations of sliding, rolling or impacting contact. Few tribometers have been designed to provide accurate simulation of wear and friction as it occurs in real machinery so that the test data obtained might be of limited value. Although recent versions of tribometers are more carefully designed to provide useful data these limitations of the experimental apparatus still remain. Older designs of tribometers were nearly always intended for tests at ambient conditions, but in current developments of the new generation of tribometers, the provision for testing in extreme environments to meet the demands of modern technology is often incorporated.

Tribological research also depends on test equipment that does not involve a moving contact. The purpose of this latter type of tribometer is to provide data on the basic mechanisms of friction, wear and lubrication. The chemistry of lubrication mechanisms is often studied in apparatus that simulate certain aspects of a dynamic contact such as exposure of the unoxidized metallic surface. Use of this type of apparatus allows experimental data to be obtained that is not obscured by complex interactions between friction and wear mechanisms.

REFERENCES

1 J.H. Chen and C.R. Ursell, Comparison of wire rope life using nylon and steel sheaves, Part 1, Off-highway Vehicle Meeting and Exposition, MECCA, Milwaukee, 10-13 Sept. 1979, SAE Publ. No. 790904.

2 J.C. Cree, Caterpillar L1 and Chevrolet L4 test process, Symposium on Engine Testing of Lubricating Oil, 1953, The Institute of Petroleum.

3 Friction and wear devices, Revised and Enlarged Report of the Sub-committee on Wear and Lubrication, Fundamentals Committee, 2nd edition, American Society of Lubrication Engineers, Illinois, USA, 1976.

4 S. Adelman, C.R. Taylor and N.C. Heglund, Sweating on paws and palms: What is its function, *American Journal of Physiology*, Vol. 229, 1975, pp. 1400-1402.

5 Standard test methods for measurement of extreme pressure properties of lubricating fluids (Four-Ball Method), ASTM Standard - D 2783-88.

6 Determination of extreme pressure and antiwear properties of lubricants - Four Ball Machine Method, IP 239/97, The Institute of Petroleum, Standards for Petroleum and Its Products.

7 Standard test methods for measurement of extreme pressure properties of fluid lubricants (Falex Pin and Vee Block Methods), ASTM Standard - D 3233-93.

8 Anti-wear, friction and corrosion protection properties of fluid lubricants: Falex Machine, The Institute of Petroleum, Standards for Petroleum and Its Products Test Methods: IP 241 Extreme Pressure Properties, 1984.

9 Standard test methods for measurement of extreme pressure properties of lubricating fluids (Timken Method), ASTM Standard - D 2782-94.

10 Determination of extreme pressure properties of lubricating fluids - Timken Method, IP 240/84, The Institute of Petroleum, Standards for Petroleum and Its Products.

11 K.R. Meckleburg, Forces in the Falex configuration, *Transactions ASLE*, Vol. 18, 1975, pp. 97-104.

12 G. Plint. Private communication, 1993.

13 F.P. Bowden and D. Tabor, The friction and lubrication of solids, Part 1, Oxford University Press, 1954.

14 T.N. Mills and A. Cameron, Basic studies on boundary, E.P. and piston ring lubrication using a special apparatus, *Transactions ASLE*, Vol. 25, 1982, pp. 117-124.

15 P.H. Shipway and I.M. Hutchings, Influence of nozzle roughness on conditions in a gas-blast erosion rig, *Wear*, Vol. 162-164, 1993, pp. 148-158.

16 L. Fang, X.L. Kong, J.Y. Su and Q.D. Zhou, Movement patterns of abrasive particles in three-body abrasion, *Wear*, Vol. 162-164, 1993, pp. 782-789.

17 K.L. Rutherford and I.M. Hutchings, Theory and application of a micro-scale abrasive wear test, *Journal of Testing and Evaluation*, Vol. 25, 1997, pp. 250-260.

18 V. Thomson, H.E. Hintermann and L. Chollet, The determination of composition depth profiles using spherical erosion and scanning Auger electron spectroscopy, *Surface Technology*, Vol. 8, 1979, pp. 421-428.

19 G.B. Stachowiak, G.W. Stachowiak, The effects of particle characteristics on three-body abrasive wear, *Wear*, Vol. 249, 2001, pp. 201-207.

20 G.W. Stachowiak, Particle angularity and its relationship to abrasive and erosive wear, *Wear*, Vol. 241, 2000, pp. 214-220.

21 M.E. Woods, W.F. Mandler, Jr. and T.L. Scofield, Designing ceramic insulated components for the adiabatic engine, *Bulletin of the American Ceramic Society, Heat Engine Issue*, Vol. 64(2), 1985, pp. 287-289 and 292-293.

22 K.F. Dufrane, Wear performance of ceramics in ring/cylinder applications, *J. Am. Ceram. Soc.*, Vol. 72, 1989, pp. 691-695.

23 J. Lankford, W. Wei and R. Kossowsky, Friction and wear behaviour of ion beam modified ceramics, *J. Mater. Sc.*, Vol. 22, 1987, pp. 2069-2078.

24 Ch. DellaCorte, Tribological properties of alumina-boria-silicate fabric from 25°C to 850°C, *Tribology Transactions*, Vol. 32, 1989, pp. 325-330.

25 M. Woydt and K.H. Habig, High temperature tribology of ceramics, *Tribology International*, Vol. 22, 1989, pp. 75-87.

26 H.E. Sliney and Ch. DellaCorte, A new test machine for measuring friction and wear in controlled atmospheres to 1200°C, *Lubrication Engineering*, Vol. 47, 1991, pp. 314-319.

27 K. Miyoshi, F.S. Honecy, P.B. Abel, S.V. Pepper, T. Spalvins and D.R. Wheeler, A vacuum (10^{-9} Torr) friction apparatus for determining friction and endurance life of MoSx films, *Tribology Transactions*, Vol. 36, 1993, pp. 351-358.

28 M.G. Gee, C.S. Matharu, E.A. Almond and T.S. Eyre, The measurement of sliding friction and wear of ceramics at high temperatures, *Wear*, Vol. 138, 1990, pp. 169-187.

29 K. Miyoshi, J.J. Pouch, S.A. Alterovitz, D.M. Pantic and G.A. Johnson, Adhesion, friction and wear of plasma-deposited thin Silicon nitride films at temperatures to 700°C, *Wear*, Vol. 133, 1989, pp. 107-123.

30 A. Iwabuchi, Y. Tatsuyanagi, K. Yoshida and M. Sugimoto, Tribological properties of cryogenic materials at 4 K in liquid helium under fretting conditions, *Wear*, Vol. 159, 1992, pp. 107-114.

31 A.J. Slifka, J.D. Siegwarth, L.L. Sparks and D.K. Chaudhuri, Apparatus for measurement of coefficient of friction, Advances in Cryogenic Engineering (Materials), editors: R.P. Reed and F.R. Fickett, Vol. 36, Plenum, New York, 1990, pp. 1119-1125.

32 D.K. Chaudhuri, A.J. Slifka and J.D. Siegwarth, Friction and oxidative wear of 440C ball bearing steels under high load and extreme bulk temperatures, *Wear*, Vol. 160, 1993, pp. 37-50.

33 D.H. Buckley, Surface Effects in Adhesion, Friction, Wear and Lubrication, Elsevier, Amsterdam, 1980.

34 M. Suzuki, M. Moriyama, M. Nishimura and M. Hasegawa, Friction and wear of self-lubricating composites at temperatures up to 450°C in vacuum, *Wear*, Vol. 162-164, 1993, pp. 471-479.

35 T.G. McDonald, D.E. Peebles, L.E. Pope and H.C. Peebles, Device to measure friction coefficients and contact resistance inside a scanning Auger microscope, *Review Science Instrum.*, Vol. 58, 1987, pp. 593-597.

36 S.C. Lim and J.H. Bruntion, A dynamic wear rig for the scanning electron microscope, *Wear*, Vol. 101, 1985, pp. 81-91.

37 J.P. Netzel, Seal technology, a control for industrial pollution, *Lubrication Engineering*, Vol. 46, 1990, pp. 483-493.

38 P.L. Ko, A. Wozniewski and P.A. Zhou, Wear-corrosion-resistant materials for mechanical components in harsh environments, *Wear*, Vol. 162-164, 1993, pp. 721-732.

39 B. Loffelbein, M. Woydt and K-H. Habig, Sliding friction and wear of ceramics in neutral acid and basic aqueous solutions, *Wear*, Vol. 162-164, 1993, pp. 220-228.

40 Z.R. Zhou and L. Vincent, Effect of external loading on wear maps of aluminium alloys, *Wear*, Vol. 162-164, 1993, pp. 619-623.

41 G.W. Stachowiak, G.B. Stachowiak and A.W. Batchelor, Suppression of fretting wear between roping wires by coatings and laser alloyed layers of molybdenum, *Wear*, Vol. 178, 1994, pp. 69-77.

42 H.H. Ghasemi, M.J. Furey and C. Kajdas, Surface temperatures and fretting corrosion of steel under conditions of fretting contact, *Wear*, Vol. 162-164, 1993, pp. 357-369.

43 F.P. Bowden and D. Tabor, The Friction and Lubrication of Solids, Part 2, Clarendon Press, Oxford, 1964.

44 F.P. Bowden and E.H. Freitag, The friction of solids at very high speeds, I. Metal on metal, II. Metal on diamond, *Proc. Royal Society of London, Series A*, Vol. 248, 1958, pp. 350-367.

45 P.A. Engels, Impact wear of materials, Elsevier, Amsterdam, 1976.

46 M. Sato, P.M. Anderson and D.A. Rigney, Rolling-sliding behaviour of rail steels, *Wear*, Vol. 162-164, 1993, pp. 159-172.

47 A.J. Perez-Unzueta and J.H. Beynon, Microstructure and wear resistance of pearlitic rail steels, *Wear*, Vol. 162-164, 1993, pp. 173-182.

48 A.W. Crook, The lubrication of rollers, *Phil. Trans. Roy. Soc., London*, Series A, Vol. 250, No. 981, 1958, pp. 387-409.

49 J.E. Garnham and J.H. Beynon, Dry rolling-sliding wear of bainitic and pearlitic steels, *Wear*, Vol. 157, 1992, pp. 81-109.

50 M. Masuko, Y. Ito, K. Akatsuka, K. Tagami and H. Okabe, Influence of sulphur-base extreme pressure additives on wear under combined sliding and rolling contact, Proc. Kyushu Conference of JSLE, Oct. 1983, pp. 273-276 (in Japanese).

51 A.P. Mouritz and I.M. Hutchings, The abrasive wear of steel during rolling-sliding contact with rock counterfaces, *Wear*, Vol. 134, 1989, pp. 99-113.

52 M. Chandrasekaran and A.W. Batchelor, In-situ observation of sliding wear tests of butyl rubber in the presence of lubricants in an X-ray microfocus instrument, *Wear*, Vol. 211, 1997, pp. 35-43.

53 A.W. Batchelor, M. Chandrasekaran, Y. Fu and H. Xing, In-situ friction and wear tests in an X-ray microscope, *The Tribologist, Transactions of JST*, Vol. 43, No. 1, 1998, pp. 8-14.

54 M. Chandrasekaran, A.W. Batchelor and N.L. Loh, In situ observations of frictional seizure by X-ray imaging, Part 1 Methods, *Journal of Materials Science*, Vol. 35, No. 7, 2000, pp. 1597-1602.

55 K.J. Wahl, M. Belin and I.L. Singer, A triboscopic investigation of the wear and friction of MoS_2 in a reciprocating sliding contact, *Wear*, Vol. 212, 1998, pp. 212-220.

56 I.L. Singer, S.D. Dvorak and K.J. Wahl, Investigation of third body processes by in vivo raman tribometry, 9th Nordic Symposium on Tribology, NORDTRIB 2000, Technical Research Centre of Finland ESPOO 2000, publisher - Julkaisija-Utgiuare, Vol. 1, 2000, pp. 31-41.

57 Raman Microscope used for in situ tribological studies, *Renishaw Spectroscopy Innovations, News from the Spectroscopy Products Division*, Issue 4, June 1999, pp. 1-2.

58 Z. Neder, K. Varadi and K. Friedrich, Quantitative evaluation of wear grooves in a polymercoating system by laser profilometry and computer aided scratch analysis, *Advanced-Composites-Letters*, Vol. 9, No. 5, 2000, pp. 315-325.

59 M.Q. Zhang, K. Friedrich, K. Batzar and P. Thomas, Abrasive wear mechanisms of fluoropolymer based composite coatings on aluminum substrates, *Wear*, Vol. 200, 1996, pp. 122-136.

60 Beauchamp Tower, First report on friction experiments, *Proc. Inst. Mech. Engrs.*, Nov. 1883, pp. 632-659.

61 R. Stribeck, Die Wesentlich Eigenschaften der Gleit-und-Rollen-lager, Zeitschrift des Vereines deutscher Ingenieure, Vol. 46, 1902, No. 38, pp. 1341-1348, pp. 1432-1438, also No. 39, pp. 1463-1470.

62 M. Akkok and C.M.McC. Ettles, The effect of load and feed pressure on whirl in a grooved journal bearing, *ASLE Transactions*, Vol. 22, 1979, pp. 175-184.

63 J.A. Cole and C.J. Hughes, Oil flow and film extent in complete journal bearings, *Proc. Inst. Mech. Engrs.*, Vol. 170, 1956, pp. 499-510.

64 A. Takeuchi, M. Sato and H. Aoki, Effect of speed on bearing characteristics in mixed lubrication, *Transactions JSLE*, Vol. 33, 1987, pp. 62-68.

65 M. Nonogaki and T. Nakahara, Frictional characteristics in mixed lubrication of journal sliding bearing, *Tribologist, Transactions of JST*, Vol. 38, 1992, pp. 345-350.

66 L.L. Ting, Development of laser fluorescence technique for measuring piston-ring oil-film thickness, *Transactions ASME, Journal of Lubrication Technology*, Vol. 102, 1980, pp. 165-171.

67 A.E. Smart and R.A.J. Ford, Measurement of thin liquid films by a fluorescence technique, *Wear*, Vol. 29, 1974, pp. 41-47.

68 R.A.J. Ford and C.A. Ford, Laser-based fluorescence techniques for measuring thin liquid films, *Wear*, Vol. 51, 1978, pp. 289-297.

69 B.O. Jacobson, Rheology and Elastohydrodynamic Lubrication, Elsevier, Amsterdam, 1991.

70 G.M. Hamilton and S.L. Moore, Deformation and pressure in an elastohydrodynamic contact, *Proc. Royal Society of London, Series A*, Vol. 322, 1971, pp. 313-330.

71 A. Dyson, H. Naylor and A.R. Wilson, The measurement of oil film thickness in elastohydrodynamic contacts, *Proc. Inst. Mech. Engrs.*, Vol. 180, Pt. 3B, 1965, pp. 119-134.

72 L.B. Sibley and A.E. Austin, An X-ray method for measuring thin lubricant films between rollers, *ISA Transactions*, Vol. 3, 1962, pp. 237-243.

73 L.B. Sibley, J.C. Bell, F.K. Orcutt and C.M. Allen, A study of the influence of lubricant properties on the performance of aircraft gas engine rolling contact bearings, WADD Technical Report, 1960, pp. 60-189.

74 J.W. Kannel and S.S. Bupara, Rheology of lubricants in real bearing contacts, *Transactions ASME, Journal of Lubrication Technology*, Vol. 97, 1975, pp. 228-235.

75 B.J. Hamrock and D. Dowson, Ball Bearing Lubrication, The Elastohydrodynamics of Elliptical Contacts, John Willey & Sons, 1981.

76 A.T. Kirk, Hydrodynamic lubrication of perspex, *Nature*, Vol. 104, 1962, pp. 965-966.

77 A. Cameron and R. Gohar, Theoretical and experimental studies of the oil film in lubricated point contacts, *Proc. Roy. Soc., London*, Series A, Vol. 291, 1966, pp. 520-536.

78 O. Marklund and L. Gustafson, Interferometry-based measurements of oil-film thickness, *Proc. Inst. Mech. Engrs, Part J, Journal of Engineering Tribology*, Vol. 215, 2001, pp. 243-259.

79 M.E. Aderin, G.H. Johnston, H.A. Spikes, T.G. Balson and M.G. Emery, The film-forming properties of polyalkylene glycols, *Synthetic Lubrication*, Vol. 10, 1993, pp. 23-45.

80 G.J. Johnston, R. Wayte and H.A. Spikes, The measurement and study of very thin lubricant films in concentrated contacts, *Tribology Transactions*, Vol. 34, 1991, pp. 187-194.

81 H.A. Spikes and P.M. Cann, The development and application of the spacer layer imaging method for measuring lubricant film thickness, *Proc. Inst. Mech. Engrs, Part J, Journal of Engineering Tribology*, Vol. 215, 2001, pp. 261-277.

82 G.R. Paul and A. Cameron, An absolute high-pressure microviscometer based on refractive index, *Proc. Royal Society, London*, Series A, Vol. 331, 1972, pp. 171-184.

83 G.R. Paul, Time dependent viscosity following a pressure rise measured on an impact viscometer, *Transactions ASLE*, Vol. 19, 1974, pp. 17-22.

84 P.L. O'Neill and G.W. Stachowiak, High shear rate impact microviscometer, *Tribology International*, Vol. 29, No. 7, 1996, pp. 547-557.

85 P.L. Wong, S. Lingard and A. Cameron, The high pressure impact viscometer, *Tribology Transactions*, Vol. 35, 1992, pp. 500-508.

86 G.R. Paul and A. Cameron, The ultimate shear stress of fluids at high pressures measured by a modified impact microviscometer, *Proc. Royal Society, London, Series A*, Vol. 365, 1979, pp. 31-41.

87 G.R. Paul and A. Cameron, Time dependent viscosity effects measured on an impact viscometer, *Nature (London)*, Vol. 248, 1974, pp. 219-220.

88 M.F. Workel, D. Dowson, P. Ehret and C.M. Taylor, Design and development of a ball impact apparatus for the direct measurement of lubricant friction under high pressure and shear rates, *Proc. Inst. Mech. Engrs, Part J, Journal of Engineering Tribology*, Vol. 215, 2001, pp. 211-222.

89 P. Frantz, A. Artsyukhovich, R.W. Carpick and M. Salmeron, Use of capacitance to measure surface forces. 2, Application to the study of contact mechanics, *Langmuir*, Vol. 13, 1997, pp. 5957-5961.

90 J.N. Israelachvili, Adhesion, friction and lubrication of molecularly smooth surfaces, Fundamentals of Friction: Macroscopic and Microscopic Processes, editors: I.L. Singer and H.M. Pollock, NATO ASI Series, Kluwer Academic Publishers, Dordrecht, 1992, pp. 351-381.

91 A.R. Savkoor, Models of friction based on contact and fracture mechanics, Fundamentals of Friction: Macroscopic and Microscopic Processes, editors: I.L. Singer and H.M. Pollock, NATO ASI Series, Kluwer Academic Publishers, Dordrecht, 1992, pp. 111-133.

92 F.P. Bowden and D. Tabor, The Friction and Lubrication of Solids, Clarendon Press: Oxford, 1950.

93 D. Tabor and R.H. Winterton, The direct measurement of normal and retarded van der Waals forces, *Proc. Royal Society, London*, Series A, Vol. 312, 1969, pp. 435-450.

94 D. Tabor, Surface forces and surface interactions, *Journal of Colloid Interface Sci.*, Vol. 58, 1977, pp. 2-13.

95 J.N. Israelachvili, Thin film studies using multiple-beam interferometry, *Journal of Colloid Interface Sci.*, Vol. 44, 1973, pp. 259-272.

96 J.N. Israelachvili, The calculation of van der Waals dispersion forces between macroscopic bodies, *Proc. Royal Society, London*, Series A, Vol. 331, 1972, pp. 39-55.

97 R.G. Horn, J.N. Israelachvili and F.J. Pribac, Measurement of the deformation and adhesion of solids in contact, *J. Colloid Interface Sci.*, Vol. 115, 1987, pp. 480-492.

98 A. Tonck, J.M. Georges and J.L. Loubet, Measurement of intermolecular forces and the rheology of dodecane between alumina surfaces, *J. Colloid Interface Sci.*, Vol. 126, 1988, pp. 150-163.

99 H.K. Christenson, Adhesion and surface energy of mica in air and water, *J. Phys. Chem.*, Vol. 97, 1993, pp. 12034-12041.

100 J.M. Georges, A. Tonck and D. Mazuyer, Interfacial friction of wetted monolayers, *Wear*, Vol. 175, 1994, p. 59-62.

101 S. Bec and A. Tonck, Nanometer Scale mechanical properties of tribochemical films, Proc. 22nd Leeds-Lyon Symposium, Lyon, France, 5-8 September 1995, editors: G. Dalmaz, D. Dowson, C. Taylor and T. Childs, 1995, pp. 173-184.

102 A. Tonck, S. Bec, D. Mazuyer, J.M. Georges and A.A. Lubrecht, The Ecole Centrale de Lyon surface force apparatus: An application overview, *Proc. Inst. Mech. Engrs, Part J, Journal of Engineering Tribology*, Vol. 213, 1999, pp. 353-361.

103 R.E. Kornbrekke, I.D. Morrison and T. Oja, Electrophoretic mobility measurements in low conductivity media, *Langmuir*, 1992, Vol. 8, pp. 1211-1217.

104 D.H. Buckley, Oxygen and sulphur interaction with a clean iron surface and the effect of rubbing contact on these interactions, *Transactions ASLE*, Vol. 17, 1974, pp. 206-212.

105 L.E. Murr, Interfacial Phenomena in Metals and Alloys, Addison Wesley, London, 1975.

106 J.M. Challen and P.L.B. Oxley, Plastic deformation of a metal surface in sliding contact with a hard wedge: Its relation to friction and wear, *Proc. Roy. Soc. London*, Series A, Vol. 394, 1984, pp. 161-181.

107 M. Masuko, N. Naganuma and H. Okabe, Acceleration of the thermal reaction of sulfur with steel surface under increased pressure, *Tribology Transactions*, Vol. 33, 1990, pp. 76-84.

108 E.D. Tingle, The importance of surface oxide films in the friction and lubrication of metals, Part 2 - The formation of lubrication films on metal surfaces, *Transactions Faraday Society*, Vol. 326, 1950, pp. 97-102.

109 A.W. Batchelor, A. Cameron and H. Okabe, An apparatus to investigate sulfur reactions on nascent steel surfaces, *ASLE Transactions*, Vol. 28, 1985, pp. 467-474.

110 S. Mori, M. Suginoya and Y. Tamai, Chemisorption of organic compounds on a clean aluminium surface prepared by cutting under high vacuum, *ASLE Transactions*, Vol. 25, 1982, pp. 261-266.

111 S. Mori and M. Yoshida, Decomposition of aromatic compounds on cut nickel surface, *Tribology Transactions*, Vol. 31, 1988, pp. 128-132.

112 S. Mori, Adsorption of benzene on the fresh steel surface formed by cutting under high vacuum, *Applied Surface Science*, Vol. 27, 1987, pp. 401-410.

113 S. Mori and Y. Imaizumi, Adsorption of model compounds of lubricant on nascent surfaces of mild and stainless steels under dynamic conditions, *Tribology Transactions*, Vol. 31, 1988, pp. 449-453.

114 S. Mori and Y. Shitara, Tribochemical activation of gold surface by scratching, *Applied Surface Science*, Vol. 78, 1994, pp. 269-273.

115 K. Nakayama and H. Hashimoto, Triboemission, tribochemical reaction and friction and wear in ceramics under various N-butane gas pressures, *Tribology International*, Vol. 29, 1996, pp. 385-393.

116 S. Mori, T. Kawada and W.-C. Xu, Tribochemical decomposition of formic acid on the nascent surfaces of steel formed by scratching, *Applied Surface Science*, Vol. 108, 1997, pp. 391-397.

117 K. Meyer, H. Berndt and B. Essiger, Interacting mechanisms of organic sulphides with metallic surfaces and their importance for problems of friction and lubrication, *Applications of Surface Science*, Vol. 4, 1980, pp. 154-161.

118 K. Date, Adsorption and lubrication of steel with oiliness additives, Ph.D. thesis, London University, September 1981.

119 A.J. Groszek, Activation of iron surfaces by chemisorption of some E.P. and anti-wear compounds, *Wear*, Vol. 18, 1971, pp. 279-289.

120 L.E. Pope and D.E. Peebles, In situ friction, wear and electrical resistance of precious metal alloys, *Tribology Transactions*, Vol. 31, No. 2, 1987, pp. 202-213.

121 M. Beltzer, Assessing adsorption of conventional friction modifying molecules by relative contact potential difference measurements, *Transactions ASME, Journal of Tribology*, Vol. 114, 1992, pp. 675-682.

122 W.A. Zisman, A new method of measuring contact potential differences in metals, *Review of Scientific Instruments*, Vol. 3, 1932, pp. 367-370.

123 B. Bhushan, Nanomechanical properties of solid surfaces and thin films, Handbook of Micro/Nano Tribology, editor: B. Bhushan, CRC Series Mechanics and Materials Science, CRC Press Inc., 1995.

124 O.S. Heavens, Some factors influencing adhesion of films produced by vacuum evaporation, *Journal de Physique ie le Radium*, Vol. 11, No. 7 July, 1950, pp. 355-360.

125 P.J. Blau, Applications of microindentation methods in tribology research, Microindentation Techniques in Materials Science and Engineering, editors: P.J. Blau and B.R. Lawn, ASTM STP 889, American Society for Testing and Materials, Philadelphia, PA, 1986, pp. 209-226.

126 B. Bhushan, Tribology and Mechanics of Magnetic Storage Devices, Springer-Verlag, New York, 1990.

127 K.L. Mittal (editor), Adhesion measurements on thin coatings, Thick Coatings and Bulk Coatings, ASTM Special Technical Paper No. 640, 1978, Symposium at ASTM Headquarters, Nov. 2-4, 1976, Philadelphia, PA, 1978.

128 B. Bhushan, Overview of coating materials, surface treatments, and screening techniques for tribological applications — Part 2: Screening techniques, Testing of Metallic and Inorganic Coatings, A Symposium sponsored by ASTM Committee B-8 on Metallic and Inorganic Coatings, Chicago IL, 14-15 April 1986, (editors: W.B. Harding and G.A. DiBari), ASTM Special Technical Paper 947, Philadelphia PA, 1987, pp. 310-319.

129 J.D.S. Campbell, Mechanical properties of thin films, Handbook of Thin Film Technology, editors: L.I. Maissel and R. Glang, McGraw-Hill, New York, 1970.

130 P.J. Burnett and D.S. Rickerby, The relationship between hardness and scratch adhesion, *Thin Solid Films*, Vol. 154, 1987, pp. 403-416.

131 S.J. Bull and D.S. Rickerby, New developments in the modeling of the hardness and scratch adhesion of thin films, *Surface Coatings Technology*, Vol. 42, 1990, pp. 149-164.

132 T.W. Wu, R.A. Burn, M.M. Chen and P.S. Alexopoulos, Micro-indentation and micro-scratch tests on sub-micron carbon films, Symp. Proc., Materials Research Society, Pittsburgh, Vol. 130, 1989, pp. 117-121.

133 T.W. Wu, A.L. Shull and R. Berriche, Microindentation Fatigue Tests on Submicron Carbon Films, *Surface Coatings Technology*, Vol. 47, 1991, pp. 696-709.

134 H.E. Hintermann, Surface treatments, Science of Hard Materials, editors: R.K. Viswanadham, D.J. Rowliffe and S. Gurland, Plenum Press, New York, 1983.

135 H.E. Hintermann, Adhesion, friction and wear of thin hard coatings, *Wear*, Vol. 100, 1984, pp. 381-397.

136 P. Benjamin and C. Weaver, Measurement of adhesion of thin films, *Proc. R. Soc. London, Series A*, Vol. A254, 1960, pp. 163-176.

137 J.E. Greene, J. Woodhouse and M. Pestes, A technique for detecting critical loads in the scratch test for thin-film adhesion, *Rev. Sci. Instrum.*, Vol. 45, 1974, pp. 747-749.

138 A.J. Perry, The adhesion of chemically vapour-deposited hard coatings on steel - the scratch test, *Thin Solid Films*, Vol. 78, 1981, pp. 77-93.

139 A.J. Perry, Scratch adhesion testing of hard coating, *Thin Solid Films*, Vol. 197, 1983, pp. 167-180.

140 M. Laugier, The development of scratch test technique for the determination of the adhesion of coating, *Thin Solid Films*, Vol. 76, 1981, pp. 289-294.

141 S. Jacobson, B. Jonsson and B. Sundquist, The use of fast heavy ions to improve thin film adhesion, *Thin Solid Films*, Vol. 107, 1983, pp. 89-98.

142 J.H. Je, E. Gyarmati and A. Naoumidis, Scratch adhesion test of reactively sputtered TiN coatings on a soft substrate, *Thin Solid Films*, Vol. 136, 1986, pp. 57-67.

143 P.A. Steinmann, Y. Tardy and H.E. Hintermann, Adhesion testing by the scratch test method: The influence of intrinsic and extrinsic parameters on the critical load, *Thin Solid Films*, Vol. 154, 1987, pp. 333-349.

144 J. Sekler, P.A. Steinmann and H.E. Hintermann, The scratch test: Different critical load determination techniques, *Surface Coatings Technology*, Vol. 36, 1988, pp. 519-529.

145 B. Bhushan, B.K. Gupta and M.H. Azarian, Nanoindentation, microscratch, friction and wear studies of coating for contact recording applications, *Wear*, Vol. 183, 1995, pp. 743-758.

146 B. Bhushan and B.K. Gupta, Micromechanical characterization of Ni-P coated aluminum-magnesium glass and glass-ceramic substrates and finished thin-film rigid disks, *Adv. Info. Storage Syst.*, Vol. 6, 1995, pp. 193-208.

147 T.W. Wu, M. Moshref and P.S. Alexopoulos, The effect of the interfacial strength on the mechanical properties of aluminum films, *Thin Solid Films*, Vol. 187, 1990, pp. 295-307.

148 S. Ramalingham and S. Kim, Tribological characteristics of arc coated hard compound films, Proc. Int. Tribology Conf., Melbourne, Australia, 2-4 December 1987, IEA Nat. Conf. Publ. No. 87/18, pp. 403-408.

149 E. Santner, D. Klaffke and G. Meier zu Kocker, Comprehensive tribological characterization of thin TiN-based coatings, Proceedings of the 4th International Tribology Conference, AUSTRIB '94, 'Frontiers in Tribology', editor: G.W. Stachowiak, 5-8th December 1994, Vol. I, 1994, pp. 271-278.

4 MEASUREMENT OF FRICTION AND WEAR

4.1 INTRODUCTION

The measurement of friction and wear appears to be a relatively simple task but becomes more complex when the exact meaning of friction or wear is considered. Does wear mean a loss of material or can it mean merely a displacement of material over the surface of a wearing body? If frictional and contact forces both vary in a continuous and random manner during wear, how can a meaningful friction coefficient be obtained? Despite many years of research, there is still no general agreement on precise definitions of friction and wear, which hinders comparison of data from different research groups.

Accurate friction and wear data requires high quality measurement technology for valid results. Early research relied on rudimentary methods such as hanging weights for friction measurements but these practices have been superseded by far superior measurement techniques. The quality of friction and wear data available influences the interpretation of friction and wear phenomena so that an understanding of the merits and limitations of the measurement technology is essential to tribological experimentation.

4.2 MEASUREMENTS OF FRICTION COEFFICIENT

In almost all tribology tests (with the exception of tests based on a block sliding down an inclined slope) the friction coefficient is deduced from measurements of the friction force. There are two uncertainties involved in the determination of friction coefficient from friction force:

- the friction force varies continuously so that the notion of a precise coefficient of friction is an approximation to reality,
- the calculation of friction coefficient is based on the nominal contact load, i.e.:

Friction coefficient = Friction force / Nominal contact load

In most cases the nominal contact load is usually applied by the hanging weight to impose a contact stress on the sliding surfaces. It is essential to provide the weight with soft suspension because with rough surfaces the contact load can vary significantly, i.e. the rougher the surface the higher the load variation. Friction force is also affected by surface roughness [1]. When the relationship between variation in contact load and friction force is not known, it has to be assumed that peaks in friction force coincide with peaks in contact load to provide an averaging effect. If there is an averaging effect, then the equation for friction coefficient given above is approximately correct. This concept is illustrated schematically in Figure 4.1.

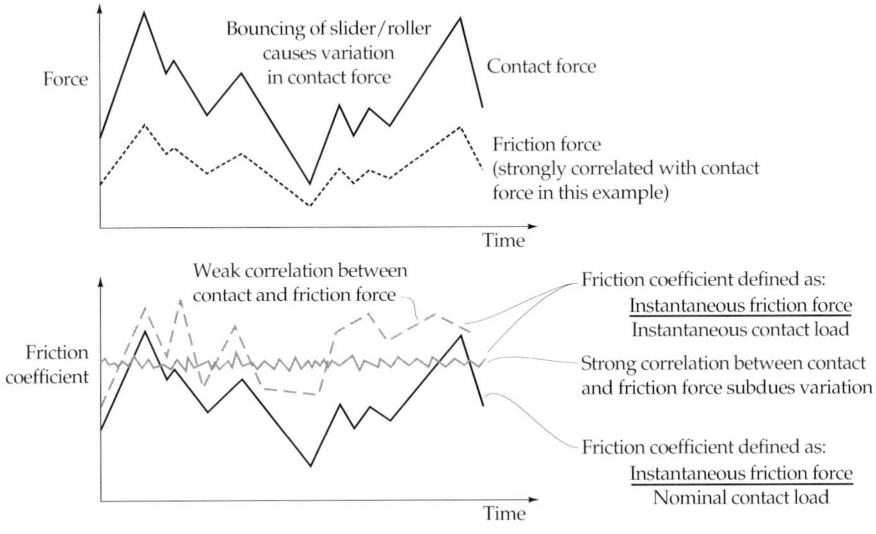

Figure 4.1 Schematic illustration of average coefficient of friction concept.

The definition of friction coefficient as friction force / nominal contact load enables the energy dissipation rate to be found directly from the friction coefficient. This feature is very useful for predicting average temperature rises. The definition of friction coefficient as instantaneous friction force / instantaneous contact load provides a measure of the intensity of asperity interaction or film shearing forces at a particular point in time. This definition of friction coefficient alone allows the effect of microscopic processes at the contacting surface to be isolated from macroscopic processes such as mechanical vibration of the sliding system. It is also possible to measure the true friction coefficient by simultaneous measurement of instantaneous contact load and frictional force [2].

Techniques of Friction Force Measurement

There are two basic types of device commonly used for the measurement of friction force and a subsequent determination of a friction coefficient. These devices are the piezoelectric force gauge and the strain gauge transducer. Where friction transducers present problems, friction coefficients can also be determined from either the angle of inclination to the vertical where sliding commences, i.e. $\mu = \tan\alpha$, where 'μ' is the coefficient of friction while 'α' is the angle of inclination, or from the use of counterbalancing weights. Another method of determining the coefficient of friction involves a pendulum and the measurement of the time necessary to extinguish the pendulum oscillations. However, none of these techniques which do not employ transducers is suitable for measuring a variable friction coefficient and are rarely used in current research.

· *Piezoelectric Force Gauges*

Piezoelectric force gauges provide a direct record of frictional force as an electrical impulse which can be recorded electronically. Earlier versions of piezoelectric force transducers originally contained a quartz crystal which has the property of emitting electric charge when it is compressed or stretched. Quartz crystals have now been largely superseded by ferro-electric ceramics or lead zirconate titanate to obtain a stronger electric signal [3]. Piezoelectric devices can be used to monitor frequencies as high as 25 [kHz] [3] so that even very rapid variations in friction force can be measured. However, it should be realized that the mass of the structure supporting the friction specimen will reduce the frequency limit of any piezoelectric device that is rigidly connected to this structure. Piezoelectric devices function by elastic deflection of a piezoelectric crystal and the operating frequency limit has to be less than the resonant frequency of the crystal and attached mass. Although the frequency limit is high when compared to other devices, i.e. strain gauge transducer, some averaging of the friction force will still occur as the following argument will show. For a typical test sliding speed of 1 [m/s], a 25 [kHz] frequency corresponds to 40 [μm] of sliding distance per cycle of oscillation. Asperity deformation and fracture of adhesive bonds between asperities are usually completed over distances shorter than 40 [μm] so that the piezoelectric force gauge will only detect friction force variations caused by individual asperity contacts at low sliding speeds. There is also a low frequency limit for piezoelectric force gauges which renders them unsuitable for measuring a steady friction force [3].

The method of mounting a piezoelectric force gauge is critical to effective measurement. Loose or excessively flexible mountings should be avoided or else the resulting vibration may affect the force gauge and subsequently the experimental data. The principle of piezoelectric force gauges and methods of mounting a piezoelectric force gauge within a tribometer are shown in Figure 4.2.

If the friction force is sufficiently small then the piezoelectric force gauge can be mounted directly to the friction specimen as shown in Figure 4.2. For tests involving larger friction forces than can be accommodated by commercially produced piezoelectric force gauges, it would be necessary to develop a lever system that imposes only a fraction of the friction force on the gauge. It is important to minimize the friction forces in lever pivots so that they will not cause a systematic error in the force measurements. A lever which is based on a strut and does not involve a pivot is shown in Figure 4.2. Vertical load on the friction specimen can be supported by flexible columns as shown in Figure 4.2 to avoid any frictional forces resulting from rolling bearing suspensions or other types of bearings. Measured values of friction coefficient can either be erroneously reduced or increased by friction forces in the specimen suspension [4]. When the force transducer is connected to the stationary specimen, a reduced friction force is measured. If the force transducer is connected to the mobile specimen, friction forces in the suspension system of the mobile specimen cause an elevated friction coefficient to be recorded. The effect of suspension friction forces is illustrated in Figure 4.3.

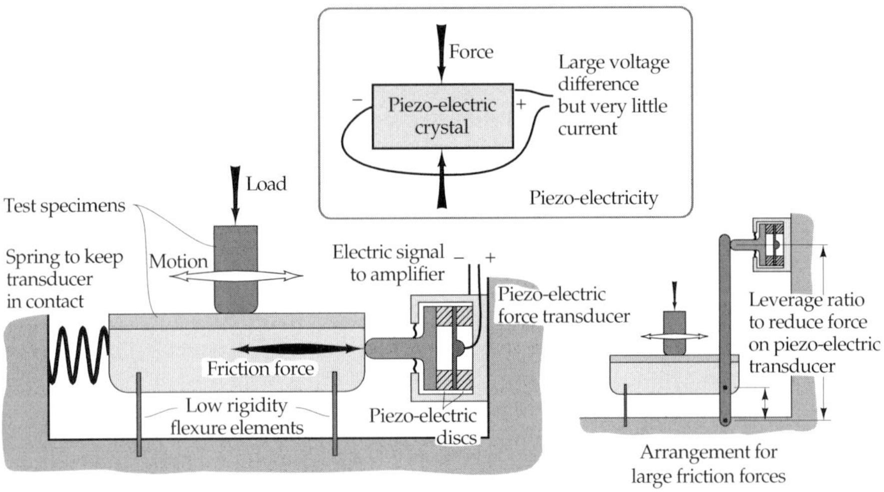

Figure 4.2 Schematic diagram of mounting of a piezoelectric force transducer in a tribometer; a) direct connection to a friction specimen, b) levered arrangement for the measurement of large friction forces.

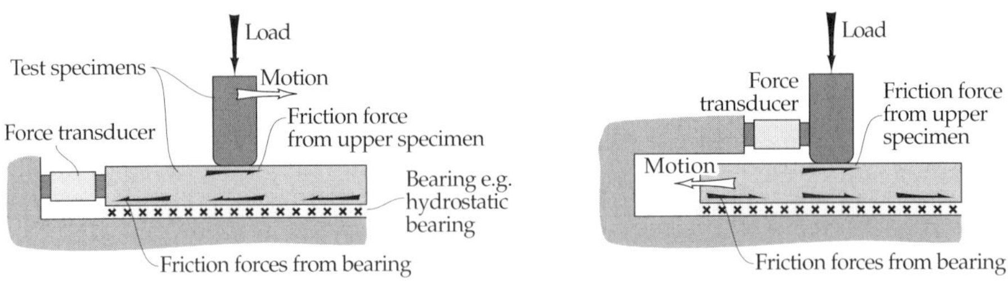

a) Force transducer records test friction force minus bearing friction forces

b) Force transducer records only test friction force

Figure 4.3 Illustration of the effect of suspension friction forces on the measurements of friction coefficient (adapted from [4]).

Piezoelectric force gauges are sensitive to temperature, vibration and corrosive agents. They are also comparatively expensive. In many instances, the use of these gauges cannot be justified despite their good force recording characteristics and other friction measuring techniques should be used instead.

- *Strain Gauged Beams*

Strain gauged beams are considerably cheaper and can be designed to suit almost any level of friction force. Friction force is usually measured from the bending of a beam arranged perpendicularly to the direction of the friction force [5]. Measurement of friction force using the principle of a flexible beam is schematically illustrated in Figure 4.4.

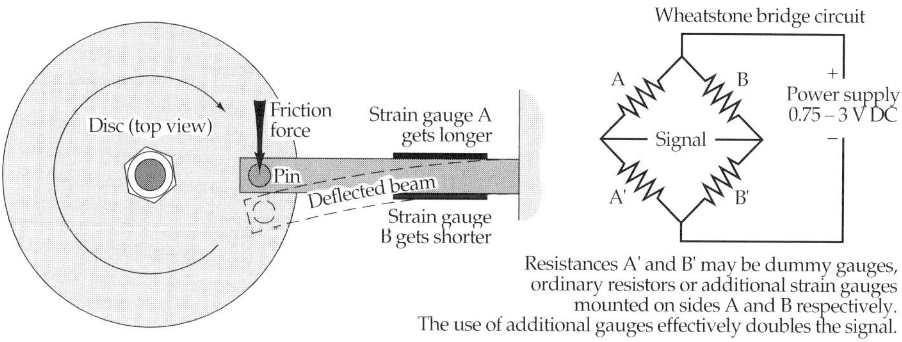

Figure 4.4 Schematic illustration of the friction transducer applying a combination of strain gauges and a flexible beam.

A similar principle is utilized to measure coefficient of friction in vacuum as illustrated in Figure 4.5. Frictional force, acting tangentially to a disc sample, is measured by recording changes in circuit resistance. From the friction force and the normal force (load) a coefficient of friction is calculated.

<u>Frequency limit of strain gauged beams</u>

In most instances, strain gauged beams provide a sensitive and accurate means of friction measurement. They are effective at recording steady friction forces where piezoelectric force gauges are unsuitable. There are, however, certain conditions where strain gauged beams are entirely unsuited for the recording of friction coefficients. The most common cause of strain gauged beam failure is excessively rapid change in friction force. Under dry sliding usually the friction coefficient can change very rapidly or else display a stick-slip characteristic. Friction changes during tests using a high frequency, short amplitude, reciprocating sliding apparatus are necessarily very rapid and strain gauges are unable to produce reliable data.

The reason for this loss of performance is that the deflection of a flexible beam is only proportional to applied force when the rate of change of the force is much slower than the resonant frequency of the beam. If the beam is subjected to a rate of change of force higher than its resonant frequency, it will merely vibrate instead of deflecting in proportion to the frictional force [6]. In extreme cases, the beam may resonate and subject the friction specimen to uncontrolled variations in sliding speed as well as failing to produce meaningful friction data. Under such conditions, the flexibility of the beam becomes a parameter that controls the friction and wear process so that the strain gauged beam can no longer be considered to function as a transducer. A solution to this problem is to raise the stiffness of the beam but this practice will reduce the sensitivity of the transducer to small friction forces.

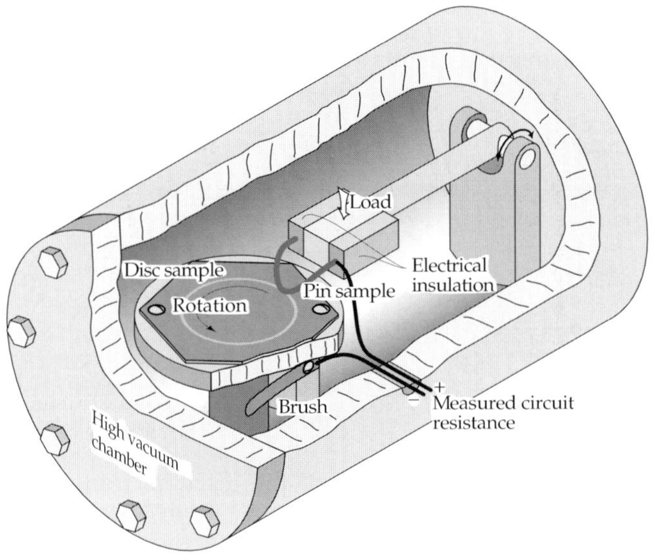

Figure 4.5 Schematic illustration of wear and friction coefficient measurement in vacuum.

For reliable experimental data, beam stiffness should be as high as possible and the small strain signals that are obtained should be electronically amplified. In some cases it is possible to obtain friction data even when there is severe transducer vibration but this requires an elaborate vibration analysis. For practical purposes, measured fluctuations in friction force should be sufficiently slow so that the strain gauged beam is only required to oscillate at frequencies less than its lowest resonant frequency of vibration. The beam stiffness may also affect the friction and wear characteristics as discussed in Chapter 3 so that the stiffness should be carefully selected to satisfy the needs of both measurement and experimental conditions.

Effect of Surface Levelling and Roughness on Measured Coefficient of Friction

Levelling of surfaces is particularly important when measuring low coefficients of friction. Even small slopes of about **1°** can result in large errors in coefficients of friction measured. This is particularly important in unidirectional sliding, e.g. during pin-on-disc experiments, where the measured friction coefficient and the true friction coefficient are not necessarily the same, especially with low friction coefficients near **0.01**. The friction coefficient is defined as the tangential force '**F**' divided by the normal force '**W**', i.e. μ = **F/W**. If the disc is tilted by a small angle 'γ' (few degrees) from the normal, then the frictional force '$\mathbf{F_m}$' (measured force) acting on a strain gauge, parallel to the table, will also be a few degrees off the 'tangential' force '**F**', as schematically illustrated in Figure 4.6.

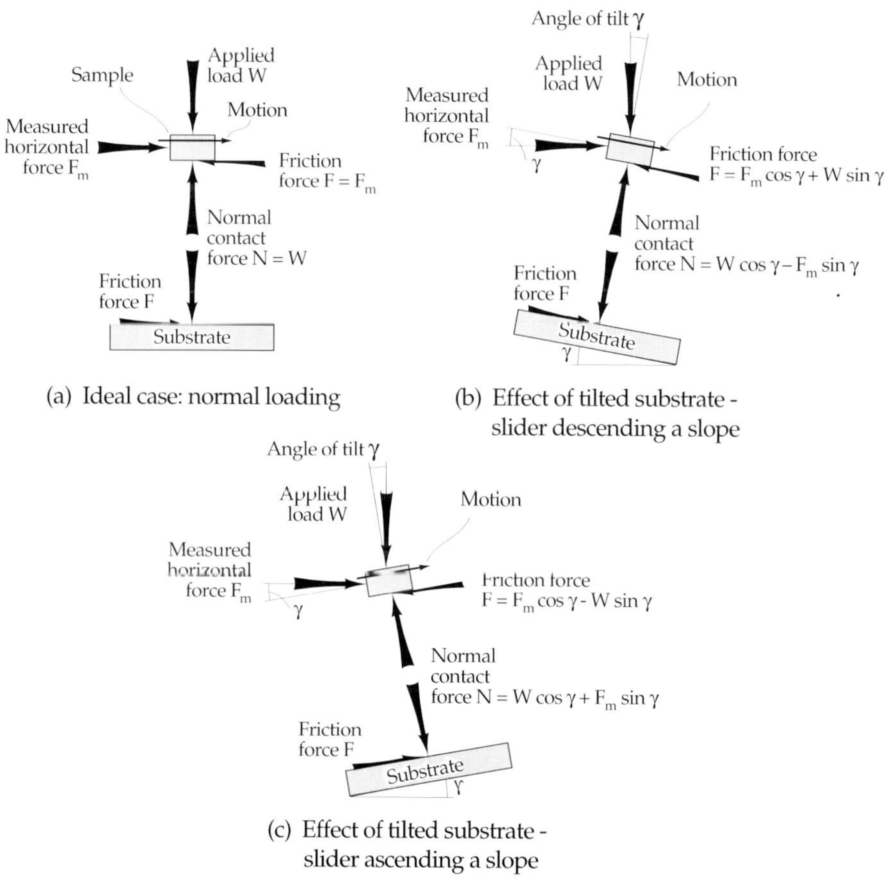

Figure 4.6 Effect of specimen tilt angle on the apparent friction force.

The arrangement shown in Figure 4.6a corresponds to the typical case where the slider aligns with the disc during sliding. If the disk is not perfectly aligned and the slider is, for example, descending then the deviation of load from perpendicular to the sliding direction results in an additional small component of '**Wsin**γ' contributing to the friction force '**F**'. As a result, the measured coefficient of friction, 'μ_m' diverges from the true friction coefficient 'μ', as it can be deduced from the free body diagram, shown in Figure 4.6b, in the following manner:

$$\mu = \frac{F}{N} = \frac{F_m \cos\gamma + W \sin\gamma}{W \cos\gamma - F_m \sin\gamma} = \frac{\frac{F_m}{W} + \tan\gamma}{1 - \frac{F_m}{W}\tan\gamma} = \frac{\mu_m + \tan\gamma}{1 - \mu_m \tan\gamma} \quad (4.1)$$

The measured coefficient of friction is then:

$$\mu_m = \frac{\mu - \tan\gamma}{\mu \tan\gamma + 1} \quad (4.2)$$

For an ascending slider the measured coefficient of friction 'μ_m' can be calculated in a similar manner from the free body diagram shown in Figure 4.6c, i.e.:

$$\mu = \frac{F}{N} = \frac{F_m \cos\gamma - W \sin\gamma}{W \cos\gamma + F_m \sin\gamma} = \frac{\frac{F_m}{W} - \tan\gamma}{1 + \frac{F_m}{W}\tan\gamma} = \frac{\mu_m - \tan\gamma}{1 + \mu_m \tan\gamma} \quad (4.3)$$

and

$$\mu_m = \frac{\mu + \tan\gamma}{1 - \mu \tan\gamma} \quad (4.4)$$

For small angles, $\mu_m = \mu + \gamma$, where 'γ' is in radians [7]. This correction is rarely made in friction measurements because 'γ' is usually very small compared to μ, i.e. for a 1° tilt, γ = **0.017** radians. However, in ultra-low friction measurements, where μ is about 0.01, this angular contribution can be significant. Therefore, to make an ultra-low friction measurement in unidirectional sliding, one must carefully level the disc, e.g., to 0.1 degree, to claim 10% accuracy for a friction coefficient of 0.017 [8].

It needs to be mentioned that this error is less of a problem in reciprocating sliding because for any angle of tilt, the deviation of measured friction from its true value is reversed when the slider changes direction of sliding. This means that the average friction force measured over a complete cycle of sliding should be very close to the true value [7]. Non-zero tilt will, however, cause the friction force in one phase of the reciprocating movement to differ from the friction force in the other phase. It is also possible that a perfectly aligned load system may lose alignment due to non-uniform wear on the slider. This would cause the slider to tip with some inevitable effect on the loading system.

It should also be noted that most but not all materials tend to display a symmetrical friction coefficient, i.e. the friction coefficient is identical whichever direction of sliding motion. However, with some materials this is not always the case. For example, materials, such as wool fibres possess a ratcheted surface where the friction coefficient is very high in one direction but not in the other. For these materials, the average friction coefficient would vary with the angle of tilt during reciprocating sliding. Thus in cases of friction tests on materials with asymmetric friction coefficients the effects of specimen tilt angle can be significant.

It has also been shown that when diamond slides on diamond coatings in air, friction coefficient can widely vary, e.g. between 0.03 and 0.5, and depends on the roughness of sliding surfaces. The coefficient of friction decreases with the decreasing surface roughness. It has been proposed by Tabor [9] that a combined adhesion/asperity climbing mechanism is responsible for the friction dependence on surface roughness. To allow for the effect of surface roughness the coefficient of friction can be calculated from the following formula [9]:

$$\mu_{av} = \frac{\mu_t(1 + \tan^2\theta)}{1 - \mu_t^2 \tan^2\theta} \qquad (4.5)$$

where:

μ_{av} is the friction coefficient between the rough surfaces [dimensionless];

μ_t is the true friction coefficient between the smooth surfaces [dimensionless];

θ is the slope angle of the surface asperities [dimensionless].

More information on the effects of surface roughness in ultra-low friction measurements can be found in [10].

Simultaneous Measurement of Load and Friction Force

The instantaneous contact force should be continuously measured for exact determinations of the friction coefficient. This contact force can be measured by fitting strain gauge flexure elements that deflect with variations of force in the load axis [2,5]. The flexure elements have a plane of flexibility, which is orthogonal to the flexure elements for friction force measurements. Simultaneous load and friction force measurement by the use of two strain gauged beams is illustrated schematically in Figure 4.7.

It should be mentioned that the system of two strain gauged beams is even more susceptible to resonant vibration than the single strain gauged beam. Thus in cases where rapid variations in friction occur during testing, it may be more effective to use two piezoelectric force gauges for load and friction force measurements. To measure the load, the piezoelectric force gauge should be connected to the stationary test specimen in order to detect the changes in contact force from normal movements, i.e. 'bouncing', of the mobile specimen.

It should also be noted that strain gauged beams are prone to twisting if the flexible beam is located too far above or below the plane of sliding. Ideally, the longitudinal axis of the beam should lie on the plane of sliding but this is usually prevented by mechanical interference between test specimens. The effect of distance between the longitudinal beam axis and the plane of sliding is shown in Figure 4.7. Twisting of the beam as a result of excessive distance between the beam axis and the plane of sliding causes tipping of the test specimen, supported by the strain gauged beam (e.g. the pin of a pin-on-disc machine), during sliding.

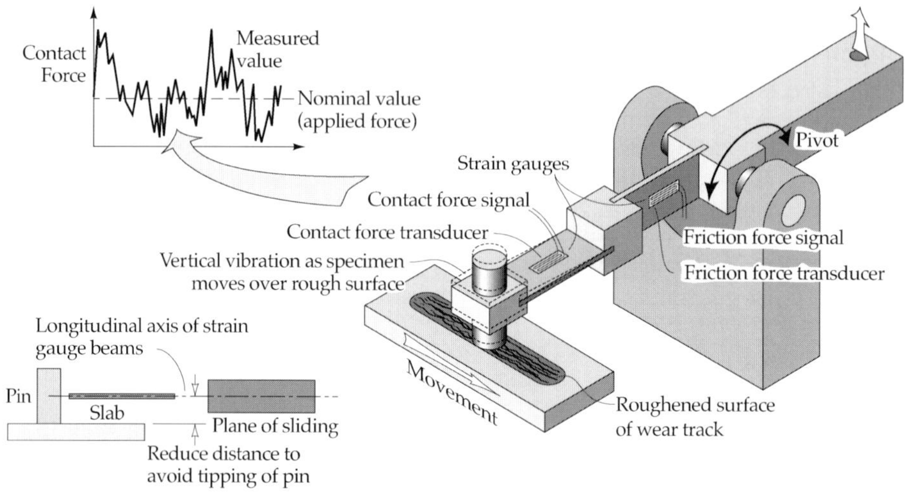

Figure 4.7 Schematic illustration of strain gauge instrumentation for simultaneous measurement of contact load and friction force.

Tipping of a specimen causes uneven wear and non-uniform distribution of contact stresses on the wearing surface, which may affect the validity of experimental data. A design of tribometer where the dynamic contact is located at the centre of four strain gauged beams placed above and below the plane of sliding contact virtually eliminates the effect of specimen tipping at the expense of heightened complexity of the design [2].

Miscellaneous methods

Friction measurement using electrical transducers is a relatively recent innovation in tribology and frictional data can also be obtained by other means. The most commonly used methods are:

- angle of inclination to commence sliding,
- tangential force to initiate sliding,
- pendulum method.

These three methods are schematically illustrated in Figure 4.8.

Early research in tribology depended on the use of either a sloping table with variable angle of inclination or a hanging weight that was attached by a pulley and cord to the test specimen [11]. These methods are rarely used in current research, apart from for one or two notable exceptions. The angle of inclination method was applied to the first experimental demonstration of high friction coefficients between clean metal surfaces under vacuum [12]. This experimental arrangement allowed the friction coefficients in a simple vacuum apparatus to be measured without the complexity involved in obtaining electrical signals from a friction transducer mounted inside the vacuum chamber. Friction experiments were performed in a small glass vessel, which could be tilted at increasing angles of inclination until a test specimen commenced sliding.

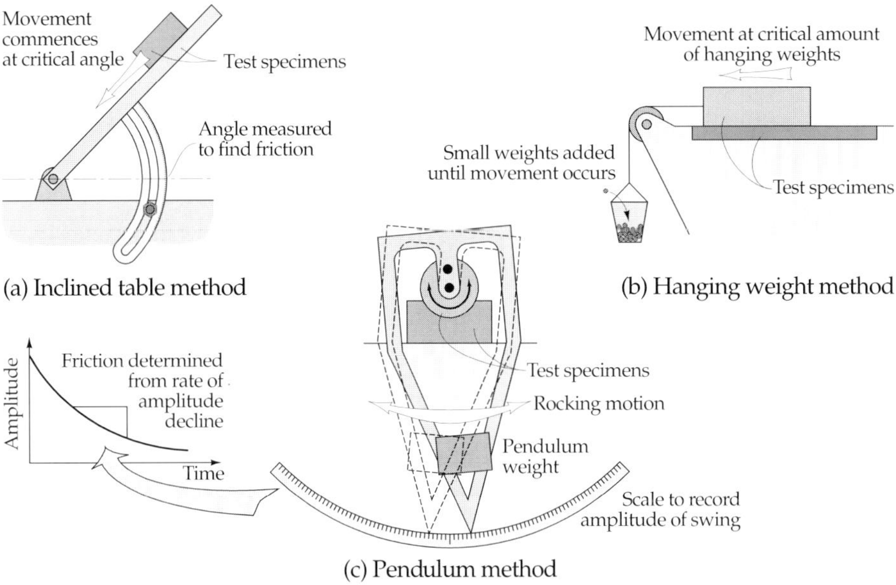

Figure 4.8 Schematic illustration of the most commonly used methods of friction coefficient measurement without the use of electronic transducers.

A more recent example of the application of inclination method in the measurement of coefficient of friction is found in biotribology. An inclined exercise table was used to measure the friction coefficient of animal footpads to the ground [13]. Although the inclined table only permits an elementary measurement of the maximum value of the friction coefficient of the footpads before slipping occurs, this method is still preferable to the complexity of applying instrumentation to the animals feet or to the exercise machine.

The oscillating pendulum method permits the measurement of very small coefficients of friction operating in dynamic contacts which consist of one rotating component and one static component [14]. A common example of this is a journal bearing, consisting of a shaft and a bush as schematically illustrated in Figure 4.8c. A basic disadvantage of the pendulum method is that reciprocating sliding can only be studied over a narrow range of sliding speeds. Perhaps the most useful and exciting application of the pendulum method is found in the measurements of coefficient of friction in synovial joints [15,16]. During these experiments it is essential to preserve the synovial joint in its original condition. An intact knee or hip with synovial joint is placed in a pendulum apparatus and the friction coefficient is measured without the problem of mounting force transducers to human tissue [15-19].

Measurement of Friction Coefficient With One Stationary Test Specimen

Most tribometers are designed with a stationary and a mobile test specimen in contact. Instrumentation to measure friction and wear reflect this general rule and in most cases the instrumentation, i.e. force transducers, load cells, etc., are connected to the stationary specimen. For the minority of tribometers that involve two moving specimens, different arrangements to measure friction and wear are required.

Friction Measurements When Both Test Specimens Are Mobile

When both test specimens are mobile, the measurement of friction becomes more complex. Any electrical connection between a transducer mounted on a specimen and a remote amplifier requires a device to allow for movement between the specimen and its surroundings. Two-disc apparatus and metal rolling experiments are typical examples of this problem. A commonly used method involves fitting a slip ring to the shaft of the test disc or roller which allows the passage of an electrical signal from strain gauges mounted on the moving shaft to a stationary amplifier. A simpler but more approximate method involves monitoring of the power consumed to drive the rotating shafts and equating this to the frictional power dissipation in the test contact plus any power dissipation in bearings (which is usually very small). Experiments involving these techniques are limited in number and new more effective methods of measurement will certainly be developed with the rapid advancement of technology.

4.3 MEASUREMENTS OF WEAR

The measurement of wear is partly determined by the exact definition of wear that is applied. The simplest definition of wear is perhaps the change in mass of an object as wear progresses. It is possible to measure this mode of wear by weighing the worn object before and after a wear test. The difficulty with this wear assessment, based on mass changes, is that no allowance is made for material displaced by wear and yet remaining attached to the worn object. Examples of displaced material are lips and trailing strands of material on the downstream side of a wear specimen, i.e. the material is not removed as a wear particle, yet it is no longer able to support the contact load so that wear can be said to have occurred. In this case, the wear would more accurately be determined from measurements of the distance between the worn surface and a datum located elsewhere on the worn specimen, i.e. depth of the wear scar.

Another problem in the experimental measurement of wear is the loss of data when wear is measured over a series of intervals as opposed to continuous monitoring. Much information about wear can be found from continuous records of the progress of wear. Sudden or periodic loss of material from a worn contact is more reliably observed from continuous wear measurements than from occasional measurements. Continuous wear records enabled a periodic release by a polymer specimen of layers of molten polymer during high speed sliding against a metal counterface to be detected [20]. Accumulation of material as transfer films and later release as wear particles also requires continuous monitoring of wear [21]. The differences between wear measurements based on mass changes and specimen dimensions and the loss of information when continuous wear recording is substituted by periodic measurements are illustrated schematically in Figure 4.9.

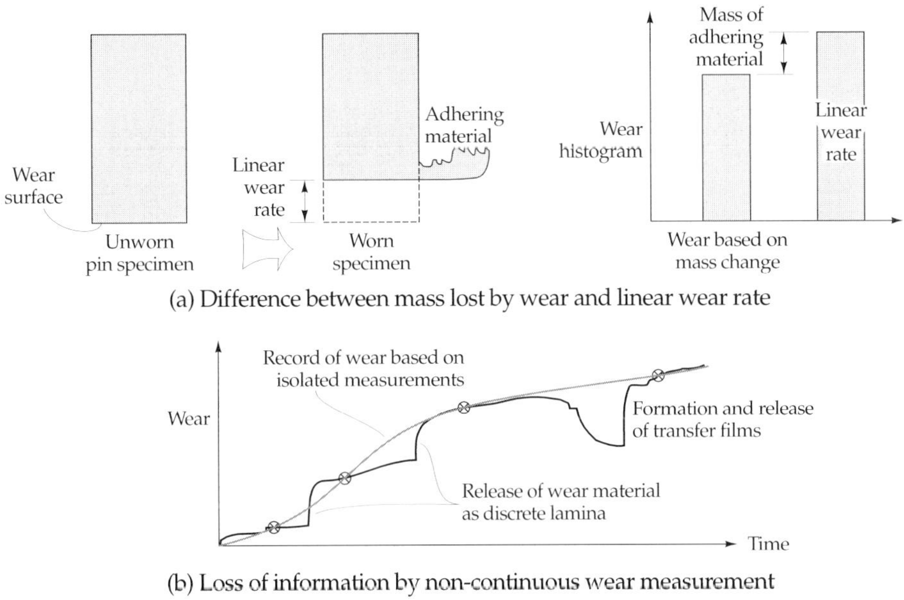

Figure 4.9 Schematic illustration of the difference between wear measurements based on mass changes and on dimensional changes (a) and loss of information when continuous wear recording is substituted by periodic measurements (b).

At present there appear to be three basic methods of measuring wear:

- detection of change in mass,
- measurement of reduction in dimension of a worn specimen and
- profilometry of the worn specimen.

There are also other methods of wear measurements, e.g. based on the detection of radioactivity from specimens irradiated in a nuclear reactor, in a particle accelerator or using an isotope source, e.g. cobalt. Often radioisotopes are used as a cheaper alternative. Measurements of the level of radioactivity in fluids or lubricants flushed through the worn contact provide an estimate of the rate of wear. This method is known as thin layer activation and is used for wear tests of inaccessible contacts such as those found inside engines [22]. In some cases small inserts of radioactive material implanted into wearing surfaces are used in order to reduce the levels of radiation. Another simple method of wear measurements involves making a hardness indentation and measuring changes in the size of the indentation imprint. A more recently developed method that is suitable for remote wearing contacts is an application of interference between ultrasonic waves. The change in dimension of a specimen can be detected from variations in interference between transmitted and reflected ultrasonic waves within the sample [23]. The characteristics of main wear measurement methods are described below.

Determination of Wear from Weight Loss

Measurement of mass change is usually performed using an analytical balance. Quite accurate data can be obtained if the specimens are cleaned before measurements and are only handed remotely using tongs or tweezers for small specimens. Mass changes in wear are usually small, e.g. of a few mg, and a sensitive analytical balance is required. Changes in mass of the specimen may not only be the result of wear. Other factors such as corrosion of the specimen and absorption of fluids need to be considered particularly when tests on materials in corrosive and wet environments are performed. Simple examples of this type of tests are corrosive-abrasive wear experiments and studies of polymers wear in organic fluids.

Measurement of Wear by Change in Component Size

Reduction in the dimensions of the worn specimen are usually monitored by connecting a displacement transducer to the surface of the worn specimen that is directly above the wear scar [e.g. 20,21]. Sensitivity of the displacement transducer should allow the detection of, for example, the release of transfer films from the worn surface, i.e. it should be of about **1** [μm]. For this purpose, linear variable differential transformer (LVDT) transducers or non-contact inductive proximity probes coupled to an electronic amplifier are typically used. In almost all experiments involving LVDTs contact between the displacement transducer and the test specimen is maintained by the compressive force of spring loading. Even very small movements of the specimen due to wear can be detected by the LVDT. However, the use of LVDT is not always convenient and often non-contact inductive proximity probes are used. An output from these displacement transducers and amplifier is transferred to either a chart recorder or a computer-based data acquisition system, e.g. based on the LabView software, allowing a continuous record of wear to be obtained.

The use of displacement transducers in wear measurements is very convenient since it provides continuous information about the progress of wear. It also provides additional information as, for example, in the case of studying the wear of polymers, whether the initial creep, occurring in specimens immediately after loading, has reduced to a sufficiently low level to allow valid wear measurements to be made. Displacement measurements of wear are most effective when almost all of the wear is confined to one specimen, e.g. the pin of a pin-on-disc test, as the technique does not provide any information about the distribution of wear between specimens. Wear measurement using a proximity or LVDT sensor and its associated problems are schematically illustrated in Figure 4.10.

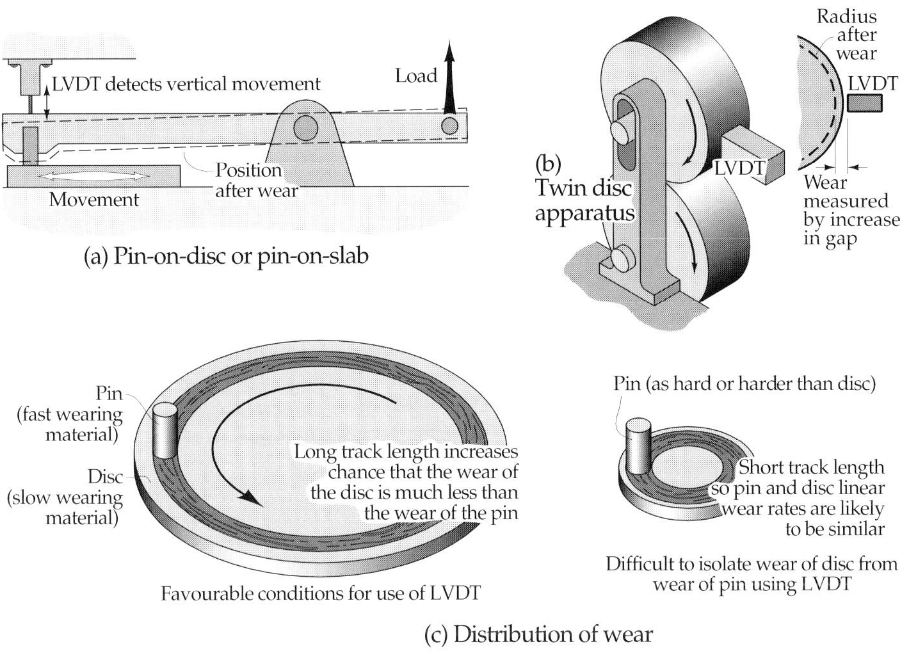

Figure 4.10 Schematic illustration of wear measurement using displacement transducers and associated problems.

Determination of Wear by Profilometry

A commonly used technique to evaluate the worn volume from the wear scar is profilometry. For this purpose an optical projector, e.g. profile projector, stylus profilometry, e.g. Talysurf, or laser scanning profilometry, e.g. UBM, are being used. Optical projector and stylus profilometry are the older techniques developed several decades ago while laser optical profilometry is a more recent technique. All of these techniques can provide much information about the topography of the wear scar and distribution of wear between specimens. Optical projector can offer a simple yet sensitive technique of measuring wear changes in specimens that have a simple shape such as a pin. The technique is based on projecting an image of the object, e.g. pin, on a screen and measuring the change in dimensions of the silhouette of the worn specimen, as schematically illustrated in Figure 4.11a. One of the limitations of this technique is that distortion of the pin by creep and attachment to the wear scar of displaced material can prevent observation of the true worn profile.

The stylus profilometry technique provides a picture of the wear scar which is compiled by making several evenly spaced traverses of the stylus or laser over the wear scar. Wear can be assessed from the deepest wear scar profile thus detected or else the volume of the wear scar can be calculated by numerical integration of the measured wear scar sections. The technique is, however, time consuming and is usually applied to record total wear at the end of a test. The technique is schematically illustrated in Figure 4.11b.

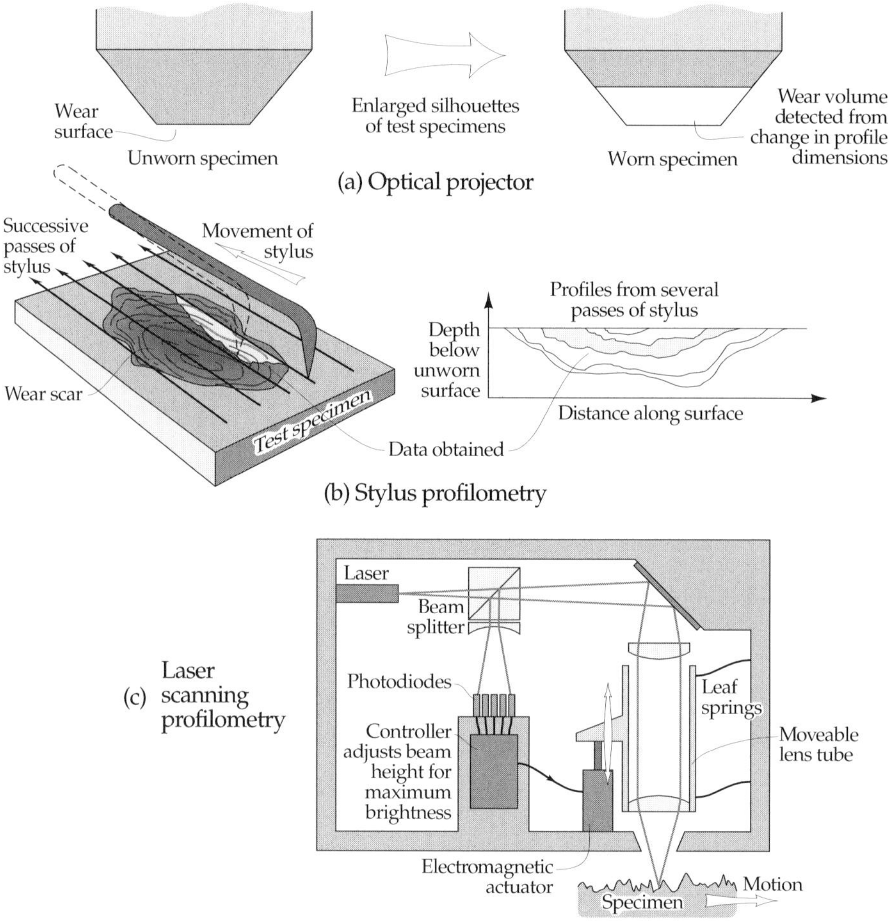

Figure 4.11 Schematic illustration of determination of wear volume by profilometry; a) optical projector, b) stylus profilometry, c) laser scanning profilometry.

In scanning profilometry light from a semiconductor laser is focused on the surface measured generating a surface incident spot of approximately **1** [μm] in diameter. The spot is then imaged onto a sensor with photodiodes via the beam splitter. Combined output from the photodiodes gives the focus error signal. This signal is then used to control the position of the moveable objective lens within the sensor. The lens is moved in real time in such manner that the signal is always maximum, i.e. the focal spot of the beam always remains coincident with object surface. Surface displacement reproduced by the movement of the lens is then measured by the light balance system attached to the lens. The focus error signal can be added to the light falling on all diodes in the sensor resulting in 'reflection intensity signal' which is used to generate microscope quality images of the surface for general inspection over large lateral areas [24]. The schematic illustration of the operating principles of the laser scanning profilometry is illustrated in Figure 4.11c. Other optical profilometry techniques are described in Chapter 6.

Specialized Techniques

There are numerous specialised techniques developed to measure wear under specific experimental conditions. The most widely known methods are thin layer activation by radioactivity and ultrasonic interference measurements of dimensional changes. Other methods include the use of proximity transducers (mentioned on page 92), e.g. to measure the clearance between a piston ring and cylinder wall [25].

The thin layer activation method involves the activation of a thin surface layer of the wearing surface by irradiation in a nuclear reactor or in a particle accelerator. Collection of the radioactive wear particles from retrieved lubricant samples and measurement of their radioactivity provides an index of the extent of wear. With an appropriate choice of source of radiation, e.g. slow neutrons, fast neutrons, protons, etc., different elements in wearing materials can be activated and hence the differences in wear ratios of two interacting components can be accurately measured. For example, with bearing materials the application of different activation methods can render either iron or chromium radioactive allowing to differentiate between the wear rates of the individual elements of the bearing. In internal combustion engine, by irradiating adjacent wearing components with different isotopes, it is possible to distinguish between, for example, wear of a piston ring and wear of a cam and tappet. The main disadvantages of thin surface layer activation are:

- the difficulty in finding a calibration curve between wear rates and wear particle density in the lubricant,
- dependence on flushing of the wearing contact by a lubricant and
- the hazards involved with the use of radioisotopes. Thus, this method appears to have fallen into disuse in recent times.

Another method of wear measurement involves the ultrasonic interference technique [23]. A direct measurement of dimensional changes caused by wear can be obtained by this method while the experimental hazards are limited. The method is sensitive to very small changes in dimensions, of about 1 [μm], and so can be used to confirm the performance of wearing contacts displaying very limited wear. The technique is, however, highly specialised and requires considerable expertise in obtaining reliable data. Wear measurements by thin layer activation and ultrasonic interference are illustrated schematically in Figure 4.12.

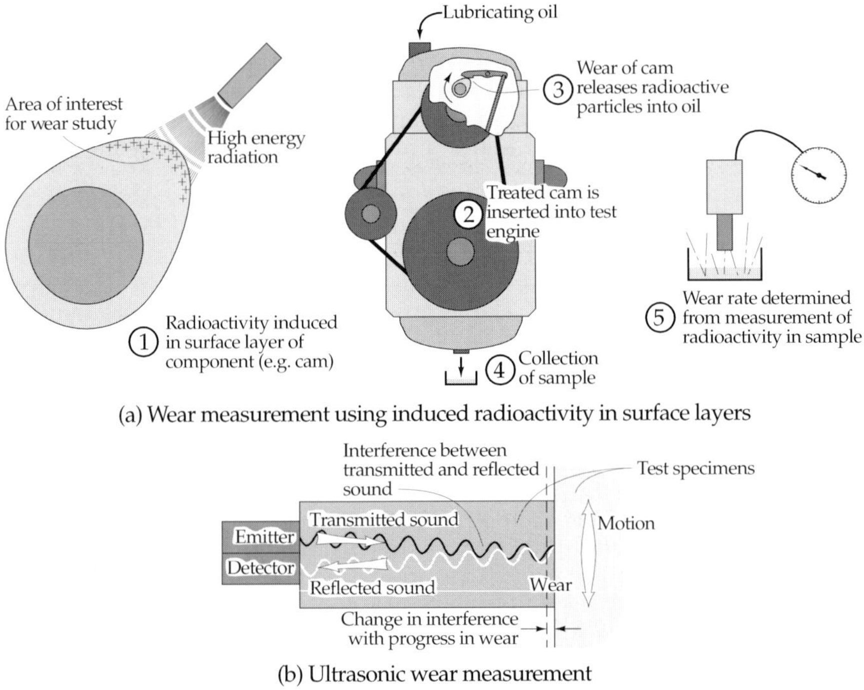

Figure 4.12 Measurement of wear by; a) induced radioactivity in surface layers, b) interference between transmitted and reflected ultrasonic waves.

Measurements of Very Small Wear Volumes in Real Equipment

The development of the Atomic Force Microscope (AFM), described in Chapter 8, has permitted the accurate measurement of very small changes in the external dimensions of solid objects. This means that a very shallow and small wear scar can now be accurately measured. Wear volumes as small as 1 [mg] in a 0.6 [kg] component were measured on steel cam rollers using an AFM to determine changes in surface topography and microhardness indentations [26]. The indentations served as reference points to locate the AFM during a surface scan. Depths of wear as shallow as 30 [nm] were measured by this technique [26]. The practical application of this technique is that measurement of subtle wear changes occurring in equipment at realistically moderate loads and speeds is now possible. The AFM enables the accurate resolution of very small wear volumes that are characteristic of a machine operating within its safe region.

The conventional method of accelerated testing involves abnormal loads and speeds to generate a large wear volume, which can be more easily measured by traditional methods. This practice incurs the risk that the wear mechanisms at the elevated loads and speeds might not be the same as those prevalent at normal operating loads and speeds. Consequently, the true relationship between wear and material or operating parameters may not be found from this conventional form of accelerated testing.

Selection of Measurement Technique for a Particular Application

The relative advantages and disadvantages of wear measurement techniques are summarised in Table 4.1.

Table 4.1 Relative merits of wear measurement techniques.

Technique	Advantages	Disadvantages
Weighing	Simple and accurate	Data corrupted by displaced or transferred material
In-situ measurement of change in length of worn specimen	Accurate and allows continuous record of wear rates	No discrimination between wear of either specimens
Stylus profilometry	Very accurate. Gives distribution of wear between specimens	Slow and mostly suitable for the end of the test. Expensive equipment required
Laser scanning profilometry	Very accurate and fast. Gives distribution of wear between specimens	Expensive equipment required
Optical profilometry	Simple and rapid	Method impossible when specimen has complex shape or its shape is distorted by wear or creep under load
Surface activation	In-situ measurements of wear in closed machinery. Possibility of simultaneous measurement of wear rates of various parts	Inaccurate and difficult to ensure safety of personnel
Ultrasonic interference	Sensitive to small changes in dimension	Specialised technique that requires expertise

As can be seen from Table 4.1 and the preceding text, no single technique can be considered as suitable for every wear measurement problem. In most cases it is necessary to consider the requirements of any particular experiment or test before deciding on the appropriate method of wear measurement. In many cases, two methods of wear measurement should be performed concurrently to check the accuracy of the data required for the particular experimental conditions.

4.4 INDIRECT TECHNIQUES OF FRICTION AND WEAR MEASUREMENT

Friction and wear are often very difficult to measure directly because of the inaccessibility of dynamic contacts. Wear measurements, in particular, often involve disturbing the test specimens which is undesirable. As a result of these problems, it is sometimes preferable to measure a quantity, which is causally related to friction and wear but does not require invasive measurement techniques.

Emissions of heat, noise and vibration are suitable sources of data on friction and wear since they have the characteristic of being transmitted away from the dynamic contact to a remote transducer. Monitoring of these 'tribo-emissions' (e.g. emission of noise [27]) is particularly convenient in machine condition monitoring. In practical machinery it is usually undesirable or impractical to modify the equipment in order to fit a wear or friction transducer. Data obtained from tribo-emissions is not only a substitute for direct friction and wear data but it can also reveal other details of a dynamic contact.

Wear Estimation by Acoustic Emission

A basic parameter describing noise is the sound energy emitted, often referred to in the technical literature as the acoustic emission. In a wearing contact, asperity interaction between opposing surfaces generates acoustic emission. Each time asperities from opposing surfaces make contact, wear is initiated or advanced and energy is released to generate vibration [28,29]. Specific causes of vibration are dislocation migration induced by plastic deformation, extension of cracks and collapse of voids [28]. It is also suggested that the elastic deformation of asperities and their subsequent 'pop-up' when the load is released may contribute to the generated vibrations of certain frequencies [e.g. 30]. Under a sufficiently limited range of experimental conditions wear and acoustic emission can both be considered proportional to frictional power dissipation and both parameters should also display a mutual proportionality [29]. However, it is necessary to treat carefully the relation between acoustic emission and wear because the latter varies radically with operating conditions and acoustic emission is not only dependent on wear but also on the structural vibration characteristics of the system. Principles involved in wear estimation by acoustic emission from a sliding contact are illustrated schematically in Figure 4.13.

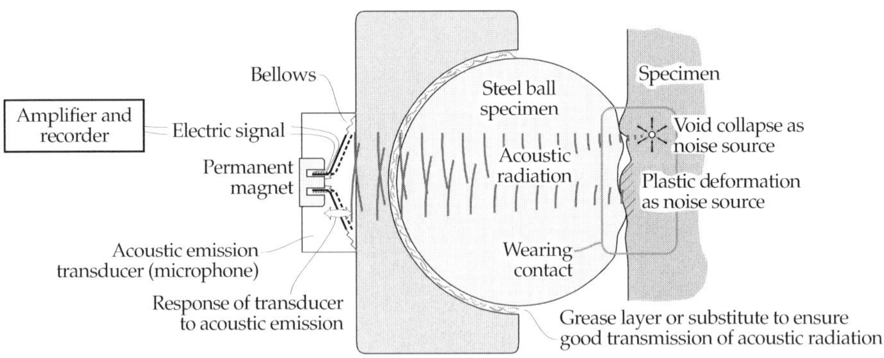

Figure 4.13 Schematic illustration of the principles involved in wear estimation by acoustic emission from a sliding contact.

Acoustic emission is conveyed from the wearing contact via a lubricating film interposed between the wear specimen and the acoustic emission transducer

[28,31]. The electrical signal generated by the transducer is then amplified and recorded or subjected to further processing such as frequency analysis.

Assuming that satisfactory acoustic emission data is obtained, the next task is to relate the data gathered to friction and wear. Although no conclusive relationship was found for frictional power dissipation and acoustic emission [32], there appears to be a better correlation between wear and acoustic emission [28,29,31,33]. An empirical correlation was found between the root mean square of the voltage of the electrical signal obtained from an acoustic emission transducer and the volume of the wear scar [31,33]. The square root of the product of wear volume and hardness of the worn material was observed to correlate well with the root mean square of the voltage of the electrical acoustic emission signal [29]. This latter relationship was found to be consistent with a common frictional energy dependence of acoustic emission and wear [29]. However, a consistent pattern or formula relating wear to acoustic emission has not yet been found.

The acoustic emission method in wear measurements may find an application in cases where a continuous record of wear is required from a wearing contact unsuited to direct wear measurements. However, a limitation of this technique is that the correlation between electrical signal root mean square voltage and wear scar volume is probably a function of the specific test apparatus. In such cases it would be necessary to check the relationship between wear and acoustic emission first by establishing a calibration curve for the particular apparatus under study.

Detection of Cracks Generated During Wear Using Thermal Emission

The progress of wear in a dynamic contact manifests itself not only by the changes in the depth of the wear track but also by the progress of other wear features such as cracks. For many studies, data on the depth of cracks may be essential to, for example, the estimation of strength loss in components due to wear, contact fatigue studies, etc. Although cracks are best observed by sectioning the material, this practice necessitates sacrificing the worn component. A non-destructive method of detecting surface cracks produced by wear is based on photothermal radiometry [34].

To detect surface cracks by photothermal radiometry, a small area of the surface is irradiated by a low power laser. The irradiated area is heated by the laser and emits thermal radiation. The thermal and optical emission from a crack is higher than from an uncracked surface because of both greater emissivity from a crack and the thermal barrier effect. A crack will block the diffusion of heat away from the area of surface irradiated by a laser so that the material immediately adjacent to the crack will be hotter than if the cracks were absent [34]. The variation in light and heat emission on the worn surface can be observed by an Infra-Red optical detector to reveal the cracks. Crack detection by photothermal radiometry is illustrated schematically in Figure 4.14.

The photothermal method allows the detection of cracks beneath layers as thick as a paint coating deposited [34] and may therefore be suitable for discernment of cracks beneath transfer layers or wear debris. The method is relatively slow since it is necessary to irradiate small areas of surface consecutively to cover a complete wear scar. Improvements in this method may remedy the slowness of surface coverage.

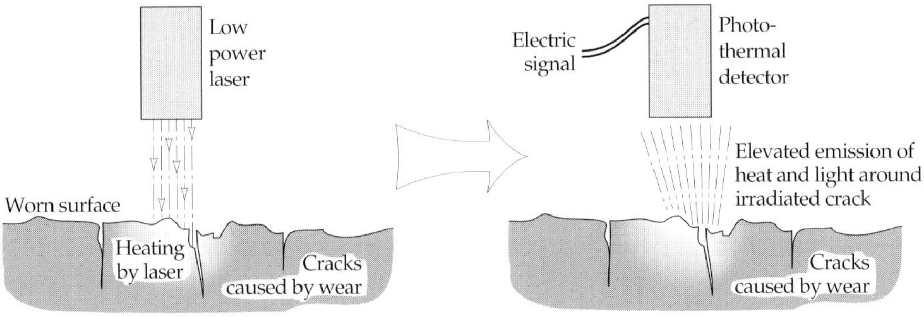

Figure 4.14 Application of photothermal radiometry to detect surface cracks.

Estimation of Friction Coefficients by Measuring the Heat Emission

A technique which provides only an approximate estimate of friction coefficients involves measurements of the temperature of the structure enclosing a dynamic contact. Temperatures in the structure around a dynamic contact are proportional to frictional power dissipation and measurements of such temperatures should, in theory, enable the determination of the friction coefficient averaged over a period of time. A major difficulty with this technique is that the temperature of the enclosing structure tends to rise progressively with time before eventually reaching a maximum value. Measurements of sudden changes in friction coefficient therefore involve the duration as well as the scale of changes in friction coefficient. This greatly complicates the estimation of friction so that the technique is only used in machine condition monitoring where a simple warning of excessive friction coefficient is required. The techniques of surface temperature monitoring are discussed further in the next chapter.

4.5 SUMMARY

The apparently simple task of measuring friction and wear is greatly complicated by practical and conceptual problems. Wear data should ideally be based on two independent methods of measurement to obtain meaningful information. Friction measurements are also subject to experimental artefacts particularly when rapid fluctuations in friction force occur. The methods of measurement of wear and friction are also influenced by the type of data required. Measurements of total wear after long periods of time and average friction coefficient require different techniques than the recording of instantaneous wear and friction coefficients. In general, assessment of wear and friction phenomena based on overall parameters such as averaged wear and friction coefficients can allow transient features of wear and friction to remain unobserved. More sensitive measurement techniques, able to record momentary changes in friction and wear, can reveal information that is fundamental to the understanding of friction and wear mechanisms. Often it is impossible to measure friction and wear directly, which necessitates reliance on measurement of other more accessible physical quantities that depend on friction and wear. Noise emission from a dynamic contact can be used to record changes in friction and wear. Valuable information on friction and wear may then be obtained with the minimum of practical difficulty by this method.

REFERENCES

1 H.M. Stanley and D.B. Bogy, Effect of surface conformity on friction, *Wear*, Vol. 168, 1993, pp. 105-108.

2 H.U. Mittman, N. Czaika and H. Czichos, A new device for simultaneous measurement of friction force, normal force and friction coefficient, *Wear*, Vol. 31, 1975, pp. 179-184.

3 A. Barber, Handbook of noise and vibration control, 6th edition, Elsevier Sci. Publ., Oxford, United Kingdom, 1992.

4 E. Rabinowicz, Friction - especially low friction, Proc. Int. Conf. on the Fundamentals of Tribology, editors: N.P. Suh and N. Saka, Publ. MIT Press, Massachusetts, USA, 1978, pp. 351-365.

5 R.G. Bayer, W.C. Clinton, C.W. Nelson and R.A. Schumacher, Engineering model for wear, *Wear*, Vol. 5, 1962, pp. 378-391.

6 E. Rabinowicz, The intrinsic variables affecting the stick-slip process, *Proc. Physical Society of London*, Vol. 71, 1958, pp. 668-675.

7 I.P. Hayward and J.E. Field, A computer-controlled friction measuring apparatus, *J. Phys. E: Sci. Instrum.*, Vol. 21, 1988, pp. 753-756.

8 I.L. Singer, Private Communication, Nordtrib 2000, Helsinki, 2000.

9 D. Tabor, Adhesion and friction, Properties of Diamond, editor: J.E. Field, Academic Press, Oxford, 1979, pp. 325-350.

10 I.P. Hayward, I.L. Singer and L.E. Seitzman, Effect of roughness on the friction of diamond on CVD diamond coatings, *Wear*, Vol. 157, 1992, pp. 215-227.

11 D. Dowson, History of Tribology, Longman Group Ltd., London, 1979.

12 F.P. Bowden and T.P. Huges, Friction of clean metals and the influence of adsorbed gases. The temperature coefficient of friction, *Proc. Roy. Soc., London*, Series A, Vol. 172, 1939, pp. 263-279.

13 S. Adelman, C.R. Taylor and N.C. Heglund, Sweating on paws and palms: What is its function, *American Journal of Physiology*, Vol. 29, 1975, pp. 1400-1402.

14 Y. Tamai, On the contact resistance between surface-oxidized metals in repeated sliding, *Wear*, Vol. 1, 1957-1958, pp. 377-383.

15 T. Sasada, Human joints - their excellent performance, *Trans. JSLE*, Vol. 31, 1986, pp. 763-768.

16 T. Takei, Structural reaction of joints to mechanical stress, *Trans. JSLE*, Vol. 31, 1986, pp. 769-774.

17 A. Unsworth, D. Dowson and V. Wright, The frictional behaviour of human synovial joints - Part I: Natural joints, *Trans. ASME, Journal of Lubrication Technology*, Ser. F, Vol. 97(3), 1975, pp. 369-376.

18 A. Unsworth, D. Dowson, V. Wright and D. Koshal, The frictional behaviour of human synovial joints - Part II: Artificial joints, *Trans. ASME, Journal of Lubrication Technology*, Ser. F, Vol. 97(3), 1975, pp. 377-382.

19 J. O'Kelly, A. Unsworth, D. Dowson, B. Robins and V. Wright, Pendulum and simulator studies of friction in hip joints, Evaluation of Artificial Joints, editors: V. Wright and D. Dowson, The Biological Engineering Society, 1977, pp. 19-29.

20 T.S. Barrett, G.W. Stachowiak and A.W. Batchelor, Effect of roughness and sliding speed on the wear and friction of ultra-high molecular weight polyethylene, *Wear*, Vol. 153, 1992, pp. 331-350.

21 T. Sasada, S. Norose and H. Mishina, The behaviour of adhered fragments interposed between sliding surfaces and the formation process of wear particles, Proc. Int. Conf. on Wear of Materials, Dearborn, Michigan, 16-18 April 1979, editors: K.C. Ludema, W.A. Glaeser and S.K. Rhee, American Society of Mechanical Engineers, New York, 1979, pp. 72-80.

22 A. Gerve, Einsatzmoglichkeiten von Radionukliden zur Untersuchung konstruktiver und schmierstoffabhangiger Einflusse auf den Verscgleiss von Maschinenteilen, VDI-Berichte, No. 196, 1973, pp. 43-47.

23 A.S. Birring and H. Kwun, Ultrasonic measurement of wear, *Tribology International*, Vol. 22, 1989, pp. 33-37.

24 A.J. Brown, High-speed optical measurement of 3-D surfaces, Proc. of SPIE (The International Society for Optical Engineering), Laser Dimensional Metrology, Recent Advances for Industrial Application, editor: M.J. Downs, Vol. 2088, 1993, pp. 190-194.

25 H. Czichos, Tribology - a systems approach to the science and technology of friction, lubrication and wear, Elsevier, Amsterdam, 1978.

26 R. Gahlin, R. Larker and S. Jacobson, Wear volume and wear distribution of hydraulic motor cam rollers studied by a novel atomic force microscope technique, *Wear*, Vol. 220, 1998, pp. 1-8.

27 V.A. Belyi, O.V. Kholodilov and A.I. Sviridyonok, Acoustic spectrometry as used for the evaluation of tribological systems, *Wear*, Vol. 69, 1981, pp. 309-319.

28 S. Lingard, C.W. Yu and C.F. Yau, Sliding wear studies using acoustic emission, *Wear*, Vol. 162-164, 1993, pp. 597-604.

29 K. Matsuoka, D. Forrest and M-K. Tse, On-line wear monitoring using acoustic emission, *Wear*, Vol. 162-164, 1993, pp. 605-610.

30 R.S. Sayles, G.M.S. deSilva, J.A. Leather, J.C. Anderson and P.B. Macpherson, Elastic conformity in Hertzian contacts, *Tribology International*, Vol. 14, 1981, pp. 315-322.

31 R.J. Boness and S.L. McBride, Wear detection and analysis using acoustic emission techniques, Proc. IEA Tribology Conf., Brisbane 3-5 December 1990, IEA Nat. Conf. Publ. No. 90/14, pp. 33-37.

32 C.L. Jia and D.A. Dornfeld, Experimental studies of sliding friction and wear via acoustic emission signal analysis, *Wear*, Vol. 139, 1990, pp. 403-424.

33 R.J. Boness, S.L. McBridge and M. Sobczyk, Wear studies using acoustic emission techniques, *Tribology International*, Vol. 23, 1990, pp. 291-295.

34 J.L. Bodnar, C. Menu, M. Egee, P. Pigeon and A. Le Blanc, Detection of wear cracks by photothermal radiometry, *Wear*, Vol. 162-164, 1993, pp. 590-592.

5 CONTROL OF THE TEST ENVIRONMENT

5.1 INTRODUCTION

Environmental factors such as temperature, humidity and oxygen exert a strong influence on most wearing contacts yet many experiments have omitted the measurement or control of these parameters. Apparent conflict of experimental data can be explained in terms of unforeseen variation in temperature, humidity or oxygen concentration. In a cooperative program of wear tests performed by various institutions, a common pattern in the wear data was found with the exception of results from one institution. Detailed investigation of the reasons for this discrepancy revealed that a cup of coffee was left close to the wearing contact during tests. The technician had used the flat surface of the tribometer as a convenient table for the coffee cup. Moist vapour from the steaming coffee reached the wearing contact and affected the progress of wear [1]. In another more extreme case, variation in test results on dry sliding of ceramics was found to occur at the same time each morning of a working day. Debriefing of the technician involved revealed that he had the habit of sitting at the other end of the laboratory and eating an orange at the same time of day as the problem with the test data was occurring. The act of breaking open an orange released an aerosol of orange oil and water which drifted over to the test machine [2]. Careful measurement and control of these environmental factors is essential for any tribology experiment and misleading results may be obtained if such precautions are ignored or omitted.

5.2 MEASUREMENT OF TEMPERATURE

The measurement of temperature is possibly one of the more intractable problems in experimental tribology. There are three temperatures that should be measured and monitored during friction and wear tests. These are:

- the bulk or ambient temperature or lubricant temperature in lubricated tests as this controls the minimum temperature in the dynamic contact,
- the localised surface temperature rise within and adjacent to the area of the dynamic contact,
- transient highly localised temperature rises on small areas of worn surface or 'hot spots' as they are known in the literature.

In terms of practical measurements, the temperatures are listed above in ascending order of difficulty. While the bulk temperature of a specimen, lubricant or process fluid is important, the 'surface temperature' of the test specimens has a controlling influence on friction and wear so that the surface temperature is of greater concern. The 'surface temperature' can be referred to the temperature of a layer about 10 [μm] thick below the surface when considering the formation of wear particles or a layer only a few atoms thick when the adsorption of lubricants is involved. Most temperature measuring systems are designed to measure bulk temperature and accurate determination of surface temperature presents considerable experimental problems. Methods of temperature measurement for tribological studies involving thermocouples, infra-red pyrometry and indirect temperature determination from metallurgical phase changes are discussed below. Glass thermometers are not discussed in detail as these are only suitable for elementary measurements of e.g. air or bulk lubricant temperature.

Thermocouples

Thermocouples are widely used in tribological studies in order to provide basic data on the temperature and hence the heat generation by friction. Typical applications of thermocouples are:

- to measure the bulk temperature of lubricant or process fluid,
- to determine air or atmospheric temperature during tests and
- to measure the temperature rise in specimens tested.

In general, thermocouples are unsuitable for surface temperature analysis without special modification. This means that most commonly available thermocouples cannot provide accurate measurements of temperatures of material close to the wearing surfaces. One of the major limitation of thermocouples is associated with the finite size of the bimetallic junctions required. Even when a fine wire is used in the thermocouple, the dimensions of the bimetallic junction remain too large for the measurement of temperature rises occurring between interacting surfaces [3]. The size of the bimetallic junction means that a thermocouple cannot detect the extremely sharp temperature gradients occurring below the surfaces of objects in dynamic contact nor the equally rapid variation of temperature with time that takes place during wear or frictional contact. There are two solutions to these limitations which are:

- a dynamic thermocouple where the thermoelectric junction is the sliding contact itself,

- a planar or lamellar thermocouple made by coating successive layers of metals and insulators onto the test specimen.

Either solution has characteristic advantages and limitations so that neither offers a comprehensive scope of measurement.

The dynamic thermocouple is comparatively simple to arrange since a thermoelectric potential can be measured directly from any wearing contact between dissimilar metals. This posses a severe restriction on the range of materials to which this technique is applicable as many important classes of materials such as polymers cannot be investigated by this technique. Perhaps the most spectacular use of dynamic thermocouples was by Bowden and Tabor to demonstrate the limiting temperature of a metallic contact [3]. A low melting point metal pin was worn against a rotating steel disc and the contact temperature was estimated from the measured thermoelectric potential. It was found that the contact temperature never exceeded the lower melting point of the two metals in contact. The experiment also showed that although the thermoelectric potential and therefore contact temperature rapidly fluctuated, it was impossible to determine whether the measured fluctuation corresponded exactly to contact temperature variations or whether it was influenced by factors such as tribo-electrification [4]. A recent study of a range of pure metals in self-mated sliding, both soft and hard, revealed that tribo-electrification is significant in dry sliding [5,6]. When one of the sliding members is significantly smaller than the other, as in a pin-on-disc test, the movement of transferred wear particles carries a large amount of electrical charge [5]. In an aqueous environment, e.g. sea-water or dilute sulphuric acid, electrochemical reactions may also influence tribo-electrification. A study by Ettles and co-workers [7] confirmed the work of Bowden and Tabor but also showed the complex nature of a dynamic thermocouple.

The problem associated with the thermocouple size can be worked around by making a planar thermocouple which can be fitted onto the surface of the specimen tested. Coatings of metal and insulator as thin as 100 [nm] can be deposited onto a surface. To form a bimetallic junction, two layers of metal and two layers of insulator are required. Assuming that all coatings are 100 [nm] thick this gives a total thickness of 0.4 [μm]. Electric terminals can be fixed to the metal coatings outside of the worn contact to complete the thermocouple. The thermocouple constructed in this manner is very sensitive and accurate to surface temperature changes and has been used in studies of elastohydrodynamic lubrication [8]. Variations in lubricating oil temperature as the oil passes through an EHL contact have been determined using this type of a lamellar thermocouple. The disadvantage of this type of thermocouple is that it is easily damaged by wear so that it cannot be used during wear tests. The advantages and limitations of conventional and special purpose thermocouples are illustrated in Figure 5.1.

When conducting bulk temperature measurements with conventional thermocouples, the location of the thermocouple should be carefully selected. It is quite easy to place a thermocouple in a wrong region, e.g. where the lubricant is stagnant, and hence to fail to determine accurately the lubricant temperature which affects wear and friction. Even when a thermocouple is placed in a suitable position, it can be moved away by air currents from rapidly rotating components in the experimental apparatus.

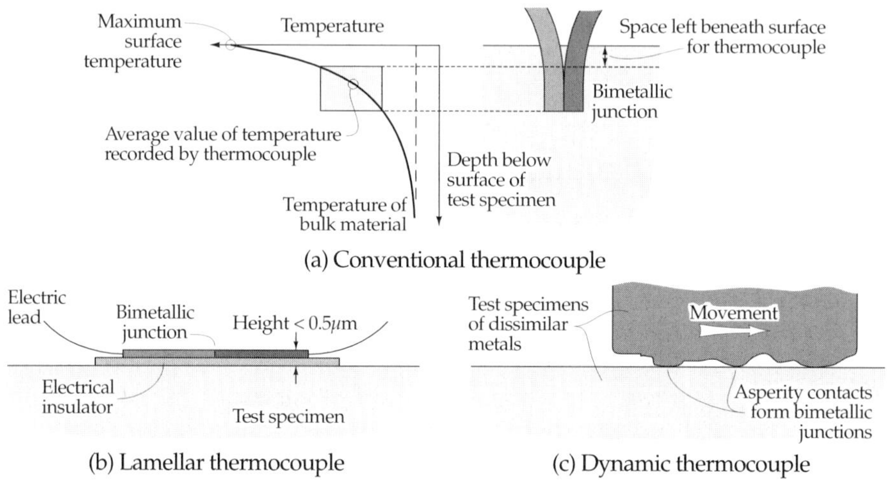

Figure 5.1 Schematic illustration of applications of conventional and special purpose thermocouples in tribological studies.

Infra-Red Pyrometry

The surface temperature of wearing materials can also be accurately measured by infra-red pyrometry. Infra-red pyrometry functions by the detection of radiation from the surface of an object. Radiation measurements are not subjected to thermal inertia limitations so that the fluctuation of frictional temperatures can accurately be determined provided that a suitable detector is available. During the measurement the intensity or power per unit area of infra-red radiation from a hot surface is recorded and the temperature is calculated from this data and a calibrated value of thermal emissivity. The relationship between temperature and infra-red emission follows the Boltzmann law of thermal radiation which states that the power per unit area of thermal emission is the product of the emissivity, the Stefan-Boltzmann constant and the fourth power of absolute temperature, i.e.:

$$\dot{Q} = \varepsilon k T^4 \tag{5.1}$$

where:

$\dot{Q}$ is the heating power (radiant energy) per unit area [W/m²];

ε is the relative emissivity [dimensionless];

k is the Stefan-Boltzmann constant, i.e. 5.6704×10^{-8} [W/m^2K^4].

Infra-red pyrometry has been used successfully to determine, for example, the temperature of a lubricant film inside an EHL contact with infra-red radiation from the film passing through a transparent sapphire window [9], the temperature of a steel counterface during a polymer wear test [10], transient temperatures in dry sliding [11]. Measurement of thermal radiation is an invaluable technique for determining the localised surface temperature rise within dynamic contacts and adjacent areas.

However, infra-red pyrometry temperature measurements of worn surfaces are not without problems. A major difficulty is to determine the emissivity of a wearing surface. While it might be possible to measure accurately the emissivity of the unworn test specimen by heating it to a known temperature and recording the level of infra-red emission, the emissivity of the same specimen may vary in an unknown fashion during wear. In effect, the wear process occurring is causing a transformation of emissivity from a pre-determined calibration constant into an experimental variable. Common causes of emissivity variation are patchy coverage of a worn surface by lubricant films during lubricated tests, chemical transformation of a worn surface by frictional heat, presence of oxygen or other reagents and the formation of transfer films particularly during sliding of dissimilar materials. Variations in emissivity are almost impossible to assess and in most cases it is necessary to assume that the scale of emissivity variation is comparatively small. Some tolerance for unknown variation in emissivity is provided by the forth power of temperature in the Boltzmann law which ensures a relatively small variation in measured temperature if the value of emissivity is wrong. For example, a 100% error in emissivity generates a 20% error in temperature. Measurement of temperature by infra-red pyrometry and the problems associated with fluctuations in emissivity are illustrated schematically in Figure 5.2.

The requirement for a separate measurement of emissivity can be achieved by measuring thermal emission (infra-red radiation) at two different wavelengths. This technique, termed two-colour thermometry, is based on the principle that the difference in thermal emission at two wavelengths is proportional to the emissivity as well as to the temperature of the radiating body. Temperature is then found from equations for thermal radiation where the emissivity has been substituted by the power levels at the two wavelengths. This technique is specialised and has the disadvantage of lowering the sensitivity of the detector since only half of the radiation collection area of the detector (known as image size) is available for each wavelength.

Infra-red pyrometry appears to be the only means of temperature measurement that can be used to assess localised temperature transients or 'hot spots' which are associated with failures in lubrication or transitions to severe wear [12]. There are two basic problems involved in the determination of 'hot spots' temperature [13], i.e. the size of the 'hot spot' is far smaller than the area of surface probed by the pyrometer and the duration of the 'hot spot' is extremely short [14]. Thermal radiation from a hot surface is emitted as a random stream of photons and the thermal emission from a 'hot spot' must be at least three times larger than the emission from the surrounding probed area to allow statistically valid discrimination of the 'hot spot' temperature from the temperature of the

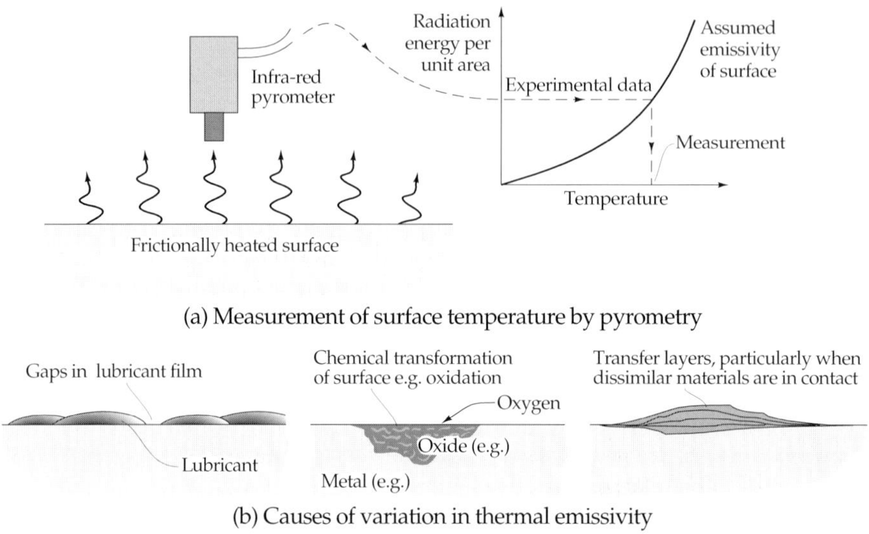

Figure 5.2 Schematic illustration of measurement of surface temperature by infra-red pyrometry during a wear test.

surrounding surface. In most cases, 'hot spot' temperatures do not exceed the melting point of metals which limits the maximum thermal radiation emitted from a 'hot spot' and as a consequence limits the minimum ratio of 'hot spot' area to probed area. The probed area or 'image area' of an infra-red pyrometer is usually greater than **0.1** [mm] by **0.1** [mm] and it can be calculated that the minimum area of a 'hot spot' cannot be less than approximately **1** [μm^2] [13] to be effectively detected. By deduction from the size of surface roughness details present on worn surfaces, it is believed that the size of a 'hot spot' is considerably less than 1 [μm^2] which means that the infra-red pyrometer cannot accurately record the 'hot spot' temperature. Another problem is associated with the extremely short existence of most 'hot spots'. It is estimated that a 'hot spot' may only persist for 200 [ns] which is much shorter than the recording period of pyrometers [14]. The problems involved in measuring 'hot spot' temperatures by infra-red pyrometry are illustrated schematically in Figure 5.3.

It has to be concluded that even with advanced instrumentation it is still not possible to measure accurately all the controlling temperatures occurring during friction and wear processes.

Indirect Temperature Evaluation from Microstructure Observations

In many instances, metallurgical examination of worn specimens after a test can provide some information about the surface temperatures occurring during the tests without the need for specialised measurements with thermocouples and pyrometers. A section of a worn surface can be examined in a microscope to determine whether in, for example, a steel specimen, frictional temperatures were sufficiently high to cause phase transformation. Metallurgical analysis of 'white layers' on steel surfaces suggests that temperatures as high as the melting point of steel can be reached [15] during sliding wear. Metallurgical changes are quite common since a basic characteristic of frictional surface temperature rises is to cause a rapid cooling in near-surface material immediately after the source of frictional heat has passed over the surface. The rapid cooling often induces microstructure change from annealed to quenched or a reduction in grain size if melting occurred. Sectioning of worn specimens allows observation of the depth of near-surface material that is affected by frictional heat. Analysis of temperature variation with depth is impossible by pyrometry and requires careful experimental preparation if thermocouples are used. The principle of deducing temperature changes by microstructural observation is illustrated schematically in Figure 5.4.

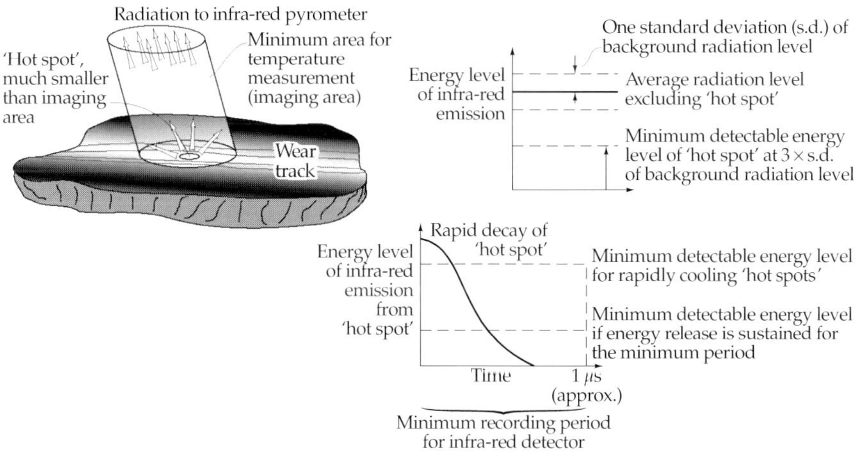

Figure 5.3 Schematic illustration of the problems associated with the measurement of transient localised surface temperatures by infra-red pyrometry.

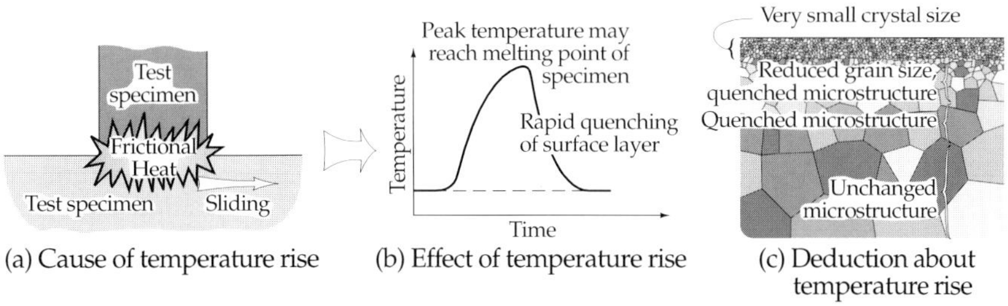

Figure 5.4 Schematic illustration of the deduction of frictional temperature rises in specimens tested based on microstructural observation.

Microstructural observations related to temperature changes are limited, however, to large temperature rises of, for example, several hundred degrees Celsius for steels and no data can be obtained about the fluctuating component of frictional temperature rise. If the temperature varies in a complex manner, i.e. shows a cyclic variation during a test, then interpretation of microstructural changes may become a complex task.

5.3 MEASUREMENT OF HUMIDITY

Humidity has a strong influence on friction and wear which necessitates its routine measurement and maintenance at a constant level during tribological experiments. The variation in humidity between various parts of the world can prevent meaningful comparison of results obtained at laboratories in different locations. For example, the relative humidity in Australia often falls as low as 15% in summer while in most tropical areas, e.g. Singapore, it rarely is less than 70% and very often exceeds 90%. There are also diurnal and seasonal variations of humidity, for example, relative humidity is usually at a maximum at dawn and at a minimum in the heat of the afternoon. This variation in relative humidity can affect wear and friction data during the course of the working day. Humidity measurements are of particular concern in studies of friction between heads and disks of computer memory systems [16] and in studies of friction and wear between ceramics [17,18]. At high humidity levels the stock holding period before the beginning of the test can have dramatic effects on wear and friction results as illustrated in Figure 5.5.

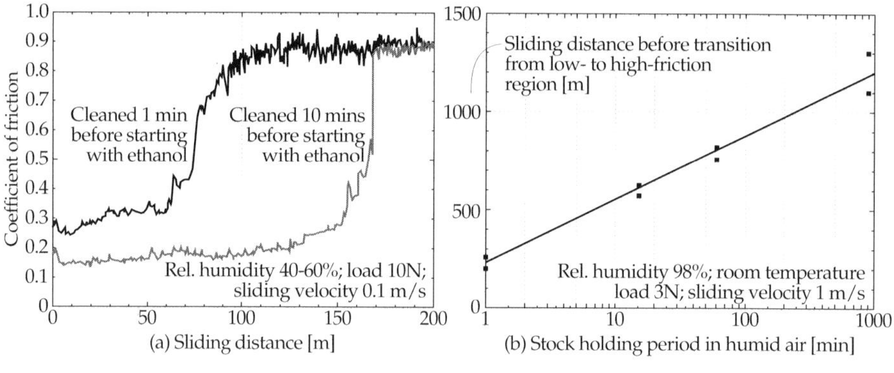

Figure 5.5 Effects of the duration between cleaning and beginning of test on the friction coefficient (a); and of stock holding period in humid air before beginning of test on the duration of low friction region for TiN sliding couples (b) [19].

Conventional techniques of humidity measurement, e.g. wet and dry bulb thermometer, are sufficient for tribological studies and are not described here. Wear and friction appear to be only affected by fairly large differences in humidity, e.g. between 20% and 70% relative humidity in tests on ceramics [17] so that precise discrimination between, for example, a relative humidity of 21% and 21.5% is not usually required.

As the test temperature declines from, for example, 30°C summer temperature to lower winter temperatures such as 5°C, the accuracy of relative humidity measurements declines too. At 5°C it is difficult to obtain an accuracy in relative humidity (R.H.) measurements that is better than ± 3% R.H [20]. The reason for this is the decline in the difference between wet bulb and dry bulb temperatures with corresponding reduction in temperature. Temperature measurements have their limitations of accuracy and this uncertainty is translated to the derived humidity values. This implies that there is little purpose in conducting the tribological tests with small differences in relative humidity when low temperatures are involved. It should also be noted that hygrometers are subject to hysteresis, i.e. time lag in measurements. If humidity changes rapidly during a test, e.g. within a few seconds due perhaps to a sudden rise in temperature, then the accuracy of the hygrometer would require confirmation by external calibration.

Another aspect of humidity measurements is departure of humidity adjacent to the wearing contact from the ambient value. In general, friction produces heat, which usually lowers the relative humidity of air surrounding a test specimen. So far there appear to have been no reported tests of humidity variations between a dynamic contact and its surroundings.

5.4 MEASUREMENT OF OXYGEN CONCENTRATION AND OTHER GASES OR AIRBORNE POLLUTANTS

Many friction and wear processes are affected by the concentration of oxygen in the atmosphere surrounding a dynamic contact or dissolved in a lubricant or process fluid [e.g. 21]. Studies where the influence of oxygen concentration on wear or friction is investigated are relatively few. Other corrosive gases, e.g. chlorine or sulphur dioxide, may also affect wear and friction but there appears to be no published data on the significance of these gases.

Most studies dedicated to the role of oxygen in friction and wear rely on a comparison between an air (or pure oxygen) and nitrogen or argon environments [e.g. 18]. Few attempts have been made to measure oxygen concentration in mixtures of air (oxygen) and nitrogen and relate the variation in oxygen concentration to changes in friction and wear [e.g. 22]. The concentration of dissolved oxygen in lubricating oil has been measured by gas chromatography in a study of the influence of oxygen on Extreme Pressure lubrication by sulphur-based additives [23]. It was found that the effectiveness of sulphur-based additives in preventing frictional seizure was closely related to the proportion of sulphur and oxygen present in the wear scar films. The relative proportion of sulphur and oxygen in the wear scar film was also found to be dependent on the concentration of dissolved oxygen in the lubricating oil as well as on the reactivity of the sulphur additive [23].

Gas chromatography is a method of separating different constituents of a mixture for the purpose of analysis by diffusing a gaseous or vaporised sample through a column of liquid. All forms of chromatography rely on the principle that different substances are transported at varying speeds through a medium. A simple example of chromatography is the separation of the colours when a droplet of ink is placed on filter paper and allowed to spread. After a period of time, a system of concentric rings of the colour components of the ink is formed on the filter paper. A schematic diagram of a gas chromatograph is shown in Figure 5.6.

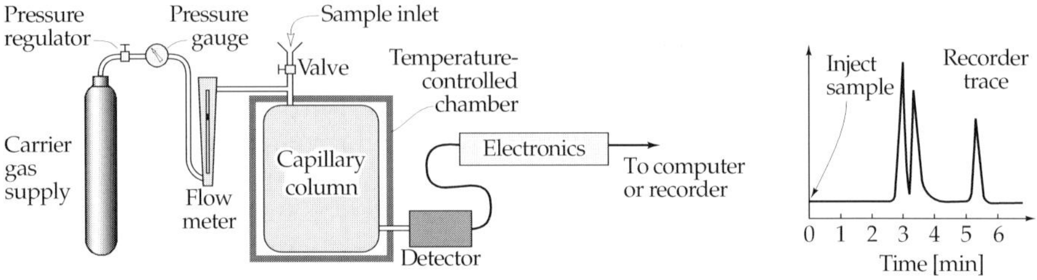

Figure 5.6 Schematic diagram of a gas chromatograph (adapted from [24]).

A carrier gas which is usually nitrogen, hydrogen or helium conveys the sample through the chromatograph. The sample of fluid or low boiling point solid is injected through a rubber membrane into a receiving chamber where the sample and carrier gas are mixed. The sample and carrier gas are often heated at this stage to ensure complete volatilization of the sample. The mixture of sample and carrier gas is then transferred to a long capillary tube filled with liquid. This liquid which acts as the separating medium is usually a hydrocarbon fluid, e.g. squalene or a silicone oil, but other liquids may also be used depending on the solubility and reactivity of the sample. The constituents of the sample pass through the capillary tubes at varying rates and are successively detected on leaving the capillary to reveal the composition of the sample.

Gas chromatography is very sensitive to small concentrations of minor constituents in a mixture and a very small sample size is required for the analysis. A typical sample size is 1 [μl] of which only 1% actually passes through the chromatograph. The remainder is released into the atmosphere as the capillary tube can only absorb a very limited quantity of sample. A complex mixture of several substances can readily be analyzed by chromatography so that this technique is suitable for the analysis of used and new lubricating oils, which contain many ingredients and impurities. The small sample size required makes this technique particularly useful in the analysis of traces of used lubricating oil or process fluid collected from test apparatus or equipment. More information on gas chromatography can be found in [24] while the application of gas chromatography to the detection of dissolved gases in lubricating oils is described in [25].

Oxygen analysis electrodes are available for direct measurements of oxygen concentration but these are apparently not suitable for lubricating oils. However, oxygen electrodes are used to measure oxygen concentration in water and find application mainly in biological and environmental studies.

5.5 SUMMARY

The measurement of basic environmental parameters such as temperature, humidity and oxygen concentration is an important part of tribological experimental methodology. If these parameters are not measured then control of friction and wear tests is compromised and the experimental results obtained corrupted. Humidity is the simplest to measure of all of these three parameters yet even so there may be some doubt about the actual level of humidity around the dynamic contact. Temperature is perhaps the most difficult parameter to measure and the problem lies not in routine measurements of overall specimen or lubricant temperature but in the determination of short-lived transient surface temperatures caused by unsteady release of frictional energy. Oxygen concentration is not often measured despite its important influence on friction and wear. In future research perhaps more consideration will be given to the role of the gaseous environment in controlling friction and wear.

REFERENCES

1 M.J. Barber, Private communication, 1993.

2 G. Plint, Private communication, 1993.

3 F.P. Bowden and D. Tabor, The Friction and Lubrication of Solids, Part II, Clarendon Press, Oxford, 1964.

4 I. Goldblatt, The importance of electron transfer processes to the wear process, Microscopic Aspects of Adhesion and Lubrication, Proceedings of the 34th International Meeting of the Societe de Chime Physique, Paris 14-18th Sept. 1981, Tribology Series 7, Amsterdam, Elsevier, 1982, pp. 521-551.

5 Y.Ch. Chiou, Y.P. Chang and R.T. Lee, Tribo-electrification mechanism for self-mated metals in dry severe wear process, Part I. Pure hard metals, *Wear*, 2003, Vol. 254, pp. 606-615.

6 Y.Ch. Chiou, Y.P. Chang and R.T. Lee, Tribo-electrification mechanism for self-mated metals in dry severe wear process, Part II. Pure soft metals, *Wear*, 2003, Vol. 254, pp. 616-624.

7 O.S. Dinc, C.M.McC. Ettles, S.J. Calabrese and H.A. Scarton, The measurement of surface temperature in dry or lubricated sliding, *Transactions of the ASME, Journal of Tribology*, Vol. 115, No. 1, 1993, pp. 78-82.

8 T.A. Dow and W. Kannel, Evaluation of rolling/sliding EHD temperatures, Proc. 6th Leeds-Lyon Symposium on Tribology, Thermal effects in Tribology, Sept., 1979, editors: D. Dowson, C.M. Taylor, M. Godet and D. Berthe, Inst. Mech. Engrs. Publ., London, 1980.

9 H.S. Nagaraj, D.M. Sanborn and W.O. Winer, Direct surface temperature measurement by infrared radiation in elastohydrodynamic contacts and the correlation with the Blok flash temperature theory, *Wear*, Vol. 49, 1978, pp. 43-59.

10 T.S. Barrett, G.W. Stachowiak and A.W. Batchelor, Effect of roughness and sliding speed on the wear and friction of ultra-high molecular weight polyethylene, *Wear*, Vol. 153, 1992, pp. 331-350.

11 S. Suzuki and F.E. Kennedy, The detection of flash temperatures in a sliding contact by the method of tribo-induced thermoluminescense, *Trans. ASME, Journal of Tribology*, Vol. 113, 1991, pp. 120-127.

12 S. Bair, I. Green and B. Bhushan, Measurements of asperity temperatures of a read/write head slider bearing in hard magnetic recording disks, *Trans. ASME, Journal of Tribology*, Vol. 113, 1991, 547-554.

13 Y.W. Chung and K.J. Wahl, Fundamental limits of flash temperature measurements of asperities using infra-red detectors, *Tribology Transactions*, Vol. 35, 1992, pp. 447-450.

14 F.E. Kennedy, discussion to Chung and Wahl [13].

15 T.S. Hong, S. Maj and D.W. Borland, The formation of white layers during sliding wear, Proc. Int. Tribology Conf. 1987, Melbourne, Australia, 2-4 Dec. 1987, pp. 193-197, IEA, Nat Conf. Publ. No. 87/18.

16 H. Tian and T. Matsudaira, The role of relative humidity, surface roughness and liquid build-up on static friction behavior of the head/disk interface, *Trans. ASME, Journal of Tribology*, Vol. 115, 1993, pp. 28-35.

17 K. Komvopoulos and H. Li, The effect of tribofilm formation and humidity on the friction and wear properties of ceramic materials, *Trans. ASME, Journal of Tribology*, Vol. 114, 1992, pp. 131-140.

18 J.M. Martin, Th. Le Mogne, C. Chassagnette and M.N. Gardos, Friction of hexagonal boron nitride in various environments, *Tribology Transactions*, Vol. 35, 1992, pp. 462-472.

19 G. Meier zu Kocker, T. Gross and E. Santner, Private communication, BAM, Berlin, 1994.

20 A.K. Sharp, Humidity: Measurement and control during the storage and transport of fruits and vegetables, CSIRO Food Research Quarterly, 1986, Vol. 46, pp. 79-85.

21 M. Masuko, Y. Ito, K. Akatsuka, K. Tagami and H. Okabe, Influence of sulphur-base Extreme Pressure additives on wear under combined sliding and rolling contact, Proc. Kyushu Conference of JSLE, October 1983, pp. 273-276 (in Japanese).

22 T. Murakami, T. Sasaki, Y. Yamamoto, H. Sakamoto and F. Hirano, Effects of oxygen dissolved in lubricating oils on oxide film formation and lubricating performance under sliding contact condition, *Journal of the Japanese Society of Lubrication Engineers (JSLE), International Edition*, Vol. 8, 1987, pp. 105-110.

23 T. Sakai, T. Murakami and Y. Yamamoto, Effect of dissolved oxygen concentration on the lubricating performance of an oil containing an organic sulphide, *Wear*, Vol. 156, 1992, pp. 175-187.

24 H.H. Willard, L.L. Merrit Jr., J.A. Dean and F.A. Settle Jr., Instrumental methods of analysis, 7th edition, Wadsworth Publishing Company, California, U.S.A., 1988.

25 J.A. Petrocelli and D.H. Lichtenfels, Determination of dissolved gasses in petroleum fractions by gas chromatography, *Analytical Chemistry*, Vol. 31, 1959, pp. 2017-2019.

6 CHARACTERIZATION OF TEST SPECIMENS

6.1 INTRODUCTION

It is well documented that tribological behaviour of a material depends on its material properties/characteristics and characterization of test specimens is an essential stage in any tribological investigation. Generic names such as mild steel, aluminium alloy or nylon do not provide sufficient specification of a test material to enable a researcher adequate control of testing, comparison of results or later replication of tests. Materials characterization in tribology involves specifications common to materials science studies as well as some characteristics that are particular to tribology. Each basic class of material, polymer, metal or ceramic, requires a slightly different specification of material properties to achieve effective control of its tribological characteristics. For example, in tests involving metals it is important to know the chemical composition of the material and its metallurgical phases, while in tests involving polymers the information on the specimen organic components, additives and degree of crystallinity is desirable. For brittle materials, such as ceramics, apart from composition details, information on grain size and material toughness is usually required. A summary of the most common materials specifications necessary for tribological testing is given in Table 6.1. Some specialized tests would require additional characteristics such as specific heat, thermal conductivity, coefficient of thermal expansion, chemical affinity, presence of lubricating layers, etc. Most of the sample characterization involves techniques and methods best described in materials science textbooks and therefore these techniques will be mentioned in this chapter only briefly. The emphasis is put on surface texture characterization and measurement of hardness.

Since wear and friction are surface-related phenomena, surface texture measurements are a necessary part of the characterization of materials in tribological testing. Surface texture results from a combined effect of roughness and waviness. Form error, although often measured on the same instruments, is usually not considered as part of surface texture.

Table 6.1 Material characteristics necessary for sample description.

Material	Sample characteristics
Metals Metallic Alloys	Name (trade name, generic name, standard designation, supplier) Chemical composition Manufacturing process and heat treatment Hardness, other mechanical properties (yield or ultimate tensile strength) Microstructure (phases, grain size, hard carbides or inclusions) Surface finish and roughness, surface anisotropy Orientation of wear surface to the direction of sliding
Polymers	Name (trade name, generic name, standard designation, supplier) Molecular weight, crystallinity, density, water absorption Manufacturing method, additional treatment (e.g. irradiation) Hardness, other mechanical properties (tensile strength, maximum elongation) Glass transition temperature, softening temperature, melting temperature Surface finish and roughness, surface anisotropy Orientation of wear surface
Ceramics Glasses	Name (trade name, generic name, standard designation, supplier) Chemical composition, impurity content, sintering aids Manufacturing process Microstructure, density, porosity, grain size, boundary glass phases Hardness, toughness, flexural strength, other mechanical properties Surface finish and roughness, surface anisotropy Orientation of wear surface
Composites	Name (trade name, generic name, standard designation, supplier) Matrix type (metal, ceramic, polymer) Filler (type, characteristics, orientation, weight or volume percentage) Manufacturing process Flexural strength, other mechanical properties Surface finish and roughness Orientation of wear surface, surface anisotropy

Contact between solid bodies is controlled on microscale by their surface roughness and on macroscale by their waviness or form error. In most tests, only the surface roughness parameters are specified for a test specimen, but the effect of waviness or form error on friction and wear cannot always be ignored. The combined effect of surface roughness and waviness on friction is schematically illustrated in Figure 6.1.

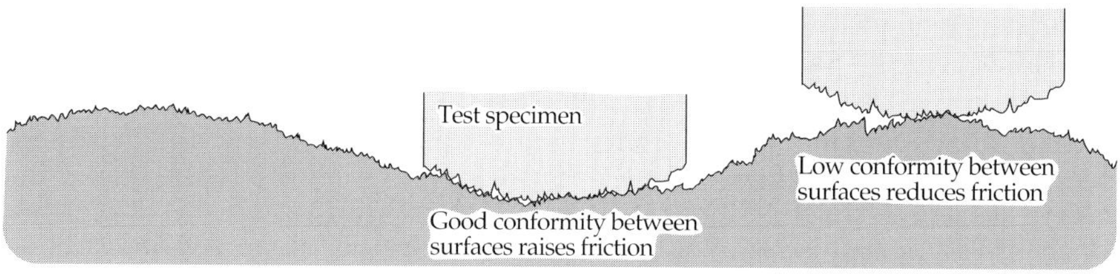

Figure 6.1 Schematic illustration of the effects of local changes in surface elevation on local friction (adapted from [1]).

As can be seen from Figure 6.1 local changes in surface elevation modifying the conformity between interacting surfaces can affect local friction. Measurements of local elevation above the mean height of the worn surface by an optical phase-shift interference microscope revealed that periodic variations in friction force correlated with gradual changes in surface elevation [1]. The cause of the friction variation, in a pin-on-disc test, was the friction dependence on the closeness or conformity of contact between the pin and the disc.

Traditionally only 2-D surface roughness parameters such as $\mathbf{R}_a$, $\mathbf{R}_q$, $\mathbf{R}_z$, $\mathbf{R}_{max}$, etc., are specified in test specimen characterization. Surface roughness measurements are usually conducted before and after testing in order to compare the original and the worn surfaces. From this comparison some information on the wear mechanisms and wear volume can be deduced, especially when 3-D surface topography is measured.

Typical methods of surface roughness measurement include mechanical stylus profilometers and noncontact techniques, which are described in this chapter. There are also other, less frequently used methods. For example, ultrasound can be used to determine surface roughness. The ultrasound techniques were originally developed to measure water surface roughness. They are now being adapted for on-line measurement of surface roughness during the machining of components. The aqueous cutting fluid is used as a medium to convey the ultrasonic sound from the machined component to the acoustic transducer [2] and the sound is then converted into surface roughness parameters. Several comparative techniques for assessing surface roughness are also available (e.g. friction and tactile tests, inductance methods, capacitance methods and pneumatic methods) [3-5]. These methods are more suitable for quality control on the shop floor rather than for laboratory measurements. Most of them are not very accurate and do not provide standard roughness parameters.

Accurate values of material's hardness and a related parameter, microhardness, are also essential data for analysis of surface contact and subsurface deformation in friction and wear. This material property is especially important in wear studies where abrasive or adhesive wear dominates and most wear damage is due to shear stresses. Microhardness measurement can reveal far more than the commonly used bulk hardness. Information on the distribution of soft phases and hard inclusions, the survival of hard coatings, quality control of surface treatments, depth profiles of hardened layers, etc., can be found from microhardness measurements. Since tribological phenomena are usually confined to the top surface layers, microhardness can often better indicate the tribological behaviour of a specimen than bulk hardness.

Recent developments of nanoindentation techniques allow for hardness measurements (nanohardness) of very thin coatings/layers where the size of the indent is not resolvable by traditional optical microscopy. Apart from nanohardness, spatial maps of other mechanical properties of surface films can be measured by nanoindentation techniques without removing the film from the substrate. In this chapter the techniques of test specimen characterization for tribological testing are described.

6.2 MEASUREMENT AND ANALYSIS OF SURFACE ROUGHNESS

Surface texture can have a strong influence on wear and friction and a general rule is to use the same texture for tribological specimens as it occurs in real application. If the objective of the testing is to compare the tribological behaviour of several materials on the same rig the specimens are prepared with very similar surface texture, which should be described in the specimens characterization. This description is especially important when testing is conducted under lubricated conditions since surface roughness affects the film thickness and lubricant retention on the surface. Also, in cases where wear is negligible the contact between the surfaces is influenced by the original surface roughness. Magnetic tapes and discs comprise one area of tribological applications where the original surface roughness affects their performance (friction and wear). When one of the surfaces in tribological contact is much softer than the other, the determination of surface roughness of the harder counterface is important as its roughness affects the abrasive wear of the softer surface.

Surface roughness on a microscale is usually measured by one of two methods:

- mechanical contact probes (stylus profilometry), and
- noncontact probes (laser or light interferometry, laser speckle analysis, SEM stereoscopy, confocal microscopy, etc.).

In order to quantify surface roughness, i.e. describe it in numerical terms, most of the above systems use similar 2-D parameters, many of them now standardised internationally (see e.g. ISO 4287, BS 1134, DIN 4768 or national standards). Detailed information on the characterization of surface roughness can be found in many textbooks [e.g. 3-6]. Standard surface roughness parameters are divided into three groups:

- amplitude parameters (e.g. R_a, R_q, R_z, R_{max}, R_{tm}, R_{pm});
- spacing parameters (e.g. peak count, peak spacing);
- hybrid parameters (e.g. average slope, average wavelength).

Some of these parameters are in common use in tribological specimen characterization ($\mathbf{R}_a$, $\mathbf{R}_q$, $\mathbf{R}_z$) and others are specified only in particular investigations. For example, peak curvature and average slope have been used in studies of contact mechanics between surfaces [7]. More detailed characterization of surface roughness would also include statistical functions describing distributions of surface features (bearing ratio curve, asperity height distribution, autocorrelation function, power spectral density, etc.) [3,5,7]. These statistical functions are seldom specified in wear and friction testing.

It has to be pointed out that there are no strict guidelines as to which parameters should be chosen for adequate surface texture description of a specimen. The choice is often dictated by tradition (e.g. '$\mathbf{R}_a$' is more popular in Europe while '$\mathbf{R}_q$' in the USA), equipment availability and industrial application. There is growing awareness, however, that a single parameter is not sufficient to characterize surface texture. Ideally, surface texture should be described by a combination of amplitude, shape and spatial parameters.

While surface roughness measures such as the '$\mathbf{R}_a$' provide a vital indication of the quality of a surface, it should also be remembered that wear and friction phenomena are usually controlled by localised abnormal regions of the surface. Small areas of the surface that are either unusually rough or smooth may induce wear particles to form or else seizure to occur. Instead of confining one's concern to the average value of roughness, it is important to consider the variability of roughness across the surface. In quantitative terms, this could be likened to the calculation of a standard deviation or a kurtosis on the '$\mathbf{R}_a$' value.

Despite providing similar surface roughness parameters, the contact and noncontact techniques differ in data collection and sampling strategies and the comparison of measurements obtained by different methods is difficult [8]. Stylus profilometry is the older and more established of the two methods. It is also the method by which various national standards for surface texture parameters are defined. A stylus is drawn with constant velocity over a surface to reveal the size and distribution of the asperities from the magnitude of vertical movement of the stylus. Laser speckle analysis examines the pattern of light reflected from a rough surface when irradiated by a laser. The analysis of the reflection pattern or 'speckle' is complex but the technique provides a noncontact measurement of surface roughness which can be invaluable when dealing with e.g. very hot or very soft test specimens. Confocal microscopy reconstructs a 3-D surface image from the set of 'optical slices', each centered about a narrow focal plane of objective lens.

Surface roughness can also be evaluated on a nanoscale using Atomic Force Microscopy (AFM) or Scanning Tunnelling Microscopy (STM) techniques. More detailed information on both the AFM and STM techniques can be found in Chapter 8.

Since real surfaces are 3-D objects full understanding of tribological processes occurring is only possible if 3-D surface topography is truly known. Recent developments in measuring systems allow for qualitative 3-D surface topography measurements, while quantitative standards are still being debated. Several 3-D parameters currently in use are derived from their 2-D equivalents while new parameters and characterization methods are continuously being developed [9-28]. Summary of the current state-of-art knowledge about 3-D surface topography measurement and description can be found in [11,28].

Measurement of Roughness by Stylus Profilometry

Surface roughness measurements using stylus profilometry have become an accepted and essential part of material surface specification in tribological studies. The technique of stylus profilometry, briefly mentioned in Chapter 4, is based on traversing the surface with a diamond stylus. This has the advantage that the stylus is in a direct contact with the surface. However, the technique is slow to use, has limited surface 'imaging' capability and tends to filter high frequency surface detail. In commonly used laboratory instruments the stylus has a fixed radius, or width, usually **1.5-2.5** [μm]. Styluses with smaller radii, of about **0.1-0.5** [μm], are also available and are used for more detailed surface measurements. The load on the stylus is in the range between **50-100** [mg] so the styluses with the very small radii could exert high contact pressures and subsequently scratch the surface. Therefore, in practice, the load on a stylus with a very fine width is greatly reduced.

In most cases, 2-D values of basic height parameters such as '$\mathbf{R}_q$' or '$\mathbf{R}_a$' are determined to characterize the surface. '$\mathbf{R}_q$' (also called 'RMS' roughness) is usually specified for optical or smooth components while '$\mathbf{R}_a$' is used for describing surface finish after machining or grinding. Surface roughness data such as '$\mathbf{R}_q$' are fundamental to, for example, concepts of the ratio of lubricant film thickness to surface roughness in elastohydrodynamic lubrication. This ratio characterizes the separation of interacting surfaces. Common surface roughness parameters '$\mathbf{R}_a$' or '$\mathbf{R}_q$' are often not sufficient to characterize the surface texture of a tribological sample. It is known that surfaces with quite different profiles, that subsequently could influence wear or friction in a different manner, may exhibit the same or very similar surface roughness average '$\mathbf{R}_a$' [4,5,29]. Since these surface profiles are generated by different finish processes, it is common to specify how the final surface was prepared (lapped, ground, turned, polished, etc.).

Surface profile graphs are often included next to the numerical roughness parameters to help visualize the shape and distribution of peaks and valleys [29]. When looking at the profile graphs it must be realised that the vertical magnification of the profile is much higher than the horizontal, often **100** times higher. In reality the peaks and valleys are not as steep as they appear on the profiles [4,5]. For surfaces with suspected pores or other irregularities such as scratches, either due to microstructure, heat treatment or surface preparation, other parameters such as maximum peak-to-valley height '$\mathbf{R}_t$' or '$\mathbf{R}_{max}$' or mean peak-to-valley height '$\mathbf{R}_{tm}$' may also be useful. Surface textures prepared by special techniques, such as for example plateau honing, often need additional parameters to best describe the symmetry of the profile about the mean line. The parameter describing this property is the skewness, '$\mathbf{R}_{sk}$' or '$\mathbf{S}_k$', which is often used when testing surfaces of engine cylinder bores or surfaces prepared for lubricated contact [4,5]. Detailed surface texture characterization (including spacing and hybrid parameters and statistical functions) is especially important in tribological tests where the influence of surface roughness on wear and friction is the test's main objective or in investigations of contact between rough surfaces. Detailed surface roughness characterization is now also being included in modelling of lubricated contacts such as EHL.

Although 2-D surface roughness data from surface contact or stylus profilometers continues to be of basic importance, it does suffer from some limitations, which cannot be overlooked. One problem is that any stylus with a finite tip radius or width is unable to record surface detail that contains concave radii smaller than the radius of the stylus tip. Therefore much of the data on fine surface characteristics is smoothed by mechanical filtering due to this finite tip size [4,5]. For example, if the tip radius is nominally **2.5** [μm] then the wavelengths of this order or smaller are significantly smoothed [5,30] as illustrated schematically in Figure 6.2. An exception to this rule is when a spherical stylus traces shallow valleys. In this case lateral resolution can be much smaller than the tip radius [31].

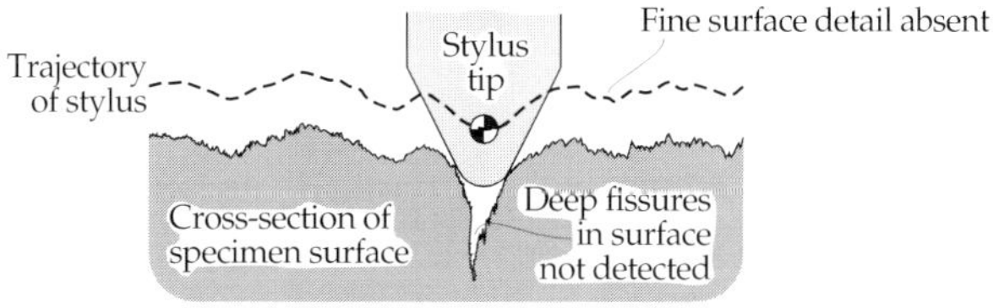

Figure 6.2 Schematic illustration of the effects of finite tip size on recorded surface roughness.

While lateral resolution depends on the stylus size, the vertical resolution can be affected by instrument vibration, electrical noise and the type of reference datum used [31]. Because of the stylus shape the angle of true sharp peaks can be rounded off and the re-entrant features and very steep slopes cannot be traced [31]. This last shortcoming can be important in characterization of cast iron, sintered or porous surfaces where the re-entrant features and holes are often present. Another source of error may be due to stylus bouncing and loosing contact with the surface. To minimize this possibility the speed of the stylus is kept sufficiently slow, usually about **1** [mm/s] or less. This contributes to the long length of time required to complete the measurement. At this point it is also necessary to mention the need for periodic checking of the accuracy of profilometer. This is done by conducting measurements on reference standards [31].

Another problem associated with stylus profilometry is that the stylus can deform or scratch the surface it passes over and consequently record a false surface profile [5,11,31]. Surface deformation is likely to occur when a stylus with a fine tip width measures a surface of low hardness using a slow traversing speed. This problem is particularly acute when the object of investigation is soft, as for example soft metals or polymers [5,32-34]. Special care has also to be taken when measuring biological surfaces such as the surface of articular cartilage [35,36]. Contemporary technical development of surface topography measurements involves the substitution of a stylus by noncontact methods to a certain extent.

It is important to realise that the conventional stylus profilometry records only a small portion of the surface detail, in the form of 2-D surface profiles. So, usually, to characterize surface roughness an average from several single surface profiles is calculated to account for small variations in surface roughness. Surface texture of specimens is rarely isotropic. Well polished or shot blasted surfaces are the exceptions where the surface roughness is nearly the same in all directions. For anisotropic surfaces the direction of surface roughness measurements should be specified (usually across the lay, or in the direction of sliding). To compare two stylus measurements, the sampling conditions such as tip radius and cut-off length should be the same. Wear processes almost always change the original surface roughness. When a surface roughness measurement is conducted on the worn surface, as schematically illustrated in Figure 6.3, the direction of the profile has to be specified (perpendicular to the sliding direction for profile depth, along the sliding direction for changes with respect to the original profile).

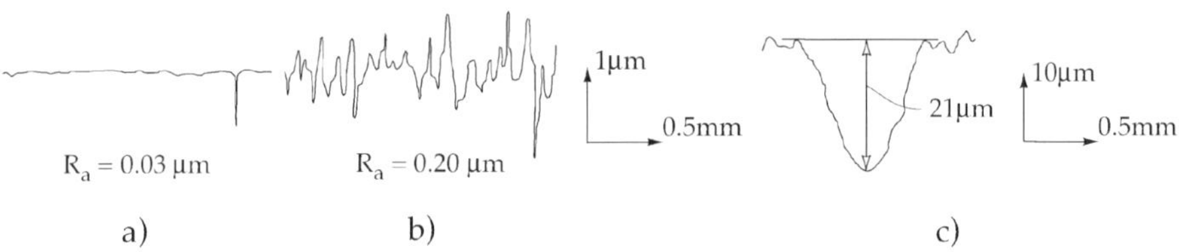

Figure 6.3 Talysurf surface profiles taken from the original (unworn) and worn surfaces; a) original polished surface, b) severely worn surface, c) surface profile across the wear track.

Surface profiles of the worn surface can provide some information on the wear mechanisms. For example, abrasive and adhesive wear generate quite different wear surface textures. This is illustrated in Figure 6.4 where a typical surface profile with ploughing grooves formed under boundary lubrication is compared with a surface profile where adhesive wear dominated under starvation lubrication. Unfiltered surface profiles across the wear scars can be used to estimate the wear volume. The application of profilometry to the characterization of wear damage is described in more detail in Chapter 4.

Figure 6.4 Surface profiles of (a) ploughing grooves under boundary lubrication; (b) adhesive wear under lubricant starvation conditions.

In stylus profilometry, 3-D surface features are acquired by taking a series of equally spaced parallel profiles from an area under investigation [4,5,11,28]. This process, however, is time consuming and requires a specialized set-up in the form of 'y' or 'x-y' stepping table. An example of 3-D surface topography acquired by equally spaced parallel surface profiles is shown in Figure 6.5.

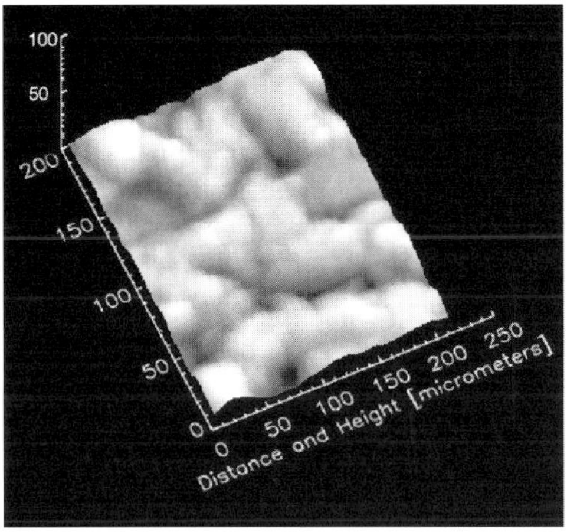

Figure 6.5 3-D surface topography image of sandblasted steel surface acquired by equally spaced parallel surface profiles.

Noncontact Surface Roughness Measurement

For most noncontact surface roughness measurements either normal light, as in light interferometry, or laser as in laser confocal microscopy, or electron radiation (SEM stereoscopy) is used to obtain the surface texture image. These techniques differ significantly in the principles applied and will be described separately. Most of them offer better lateral resolutions and some of them better vertical resolutions than stylus profilometry [37,38]. An additional advantage of noncontact techniques is the ability to measure soft surfaces without scratching. Also, most of the noncontact techniques can provide 3-D surface topography data instead of usual line profiles of stylus profilometry.

The majority of noncontact methods use light (laser or nonlaser) to obtain the surface image. From the numerous optical instruments currently available for surface roughness measurements three techniques will be described in this section in more detail: interferometric microscopy, laser speckle analysis and confocal microscopy. Information on other optical techniques such as 'Foucault knife probe' and 'optical followers' can be found in [3,5,39,40]. There are also much simpler optical methods such as light reflection (gloss meter) and light sectioning but they are not accurate enough for the characterization of tribological specimens [4,5].

- *Optical Profilometry Based on Light Interferometry*

Light interferometry is the best known noncontact surface roughness measurement technique. Its principle is based on interference of light beams reflected from two inclined surfaces [3,4]. In its simplest form the conventional surface texture interferometer produces a surface image consisting of alternate dark and light fringes by first splitting and then recombining a light beam reflected from a reference and a tested surface [3,4]. Contour fringes are produced when the two surfaces are normal to each other and rows of fringe profiles when the surfaces are inclined at a small angle [3]. Any irregularities on the surface are represented as irregularities in the fringe pattern. At first, light interferometry was restricted to measurements of steps, dents or grooves on the surface [4]. Developments in the past 25 years in electronics and computing have allowed the measurements of fringe pattern intensities to obtain standard surface texture parameters [37].

Modern interferometric microscopes can acquire quite accurate 3-D topography data of surfaces [e.g. 37,41,42]. In these instruments a light beam is split into an internal reference beam whose path is precisely monitored by piezoelectric drives and a probe beam whose optical path length varies with the change in sample surface height. The two beams recombine and form the interference fringes. The intensities of the interference fringe signal are used to measure surface heights.

The modern interferometric microscope can be operated in two modes:

- phase shifting interferometry (PSI) and
- vertical scanning interferometry (VSI).

In the PSI mode, the sample is first brought in focus and the reference mirror is then moved by a small distance at constant speed (this method is also known as integrating bucket technique) or step-by-step. This causes a phase shift between the reference beam and the probe beam. The intensities at each surface point are measured at three reference mirror positions, which vary by 90°, and the phase shift is determined. The surface heights are calculated at each point from the phase shift data obtained [37]. An alternative method of determining surface roughness, the VSI mode, involves the use of white light and locating the maximum amplitude of intensities (the maximum fringe contrast) instead of the phase shift. In the VSI mode, the magnification objective is moved vertically using a piezoelectric actuator and images are recorded at different heights above the focus point on the surface. An irradiance, defined as the ratio of the total amount of light flux reflected from the focal spot and the area of that spot, is then calculated from each image recorded. Next, irradiances calculated are used to construct an intensity signal, i.e. irradiance as a function of height. From the resulting signal low frequency components are removed using a highpass filter and the signal is further processed by a square-law detection routine and a lowpass filter. A position of the highest peak of the lowpass filter output is detected and used to calculate a relative height of the surface. At this position, a path length difference between the reference beam and the probe beam is almost equal to zero and the fringe contrast (the intensity signal) is at maximum. This procedure is repeated for each point on a surface and a 3-D surface map is then obtained. The scan size and sampling interval are fixed by the magnification of the optical system used. Vertical resolution of this instrument in the VSI mode is between **3** [nm] and **500** [μm] while in the PSI mode is between 0.3 [nm] to 160 [nm]. The lateral resolution depends on the wavelength of the light used, i.e. is equal to half of the light wavelength. The operating principles of phase shifting and vertical scanning interferometry are schematically illustrated in Figures 6.6 and 6.7.

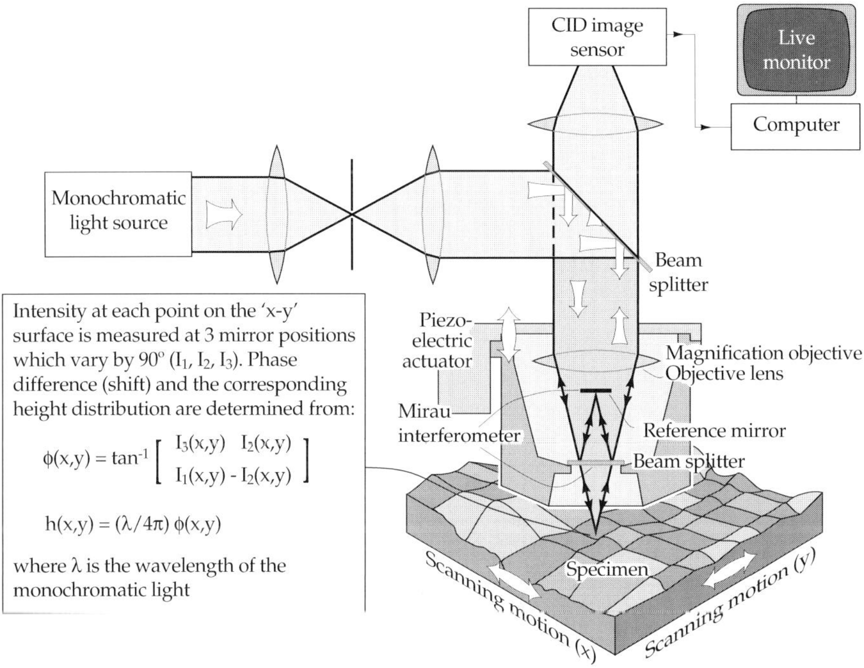

Figure 6.6 Schematic illustration of surface roughness measurements by phase shifting interferometry (adapted from [37]).

Several other designs of interferometric microscopes are described in [3,39]. Interference microscopy in the PSI mode is most suitable for measuring surface texture of low roughness and low slopes. Therefore it has found applications in measuring surface texture of computer recording systems (tapes, discs, recording heads), glass, gauge blocks and silicon wafers [37,39]. The VSI mode, on the other hand, can be used to measure surface topography with higher roughness and higher slopes [42].

As 3-D quantitative surface texture assessment is necessary for complete surface characterization most modern interferometric microscopy methods provide qualitative and quantitative 3-D surface images [37,42].

One of the more exotic interferometry techniques is holographic interferometry [39]. This technique has, for example, been used to measure wear and dimensional changes occurring in worn acetabular cups of hip prostheses and tibial components of total knee prostheses [43]. Accurate results are obtained when the acetabular cup is scanned from more than one position to compensate for the tendency of interference fringes to become very crowded over parts of the hologram. It was found that although the accuracy of the technique was insufficient for useful estimates of wear volume, the area distribution of wear or creep could be determined accurately [43].

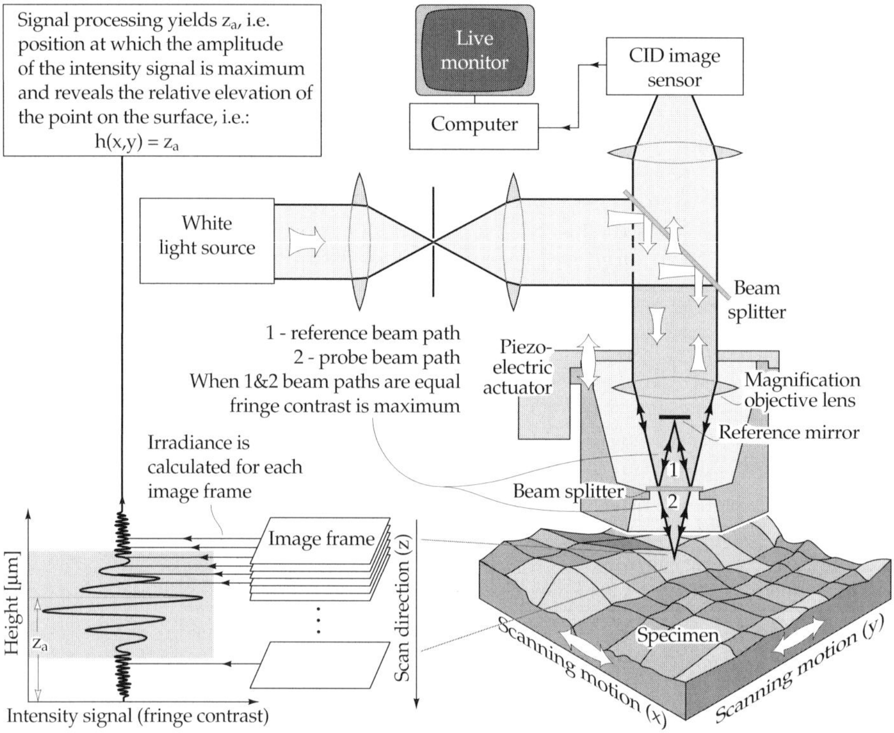

Figure 6.7 Schematic illustration of surface roughness measurements by vertical scanning interferometry (CID - Charge Injection Device).

· *Optical Profilometry Based on Light Reflection*

Another group of optical surface topography measurement techniques is based on light reflection from the measured surface. Three optical phenomena are used to quantify surface topography: speckle [44], scattering [45] and polarization [46]. These techniques have been mostly directed towards industrial applications since they offer real-time assessment of an area of a surface [39,40,47,48].

Most theoretical analyses of light scattering relate to Gaussian surface roughness and experimental calibrations between scattered light and roughness are also usually based on this type of roughness [47]. When the surface roughness becomes non-Gaussian, then the relationship between scatter and roughness changes in an unknown manner. For this reason, external calibration of roughness with, for example, a stylus profilometer is required in order to obtain accurate measurements. Light scatter analysis is therefore more suited for detecting changes in roughness or differences in roughness between specimens. Optical profilometer based on scattered light and triangulation principles has been proposed for 3-D surface topography measurements [49]. Two optical detecting systems were employed to minimize the response distortion. Since the vertical sensitivity of this system is about **3** [μm] it is suitable for measuring topography of rough surfaces.

Surface roughness measurement by speckle pattern analysis is based on the fact that when a rough surface is illuminated with coherent laser light, a random pattern of bright and dark regions, or speckle, forms. The spatial distribution and contrast of speckle patterns have been found to depend on surface roughness. The existing designs are based on three forms of speckle pattern measurements: monochromatic or polychromatic speckle contrast and speckle pattern correlation. The principle of the laser speckle roughness measuring system is schematically illustrated in Figure 6.8.

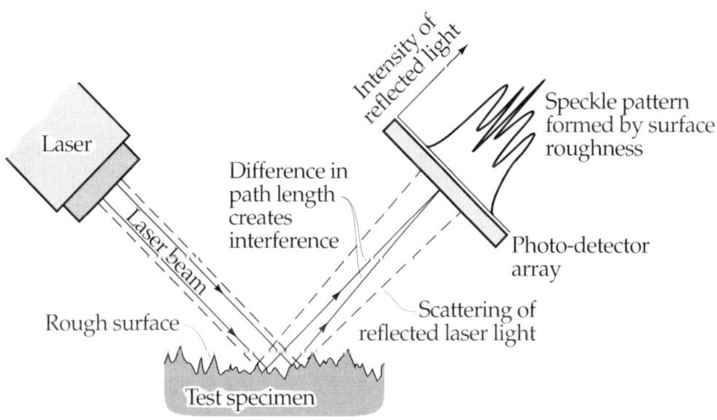

Figure 6.8 Schematic illustration of the principles of the laser speckle roughness measuring system.

The speckle pattern is generated from differing angles of reflection from a rough surface, which allow portions of a laser beam to be reflected from different locations on the rough surface and to merge at some distance from the surface. The optical interference is strong enough to generate a pattern of light and dark areas or merely cause a variation in light intensity.

A limited capacity to measure surface roughness is obtained by the use of a single wavelength of laser light (monochromatic speckle contrast) where a roughness parameter is correlated to the ratio of light intensity variation to mean light intensity (contrast) [50]. More advanced techniques employ a comparison between speckles generated at different operating conditions of the laser. One technique, known as angular spectral correlation, involves recording speckles at two different inclination angles of the laser beam to the surface for later comparison [48,51]. The advantage of this more complex technique is that the range of roughness measurement is extended from **0.1 - 0.5** [μm] '$\mathbf{R_a}$' to **0.1 - 5** [μm] [52]. In another method the speckles produced by two different wavelengths of laser light are measured and the surface roughness is calibrated against the degree of correlation between the two speckles [48]. This technique, known as spectral speckle correlation or 'SSC' [53], avoids the need for accurate positioning equipment of the laser when performing angular spectral correlation. 'SSC' is, however, very sensitive to vibrations so that it can only be used under carefully controlled conditions. The calibration between roughness and speckle parameters is influenced by the type of surface, i.e. a shot-peened, ground surface, etc. [48].

· *Light Intensity Feedback Profilometry (Focus-Detect)*

In the systems utilizing this principle the surface roughness is recorded by maintaining the laser beam focal point always coincident with the object surface. This is achieved either by a set of moveable objective lens [54], as illustrated schematically in Figure 4.11c, or by a feedback arrangement which moves the specimen in both vertical and horizontal planes [55], as shown in Figure 6.9. At the focal point the intensity of the reflected beam is maximum. Attempts have also been made to design the systems such that they can operate in both modes, either 'fixed specimen' or 'fixed lens' mode [39,56,57].

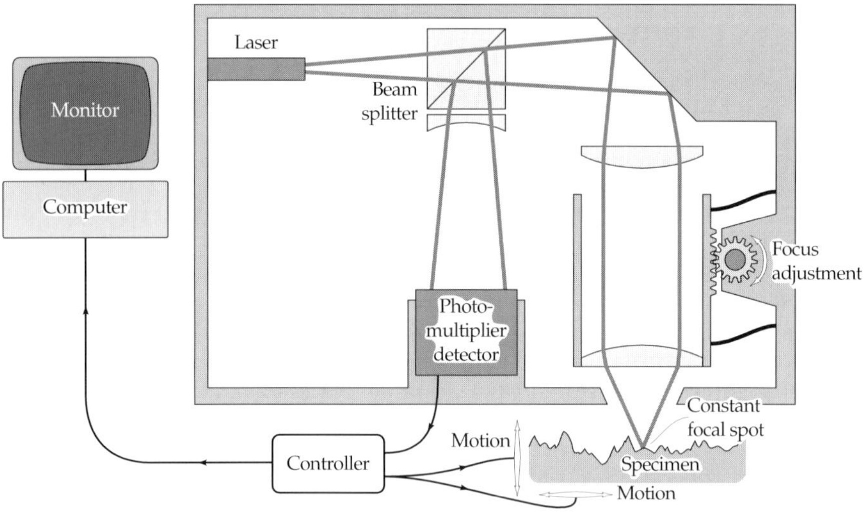

Figure 6.9 Schematic illustration of surface roughness measurements by a light intensity feedback profilometry.

· *Confocal microscopy*

Confocal microscopy (CM) is a rather new technique that can be applied to characterize surface texture [3,58]. The technique is known better in biological sciences where confocal microscopy allows the generation of sharp images of translucent tissue structures, often with the aid of fluorescent dyes [58]. In engineering confocal microscopy has been used to obtain images of, for example, sandblasted [30], ground, lapped and milled [59] and worn or damaged surfaces [58]. Confocal microscopy uses either laser or nonlaser illumination as a light source and several scanning techniques to obtain the images [58]. The operating principles of the laser scanning confocal microscopy (LSCM) are schematically illustrated in Figure 6.10.

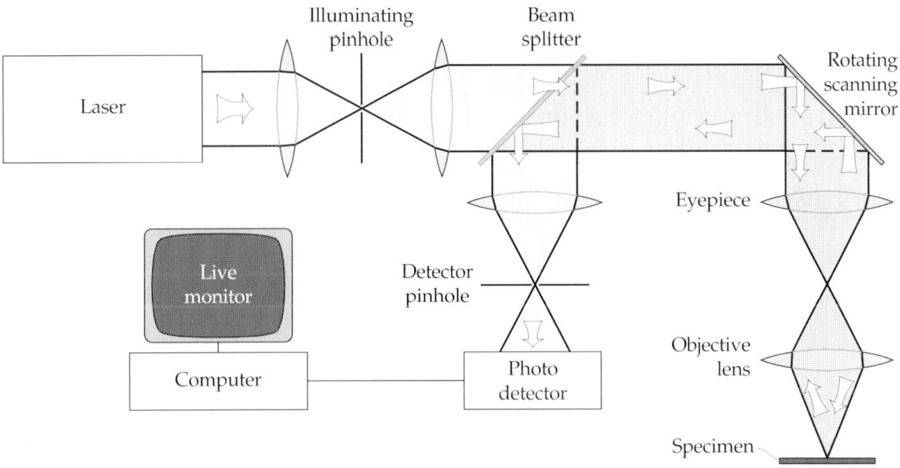

Figure 6.10 Schematic illustration of surface roughness measurements by laser scanning confocal microscopy.

The operating principle of LSCM involves focusing a light from a semiconductor laser, restricted by a pin-hole, on the surface measured. A surface incident spot of approximately 1 [μm] in diameter is generated. Light reflected from the specimen is collected by an objective lens and then via the beam splitter passes through another pin-hole before reaching a photo multiplier where the image is formed. A second pin-hole in this arrangement allows a large proportion of out-of-focus light to be rejected. The pin-hole acts as a filter allowing only light reflected from the focal spot on the surface to pass through it. All the other light reflected from the surface is cut off. In LSCM the laser beam is scanned across the surface in a raster pattern. In some devices a specimen is moved both horizontally in x-y directions and vertically. At each horizontal pass a focused image is taken and stored in the computer. The specimen is then moved vertically at the increments of 0.5 [μm] or less depending on the instrument design and the image collection process is repeated. The collected 2-D images (optical sections) at various heights of the specimen are then combined forming a 3-D image of the surface.

Confocal microscopes allow two kinds of images to be acquired: 'maximum brightness' (also called through-focus) and 'height encoded' (topography) [30,58,59]. Common 2-D surface roughness parameters can then be calculated from line scans obtained from the height encoded or topographic images [59]. However, when surface roughness parameters calculated from confocal topographic images were compared with the parameters obtained from stylus profilometry, using the same surfaces, differences in the absolute values were found [59]. Line scans from topographic images can also be used to measure the depth of wear tracks or surface cracks [58]. The surface can be viewed in 3-D (stereo view) by using two tilted confocal images [58].

Confocal microscopy techniques with real time imaging are also available [58]. The vertical and horizontal resolutions depend on the type of objective lens used and the limit of about **0.4-0.5** [μm] is often quoted [58,59]. The main disadvantage of CM is that the images can suffer from scattered background noise that may affect the determination of surface texture parameters [30,59]. In opaque samples re-entrant features cannot be imaged as the light cannot reach them. Also holes, pits, thin microcracks and nearly vertical edges are very difficult to image. These limitations of confocal microscopy are similar to those found for other surface topography measurement techniques described in this chapter. Other drawbacks of CM are that it is often slow and, as CM is based on the light reflected from the surface under examination, a moderate reflectivity of the specimen surface is required.

· *SEM Stereoscopy*

Scanning electron microscopy (SEM) is widely used to qualitatively examine surface texture of tribological specimens using either secondary electron (SE) or topography modes. The SEM images are characterized by high resolution (up to 10 [nm]) and large depth of field. Some information about relative heights in surface texture can be obtained from single SEM images, but some impressions can also be misleading [60]. Quantitative assessments of surface texture using line scans of SEM image brightness, however, have proved to be inaccurate. The main reason is that the exact correlation between image brightness and surface height is not known. Image signal amplitude and frequency can vary with magnification and some surface features are susceptible to highlighting.

It is now recognized that stereo-based image analysis methods are necessary for quantitative assessment of surface texture from high resolution SEM images [60]. The technique is called SEM photogrammetry [60] or SEM stereoscopy [38]. In SEM stereoscopy 3-D images of the surface texture are obtained from two sequential SEM images.

The surface images used in the analysis are obtained either by tilting the specimen in the SEM by a known angle (high magnification images) or by translating the specimen by a known distance (low magnification images). As the two images obtained are displaced relative to each other the position of corresponding points on these images is shifted. The distance between these corresponding points is then calculated and used to obtain the surface elevation values. The tilt angles can vary between **5°** and **15°** depending on the surface topography and are chosen to ensure that the corresponding points could be clearly recognized on both images. The accuracy of the SEM stereoscopy is mostly limited by errors intrinsic in the taking of stereo-pair micrographs [60]. More details about this technique can be found in [38,60].

The SEM stereoscopy technique, although well known, has not been widely used because of the inadequate image matching programs and lengthy computation time required to obtain the 3-D image of the surface. This, however, has recently been rectified by the development of the special software enabling the accurate acquisition of 3-D wear particle surface texture data [26,38,61]. Although surface roughness parameters can be calculated from the SEM stereo images, this technique is very rarely used for this purpose. The example of the application of SEM stereoscopy to surface texture imaging is shown in Figure 6.11.

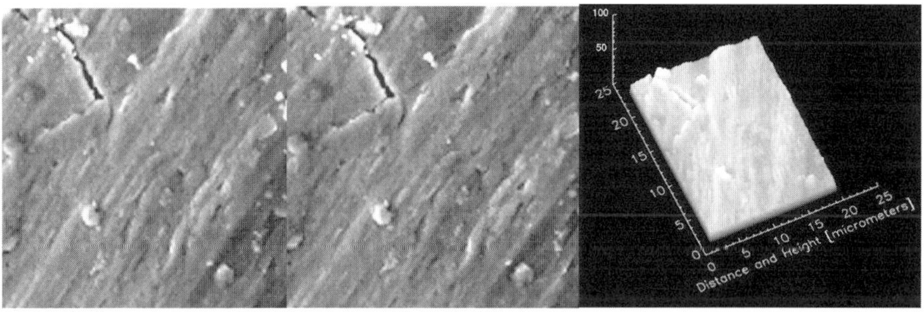

Figure 6.11 Two SEM images of worn surface obtained under two tilt angles of 0° and 10° and the resulting shaded image of the 3-D surface topography.

The major differences between various surface topography measurement techniques described in this section are data acquisition speed and lateral resolution and range. These differences are summarized in Figure 6.12.

6.3 SURFACE HARDNESS, MICROHARDNESS AND NANOHARDNESS MEASUREMENTS

Hardness measurements have long been used as an indication of materials wear resistance. Although this view of hardness has been disproved for most forms of wear, except abrasive and erosive wear, hardness is a material property that is related to wear and friction. Measurement of hardness in tribology involves the standardized methods employed in materials science studies and is not described in this book in detail. ASTM standards for measuring hardness (and microhardness) of various materials are listed in Table 6.2. The Vickers hardness test is preferred for research purposes in tribology as it is suitable for a very wide range of hardnesses and is directly related to the compressive flow stress of a material which controls the area of true contact between solids. If commercially available materials are used for tribological testing their bulk hardness is usually specified by the manufacturer. However, if any additional heat or surface treatment has been applied, the hardness should be measured again.

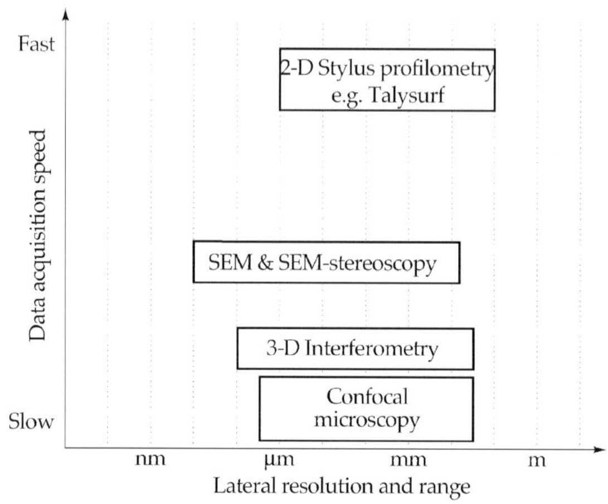

Figure 6.12 Schematic comparison of the influence of data acquisition speed and typical lateral resolution and range on the efficiency of surface topography measurement techniques.

Table 6.2. ASTM standards for hardness tests.

Metals	**Ceramics Glasses**	**Polymers Rubbers**
A 370	C 730	D 785 (Rockwell)
E 10 Brinell	C 748	D 2583 (Barcol)
E 18 Rockwell	C 849	D 1415
E 92 Vickers	C 886	D 2240 (Durometer)
E 384 Vickers Microhardness		
E 448		

Microhardness Measurements

While hardness primarily relates to a bulk property of the material, microhardness data is useful in describing not only material characteristics such as the presence of soft phases and hard inclusions but also modifications done to test specimens such as wear resistant coatings or mechanically worked surface layers. Since engineered surfaces are found in ever increasing numbers in tribological applications, specimens with thin wear or friction resistant coatings have to be characterized. The microhardness measurements are conducted on the unworn coatings before tribological testing and on the worn coatings or wear surfaces after testing. Examples of tribological surfaces where microhardness was used for their characterization include laser-alloyed layers [62], laser clad composite coatings [63], white layers [64], and mechanically mixed layers [65].

There is usually a significant hardness difference between the coating and substrate. In such cases the microhardness tests over the surface of the coated specimen can be used to check the uniformity of the coating. Even in the case of thin coatings, where microhardness measurement is compromised by substrate deformation occurring during the indentation, the microhardness measurement can still be used for comparative purposes to detect flaws or gaps in a coating. It has to be pointed out, however, that unless the thickness of the coating is already known, the accurate measurement of hardness of a thin coating is quite difficult [66-68].

Provided that the coating is not too thin, the thickness of a coating and variation of hardness with depth inside the coating can be determined from the specimen wedge sections. Some problems may arise with coatings of thickness less than 10 [μm] as the indentation marks become comparable in size to the coating thickness and it might be difficult to locate the indent with sufficient precision. Ion implanted coatings or coatings deposited by physical vapour deposition are therefore usually too thin for sectional microhardness measurements. The application of microhardness measurements to specimen characterisation is schematically illustrated in Figure 6.13.

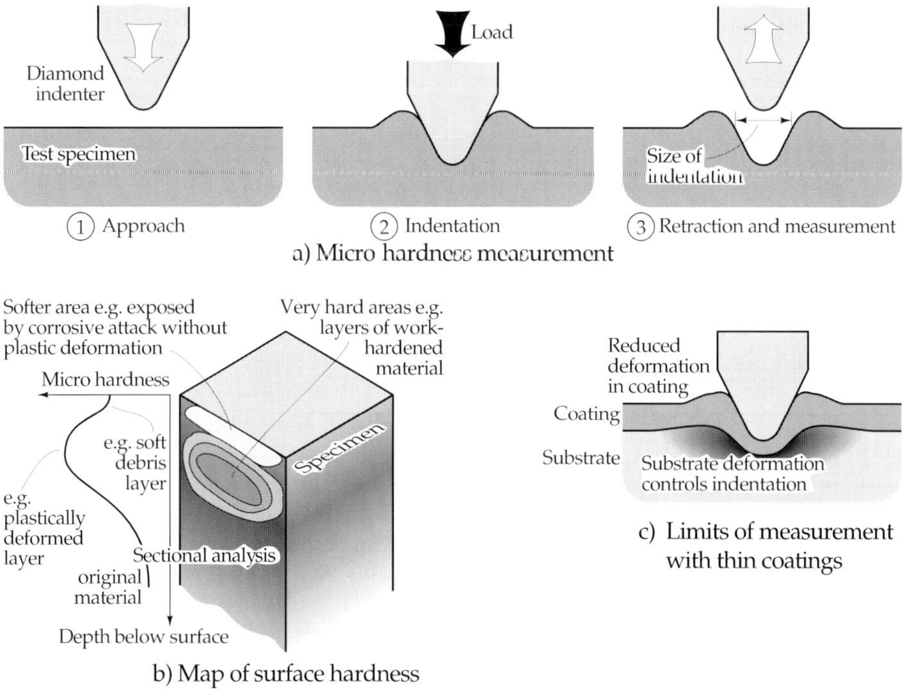

Figure 6.13 Schematic illustration of the application of microhardness measurements to the characterization of surface and subsurface.

The principle of microhardness measurement is the same as the principle involved in bulk hardness testing (as described in ASTM E-384). A hard object in a shape of a pyramid (known as the indenter) is slowly lowered onto the test specimen and then loaded with a prescribed weight. The main difference is the value of load which in the case of microhardness testers is between 1 and 1000 [g]. Hardness is calculated from the size of the plastic indentation left on the specimen surface after the indenter has been retracted.

The conventional forms of microhardness testing are known as the Vickers and Knoop tests. The smallest areas to be measured by the Vickers indenter are about **5-10** [μm] in diameter and by the Knoop indenter about **20** [μm] in length. Newer tests continue to be developed in order to investigate surfaces at ever smaller scales. These new techniques are referred to as nanohardness testing or nanoindentation techniques.

In conventional microhardness testers the area of plastic indentation is measured with an optical microscope. There is a little change in size of the indentation imprint for metals on unloading. However, small errors can occur due to the elastic contraction of the imprint [69]. To avoid these problems microhardness testers based on measuring the penetration depth rather than the imprint area have been developed [e.g. 70-77]. This method in its refined form is now mainly used in the nanohardness measurements.

Nanohardness Measurements

There are major technological and methodological differences between bulk hardness testing and nanoindentation due to the extreme degree of miniaturisation involved and additional materials properties, e.g. Young's modulus, measured at this, nano, level. Nanoindentation testers are intended for hardness measurements of thin surface coatings, such as are produced by ion implantation [78,79], physical vapour deposition [80], heat treatment [e.g. 81], polymer coatings [e.g. 82] and thin surface films [e.g. 70,83-89]. Nanoindentation tests allow a wide variety of mechanical properties of the surface films to be determined without removing the film from the substrate. Because of the extremely small size of the indentation imprint, these measurements allow the surface at various points to be probed and spatially map its mechanical properties.

Unlike conventional microhardness tests, nanoindenters not only enable measurement of the hardness of very thin films, but they can also be used to determine the elastic modulus. In conventional microhardness testing, the elastic modulus cannot be measured while the hardness of very thin films is difficult to measure at the best. The reason for that difficulty is that in order to make sure that the substrate does not affect the measurements, the size of the impression must be kept small relative to the film thickness. A conservative rule is that the depth of the imprint should be less than **10%** of the film thickness [90,91]. However, it has also been reported that for some materials the microhardness was not affected when this depth was up to **25%** of the film thickness [91-94].

In nanoindentation, the minimum depth from which the nanohardness values can be obtained can be made as small as **100** [nm] [95]. However, it seems that below **25** [nm] elastic deformation of both the indenter and indented material may prevent the accurate evaluation of hardness [96]. At a scale of indentation comparable to the size of atoms, evaluation of indentation depth must allow for the electron tunnelling distance as well as elastic and plastic deformation of the specimen and indenter [96]. The extremely small displacements in nanoindentation pose a severe limitation on techniques that could be used to accurately determine the impression area. For example, for the surface film of 1 [μm] thickness, the depth of indentation is about 100 [nm] and the area of impression is too small to be accurately determined by optical microscopes. In nanoindentation this difficulty is overcome by accurately measuring the load-displacement characteristics.

The basic operating principle of nanoindenters involves the ability to produce and measure very small loads and displacements. From the simple measurement of load, displacement and time, a material's mechanical properties are derived. Loads are typically applied via coil and magnet assembly or by piezoelectric drives. In the first case, the load is determined directly from the current flowing to the coil, while in the second specially designed load cells are employed. Displacement is almost always measured using capacitance sensors. As the indenter is driven into the film both elastic and plastic deformations occur. During the plastic deformation a permanent hardness impression, that conforms to the shape of the indenter, forms at some contact depth '$\mathbf{h}_c$' while elastic deformation produces an additional component '$\mathbf{h}_s$' of the displacement. This is schematically illustrated in Figure 6.14.

The major problems associated with the nanoindentation technique are: the extreme susceptibility to vibration, the difficulty of controlling the motion of components to within very small tolerances and the higher likelihood of strong adhesive forces impeding retraction of the indenter after unloading. To obtain reliable data without interference from vibration, etc., nanoindenters are mounted on a vibration free system and indenter movement is controlled by piezoelectric transducers capable of moving the indenter at speeds of 1 to 10 [nm/s] with a minimum detectable force of 1 [μN] [87,97]. However, load and displacement resolutions better than 0.01 [μN] and 0.1 [nm] have also been achieved [91]. Diamond tip radii typically in the range of **0.1** to **1** [μm] and an apical angle of **80°** are usually used for an indenter [87,91]. The values of these radii must be accounted for in order to obtain accurate results; this is especially true for small indentations [91].

Another major weakness of this technique is the effect of 'piling-up' or 'sinking-in' of material around the indentation on the indentation depth measured [77]. This effect can be corrected and reproducible nanohardness measurements are provided by these instruments [e.g. 70,77,91,92,98,99]. In addition this technique also allows a Young's modulus to be determined which can be directly measured from the slope of stress-strain curve in the elastic regime obtained during the unloading cycle [e.g. 77,91,99,100].

The most common method of measuring hardness and elastic modulus of the thin film involves making small indentations in the film, usually with the Berkovich indenter, while continuously recording the indentation load '**P**' and displacement 'h', during one complete cycle of loading and unloading [91,98,99,101]. The Berkovich indenter has a perfect three-sided pyramidal geometry with the same depth-to-area relation as a Vickers indenter [102,103]. This particular shape of the indenter, i.e. three-sided, is much easier to manufacture than a four-sided indenter, such as the Vickers indenter. It is quite difficult to grind the four-sided Vickers indenter to terminate at a point. A disadvantage of sharply pointed indenters is that they produce almost constant plastic strain impressions and there are difficulties in assessing the elastic modulus from a continuously varying unloading curve. These problems are eliminated with spherical indenters as the transition from elastic to plastic behaviour can easily be determined and hence the yield stress found [104]. However, sharply pointed indenters are very useful in studies of thin films requiring shallow indentation and hence the Berkovich tip is the most commonly used in the evaluation of nanomechanical properties of materials. Experimental procedures have also been developed to compensate for the tip shape [71].

During the nanoindentation test the measured quantities are the peak load '$\mathbf{P}_{max}$', the displacement at peak load '$\mathbf{h}_{max}$', the initial unloading contact stiffness $\mathbf{S} = \mathbf{dP/dh}$, i.e. the slope of the initial portion of the unloading curve, and the displacement found by linearly extrapolating the initial portion of the unloading curve to zero load '$\mathbf{h}_0$'. The typical load-displacement data is illustrated in Figure 6.14 [91].

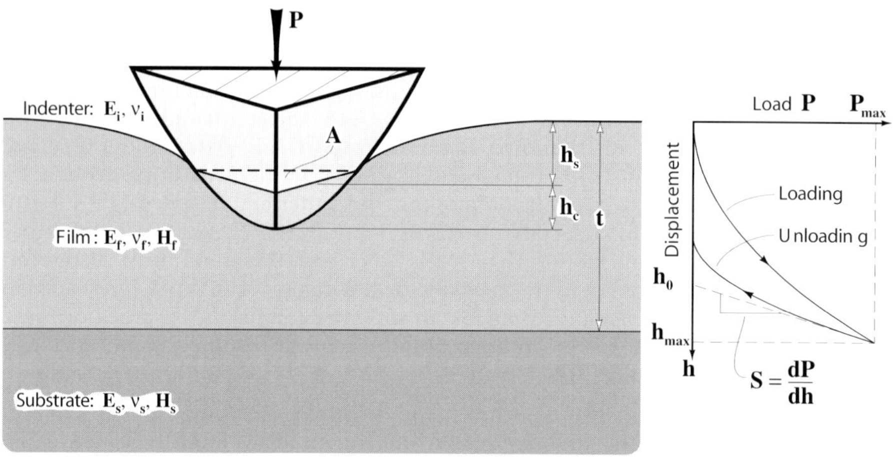

Figure 6.14 Schematic illustration of the nanoindentation process with the example of a typical indentation load-displacement data, where '$\mathbf{h}_c$' is a penetration depth (contact depth) and '$\mathbf{h}_s$' is the elastic displacement, (adapted from [91]).

As the indenter is withdrawn, load-displacement data is recorded and the elastic unloading data is used to relate experimentally measured quantities to the projected contact area and effective elastic modulus [69,91,105,106]. For example, the reduced elastic modulus of the film 'E_r' is determined from the value of the unloading contact stiffness '**S**', i.e.:

$$\boxed{\mathbf{S} = \frac{\mathbf{dP}}{\mathbf{dh}} = \frac{2}{\sqrt{\pi}}\mathbf{E_r}\sqrt{\mathbf{A}}} \tag{6.1}$$

where:

S	is the contact stiffness [N/m];
P	is the load [N];
h	is the displacement [m];
A	is the contact area [m^2];
$\mathbf{E_r}$	is the reduced elasticity modulus [Pa], i.e.:

$$\frac{1}{\mathbf{E_r}} = \frac{1}{2}\left[\frac{1-\upsilon_f^2}{\mathbf{E_f}} + \frac{1-\upsilon_i^2}{\mathbf{E_i}}\right]$$

where:

$\mathbf{E_f}$	is the elasticity modulus of the film [Pa];
ν_f	is the Poisson's ratio of the film;
$\mathbf{E_i}$	is the elasticity modulus of the indenter [Pa];
ν_i	is the Poisson's ratio of the indenter, e.g. for diamond $E_i = 1141$ [GPa] and $\nu_i = 0.07$ [107].

Hardness of the film is determined from its normal definition, i.e.

$$\boxed{\mathbf{H} = \frac{\mathbf{P_{max}}}{\mathbf{A}}} \tag{6.2}$$

Contact area at peak load is estimated from the contact depth 'h_c', i.e. $\mathbf{A} = \mathbf{f(h_c)}$ and depends on the shape of the indenter. The contact depth is evaluated from the following formula [99]:

$$\mathbf{h_c = h_0 + 0.25(h_{max} - h_0) = h_{max} - 0.75P_{max}/S} \tag{6.3}$$

where:

$\mathbf{h_0}$	is found by linearly extrapolating the initial proportion of the unloading curve [m], shown in Figure 6.14, to zero load [91];
$\mathbf{h_{max}}$	is the displacement at peak load '$\mathbf{P_{max}}$' [m];
$\mathbf{P_{max}}$	is the maximum load [N].

More advanced instruments allow a continuous measurement of contact stiffness '**S**' during the nanoindentation test [91,108,109]. When used in conjunction with equation (6.1) data obtained provides information on changes of the elastic modulus or contact area. The continuous stiffness measurement technique offers a way of continuous measurement of the contact areas of nanometre size contacts during the nanoindentation test [91,110].

The tiny indentation imprint produced by a nanoindenter is easily affected by surface roughness so care must be taken to locate the indentation mark on a flat part of the surface. Nanohardness testing is often performed with the aid of a microscope in order to facilitate precise control of the indenter and to enable the simultaneous observation of microscopic surface features together with variations in nanohardness and mechanical properties of the film. An optical microscope does not provide sufficient resolution for the indentations which are approximately **0.1** [μm] in size so an AFM (atomic force microscope) is usually used to obtain the image of the indentation area. Other specialized high magnification microscopes, described in Chapter 8, can also be used for this purpose. The nanohardness test apparatus (nanoindenter) is illustrated schematically in Figure 6.15.

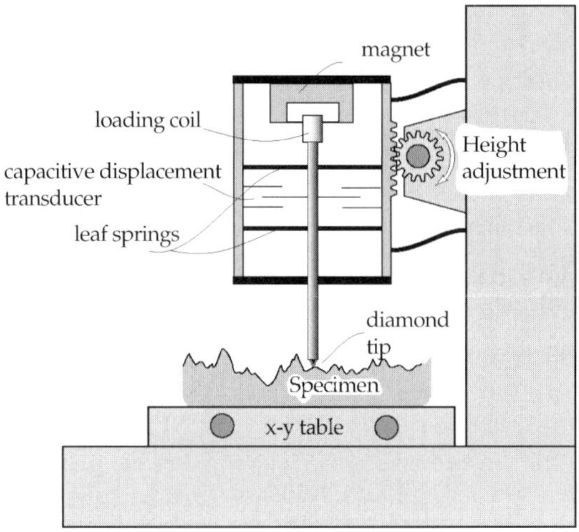

Figure 6.15 Schematic illustration of a nanoindenter.

The indentation procedure may vary slightly depending on the type of the indenter used but the fundamental steps are similar for all the indenters. The indenter is lowered first at constant load rate against the suspending springs until the contact with the sample is established. After the area to be tested has been located, the indenter load is gradually incremented and either constant load or constant displacement rate is maintained. The load and the indentation depth are monitored during both loading and unloading cycles. The resisting force of the suspending springs and the displacement associated with the measured compliance of the instrument are compensated for [77].

In one of the early designs the displacement was determined from the measurements of intensity of light reflected from the mirror on a sample dish as schematically illustrated in Figure 6.16 [87].

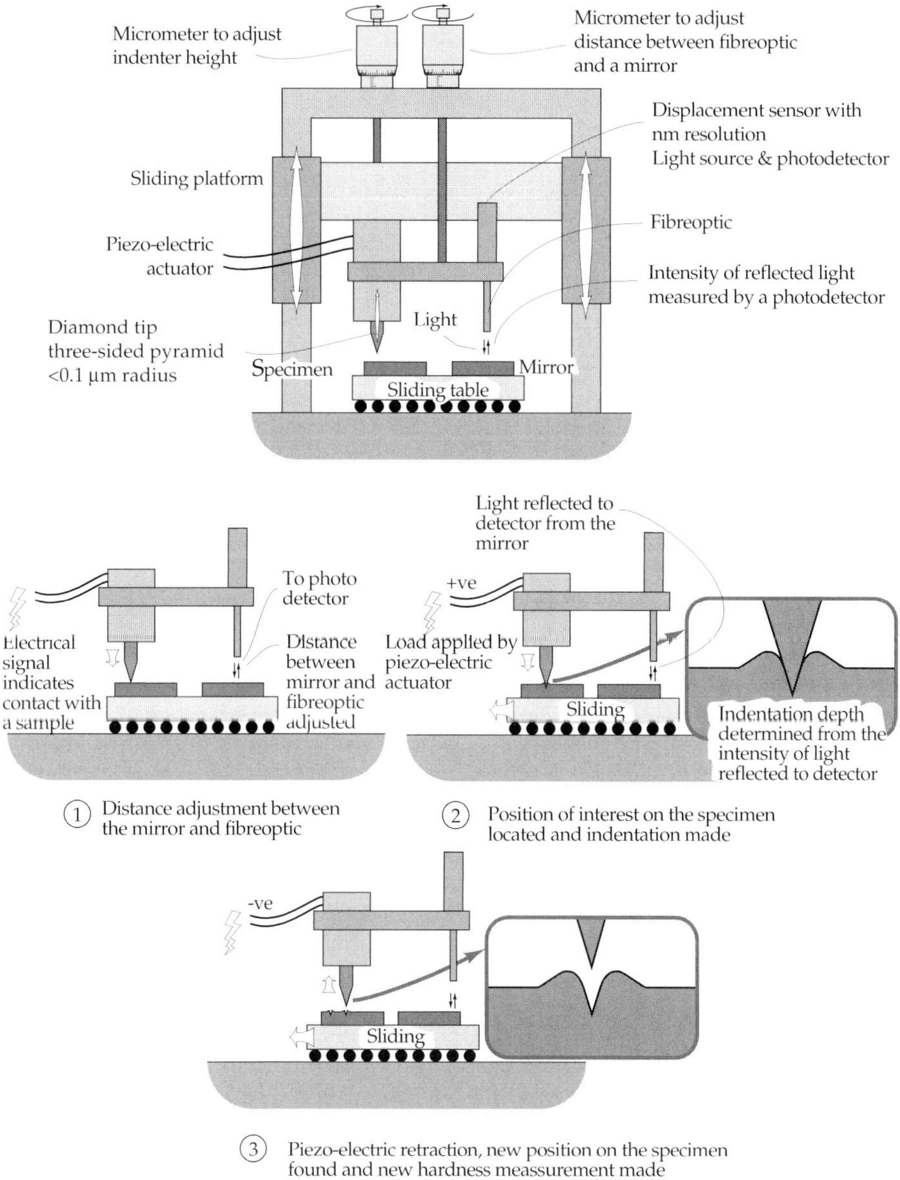

Figure 6.16 Schematic illustration of the nanohardness test apparatus (nanoindenter), (adapted from [87]).

The light was irradiated onto the mirror through the fibreoptic and on its return measured by a photodetector. After a contact has been established between the sample and an indenter the distance between the fibreoptic and a mirror was adjusted by a micrometer screw [87]. The fibreoptic position was then secured by the screw and the load applied to the indenter via piezoelectric actuator. During the loading phase the voltage to the piezoelectric actuator was increased causing the indenter to penetrate the sample, while during unloading this voltage was reduced. With this instrument a displacement resolution of 4 [nm] was reported [87].

The technological developments in this area are rapid and new, improved designs of nanoindenters are becoming available from time to time. Although the operating principles of these instruments remain unchanged the developments in electronics allow for more accurate control and better displacement and load resolutions.

The displacement sensors used vary depending on the instrument design and manufacturer. Displacement sensors include: polarization interferometers [77,111,112] (indentation depth measured with the resolution of 0.2 [nm] [77]), inductive displacement transducers [e.g. 113] (resolution 5 [nm]) or capacitive transducers [e.g. 70,114-116]. With capacitive transducers displacement resolutions as high as 0.015 [nm] [76,116] and even higher [117] have been reported.

Nanoindenters that allow friction, wear, nano scratch resistance, fracture toughness, elastic modulus measurements, etc. to be conducted are also available [e.g. 77,117].

6.4 OTHER MATERIAL CHARACTERISTICS

Apart from surface roughness and hardness/microhardness measurements other material characteristics and properties are necessary for tribological specimen characterization, as shown in Table 6.1. Typical mechanical properties such as, for example, yield strength, flexural strength, fracture toughness etc., are determined according to standardised test methods. The most common ASTM standards (USA) for basic mechanical tests are listed in Table 6.3 (the standards are updated every three years and are included in the Annual Book of ASTM Standards). Similar standards developed by different nations or organisations are also available, e.g. BSI (UK) and ISO (EU). Detailed descriptions of the mechanical testing methods can be found in many materials science books [118-120].

Some information on the specimen chemical composition is essential in any tribological experiment. For most experimental purposes, a standard composition analysis provides sufficient information. A comprehensive list of methods used for the chemical analysis of various materials can be found in [e.g. 121]. However, it is often not necessary to chemically analyse all samples used for tribological testing. Specimens made of steels, cast irons and other metal alloys are usually specified according to an engineering standard, e.g. AISI, SAE, ASTM, UNS. Materials handbooks give a good introduction to various naming systems including the explanation of each designation and comparison between the systems [121,122]. Apart from chemical composition, standard specifications provide many of the material's mechanical properties listed in Table 6.3. Pure metals are also supplied with a composition analysis and necessary information on metallurgical phases or mechanical properties.

Table 6.3 Common ASTM mechanical tests.

Property	Metals	Ceramics Glasses	Polymers Rubbers	Composites
Tensile strength Elastic modulus Yield strength Elongation	A 370 B 557 E 8 E 602 E 646	C 565 C 749 C 1273	D 638 D 882 D1623 D 412	C 1275 D 3039 D 3379 D 3552 D 4018
Notch impact	A370 E 23 E 208 E 436 E 604		D 256 D 1822 D 4812 D 5420 D 1054	
Flexural strength	E 290 E 812 E 855	C 158 C 674 C 1161 C 1211	D 648 D 747 D 790	C 393 C 1341
Fracture toughness	E 399 E 1737	E 399 E 1304 E 1421	D 5045 D 6068 E 1737	D 5528
Compressive strength	E 9 E 209	C 133 C 695 C 773	D 695 D 1621 D 395 D 575	C 364 C 365 D 3410

Ceramics are usually specified by their chemical name such as e.g. alumina, silicon carbide or silicon nitride. As the ceramics mechanical properties strongly depend on the amount of impurities and glass phases present their chemical composition should be included in the specimen description. The exact composition is often difficult to determine since manufacturers use various sintering aids which are proprietary information. The properties of the same type of ceramics can also vary depending on the manufacturing process used (sintering, hot pressing, reaction sintering, etc.) and this process should be specified. If commercial ceramics are tested their chemical and mechanical characteristics are often supplied by the producers.

Polymers are either specified according to the conventions of organic chemistry, with names often abbreviated (e.g. polyethylene, PE, polyvinyl chloride, PVC) or by the exact name given by the manufacturer (e.g. Teflon, Delrin, Plexiglas). The detailed specification of polymers can be found from handbooks or directly from a manufacturer.

Beyond this basic data, the degree of materials specification varies according to experimental needs. For example, if a study involves comparing the wear of alloys of slightly different composition or subject to different heat treatments, then a careful specification of the metallurgical phases present in the material becomes important. In such cases, metallographical images of specimen original microstructures are often included in the material characterization to better visualise the differences in, for example, the size and the distribution of phases or inclusions [123], microcracks distribution [124], grain size, etc. Metallographical micrographs, either optical or SEM, are often included in cases where even one kind of specimen is the object of tribological study. The purpose here, again, is to help in the interpretation of the wear behaviour with respect to the original material microstructure.

Various wear mechanisms are affected by material characteristics to a different degree. For example, interpretation of abrasive and erosive wear tests of metallic materials will require detailed information on the composition, size and shape of hard phases or precipitates present in the metal. Investigations of delamination wear or contact fatigue may require detailed data on the number of small inclusions present in the material. At elevated temperatures, often present during tribological tests, segregation of material impurities to the worn surface can occur so that materials specification of specimens has to include data on the impurities and minor constituents. The impurities are driven to the surface of the bulk material under the influence of heat and plastic deformation. The minority constituents, i.e. impurities, move as isolated atoms or ions through the crystalline matrix and their movement is facilitated by dislocations and thermal vibrations. Impurity migration is believed to strongly influence dry sliding under heavy loads, where high surface concentrations of, for example sulphur, help prevent seizure of the metal. Impurity migration has been extensively discussed in [125]. It has been shown that small amounts of sulphur in steels can affect the likelihood of adhesive wear so that apparently similar steels can reveal different wear characteristics if their sulphur content is not the same [125]. Even a tiny amount of surface processing such as polishing can cause impurity movement within a material and frictional temperature rises as low as 100°C can initiate surface concentration of minor elements in an alloy [126]. The mechanism of surface concentration of impurities and minor constituents due to temperature increase during wear is illustrated schematically in Figure 6.17.

Amongst wear mechanisms only fretting, which usually involves small frictional temperature rises, may not be affected by changes in surface composition, but almost all other wear modes would be affected.

According to Table 6.1 slightly different material characteristics should be supplied when testing polymers. Wear and friction of polymers is often affected by the degree of crystallinity, molecular weight and imperfections such as fusion voids. For example, it is well documented that the wear resistance of polyethylene improves with increase in molecular weight [127,128]. Glass transition temperature is important when testing amorphous polymers, while softening and melting temperatures are essential when testing thermoplastics and elastomers. Additional treatments, such as for example irradiation used for polymers in biomedical applications, can affect their subsequent wear behaviour and such treatments should be included in the description of the specimens [129].

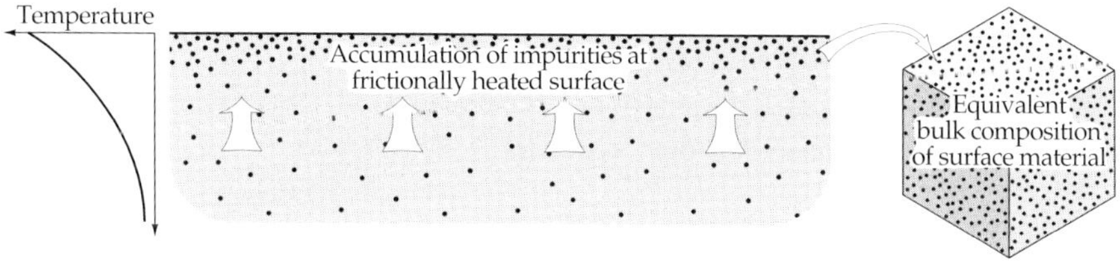

Figure 6.17 Schematic illustration of mechanism involved in the accumulation of impurities and minor constituents at worn metallic surface.

Polymers contain additives to facilitate their processing, i.e. colouring agents and anti-oxidants. Colouring agents such as cadmium oxide are present in the polymer matrix as discrete particles and these can act as crack initiators. On the other hand anti-oxidants can affect the resistance of polymers to wear in corrosive or oxidative fluids. Any information about the crystalline state of polymers and number of polymer types present should also be available.

For ceramic tribological specimens the two important characteristics are their fracture toughness and grain size. Since microfracture is often the dominating wear mechanisms of ceramics (especially in dry environments) fracture toughness can affect the wear behaviour of ceramics more than their hardness [130,131]. The wear of ceramics also changes with the change in the grain size and this property is usually included in the description of ceramic specimens [130,132]. Apart from the actual grain size, the composition and strength of the grain boundary phases also affect the tribological behaviour of the ceramics [133,134]. Different manufacturing processes of ceramics result in various amounts of residual porosity. The presence of pores usually increases the wear of ceramics.

6.5 SUMMARY

Although the characterization of specimens for tribological tests may appear to be a relatively simple task, it may in fact involve quite complex procedures. Most surface characteristics of interest relate to surface roughness and hardness. Information on the composition of a specimen is also of critical importance and there is often a distinction between bulk and surface composition of a specimen. It is difficult to prescribe a comprehensive scheme for specimen characterization because the level of information required varies depending on the type of investigation pursued. The problem is that if insufficient data is provided to characterize a specimen then much time might be wasted on experiments that are subject to unknown variations in data, rather than resolving the tribological problem of direct interest.

REFERENCES

1 H.M. Stanley and D.B. Bogy, Effect of surface conformity on friction, *Wear*, Vol. 168, 1993, pp. 105-108.

2 G.V. Blessing, J.A. Slotwinski, D.G. Eitzen and H.M. Ryan, Ultrasonic measurements of surface roughness, *Applied Optics*, Vol. 32, 1993, pp. 3433-3437.

3 D.J. Whitehouse, Handbook of Surface Metrology, Institute of Physics Publishing, Bristol and Philadelphia, 1994.

4 H. Dagnall, Exploring Surface Texture, Rank Taylor Hobson Ltd, London 1986.

5 T.R. Thomas (editor), Rough Surfaces, Longman Inc., New York, 1982.

6 E. Mainsah, J.A. Greenwood and D.G. Chetwynd (editors), Metrology and properties of engineering surfaces, Boston, Mass.; London: Kluwer Academic Publishers, 2001.

7 J.A. Greenwood and J.B.P. Williamson, Contact of nominally flat surfaces, *Proc. Roy. Soc., London*, Series A, Vol. 295, 1966, pp. 300-319.

8 B. Bhushan, Micro/nanotribology using atomic force/friction force microscopy: State of the art, New Directions in Tribology, editor: I.M. Hutchings, Mechanical Engineering Publications Ltd, 1997, pp. 141-158.

9 Surface roughness terminology - Part 1: Surface and its parameters, ISO Standard 4287/1, 1984 (International Standards Organization).

10 W.P. Dong, P.J. Sullivan and K.J. Stout, Comprehensive study of parameters for characterising three-dimensional topography. III: Parameters for characterising amplitude and some functional properties, *Wear*, Vol. 178, 1994, pp. 29-43.

11 K.J. Stout, P.J. Sullivan, W.P. Dong, E. Mainsah, N. Luo, T. Mathia and H. Zahouani, The development of methods for the characterisation of roughness in three dimensions, Commission of the European Communities, Directorate General XII, Science, Research and Development, Report EUR 15178 EN, 1993.

12 W.P. Dong, P.J. Sullivan and K.J. Stout, Comprehensive study of parameters for characterising three-dimensional topography. IV: Parameters for characterising spatial and hybrid properties, *Wear*, Vol. 178, 1994, pp. 45-60.

13 M. Hasegawa, J. Liu, K. Okuda and M. Nunobiki, Calculation of the fractal dimensions of machined surface profiles, *Wear*, Vol. 192, 1996, pp. 40-45.

14 J. Lopez, G. Hansali, H. Zahouani, J.C. Le Bosse and T. Mathia, 3D fractal-based characterisation for engineered surface topography, *International Journal of Machine Tools and Manufacture*, Vol. 35, 1995, pp. 211-217.

15 S. Ganti and B. Bhushan, Generalized fractal analysis and its applications to engineering surfaces, *Wear*, Vol. 180, 1995, pp. 17-34.

16 Y. Wang and K.S. Moon, A methodology for the multi-resolution simulation of grinding wheel surface, *Wear*, Vol. 211, 1997, pp. 218-225.

17 X.Q. Jiang, L. Blunt and K.J. Stout, Three-dimensional surface characterization for orthopaedic joint prostheses, *Proceedings of Institute of Mechanical Engineers Part H, Journal of Engineering in Medicine*, Vol. 213, 1999, pp. 49-68.

18 P. Podsiadlo and G.W. Stachowiak, The development of modified Hurst orientation transform for the characterization of surface topography of wear particles, *Tribology Letters*, Vol. 4, 1998, pp. 215-229.

19 G.W. Stachowiak and P. Podsiadlo, Surface characterization of wear particles, *Wear*, Vol. 225-229 Part II, 1999, pp. 1171-1185.

20 P. Podsiadlo and G.W. Stachowiak, Applications of Hurst orientation transform to the characterization of surface anisotropy, *Tribology International*, Vol. 32, 2000, pp. 387-392.

21 P. Podsiadlo and G.W. Stachowiak, Scale-invariant analysis of wear particle morphology - a preliminary study, *Tribology International*, Vol. 33, 2000, pp. 289-295.

22 P. Podsiadlo and G.W. Stachowiak, Scale-invariant analysis of wear particle surface morphology. I: Theoretical background, Computer Implementation and Technique Testing, *Wear*, Vol. 242, 2000, pp. 160-179.

23 P. Podsiadlo and G.W. Stachowiak, Scale-invariant analysis of wear particle surface morphology. II: Fractal dimension, *Wear*, Vol. 242, 2000, pp. 180-188.

24 P. Podsiadlo and G.W. Stachowiak, Scale-invariant analysis of wear particle surface morphology. III: Pattern recognition, *Wear*, Vol. 242, 2000, pp. 189-201.

25 P. Podsiadlo and G.W. Stachowiak, Scale-invariant analysis of tribological surfaces, Thinning Films and Tribological Interfaces, editor: D. Dowson et al, Tribology Series 38, pp. 546-557, Elsevier, 2000.

26 G.W. Stachowiak and P. Podsiadlo, Characterization and classification of wear particles and surfaces, *Wear*, Vol. 249, 2001, pp. 194-200.

27 Y.M. Niu, Y.S. Wong, G.S. Hong and T.I. Liu, Multi-category classification of tool conditions using wavelet packets and ART2 network, *Journal of Manufacturing Science and Engineering, Transactions of the ASME*, Vol. 120, 1998, pp. 807-816.

28 K.J. Stout and L. Blunt, Three-dimensional surface topography, 2nd edition, London, Penton, 2000.

29 L.Y. Prujanski, A study of surface wearing ability in relation to some technological factors, II. Effect of the method of surface finishing and running-in, *Wear*, Vol. 54, 1979, pp. 355-369.

30 M.G. Hamblin and G.W. Stachowiak, Measurement of fractal surface profiles obtained from scanning electron and laser scanning microscope images and contact profile meter, *Journal of Computer Assisted Microscopy*, Vol. 6, 1994, pp. 181-194.

31 J.F. Song and T.V. Vorburger, Surface Texture, ASM Handbook, Vol. 18, Friction, Wear and Lubrication, ASM International, 1992, pp. 334-345.

32 J.F. Song and T.V. Vorburger, Stylus profiling at high resolution and low force, *Appl. Opt.*, Vol 30, 1991, pp. 42-50.

33 W.B. Estill and J.C. Moody, Deformation caused by stylus tracking on thin gold film, *ISA Trans.*, Vol. 5, 1966, pp. 373-378.

34 R.C. Tucker and R.W. Meyerhoff, An SEM study of surface roughness measurement, Proc. 2nd Annual Scanning Electron Microscope Symposium, 1969, pp. 389-396, IIT Research Institute, Chicago.

35 R.S. Sayles, T.R. Thomas, J. Anderson, I. Haslock and A. Unsworth, Measurement of the surface microgeometry of articular cartilage, *Journal of Biomechanics*, Vol. 12, 1979, pp. 257-267.

36 T.B. Kirk, G.W. Stachowiak and A.S. Wilson, The morphology of the surface of articular cartilage, Abstracts of the Second World Congress of Biomechanics, 10-15 July, 1994, Amsterdam, editors: Leendert Blankevoort, Jan G.M. Kooloos, 1994, Vol. 1, pp. 208.

37 J.C. Wyant, C.L. Koliopoulos, B. Bhushan and D. Basila, Development of a three-dimensional noncontact digital optical profiler, *Journal of Tribology*, Vol. 108, 1986, pp. 1-8.

38 P. Podsiadlo and G.W. Stachowiak, Characterization of surface topography of wear particles by SEM stereoscopy, *Wear*, Vol. 206, 1997, pp. 39-52.

39 I. Sherrington and E.H. Smith, Modern measurement techniques in surface metrology: Part II; Optical instruments, *Wear*, Vol. 125, 1988, pp. 289-308.

40 K. Mitsui, In-process sensors for surface roughness and their applications, *Precis. Eng.*, Vol. 8, No. 4, 1986, pp. 212-220.

41 K. Creath, Phase-shifting interferometry techniques, Progress in Optics XXVI, editor: E. Wolf, Elsevier Science, 1988, pp. 357-373.

42 P.J. Caber, Interferometric profiler for rough surfaces, *Applied Optics*, Vol. 32, 1993, pp. 3438-3441.

43 S. Kakunai, T. Sakamoto and T. Matsumoto, Profile measurement of worn acetabular cup by holographic interferometry, 10th International Conference on Biomedical Engineering, 6-9 December 2000, Singapore, editor: J.C.H. Goh, National University of Singapore, pp. 165-166.

44 R.K. Erf, ed., Speckle Metrology, Academic Press, New York, 1978.

45 J.M. Bennett and L. Mattsson, Introduction to surface roughness and scattering, Optical Society of America, Washington, DC, 1989.

46 R.M. Azzam and N.M. Bashara, Polarization characteristics of scattered radiation from diffraction gratings by ellipsometry with application to surface roughness, *Phys. Rev.* B5, 1972, pp. 4721-4729.

47 T.V. Vorburger, E. Marx and T.R. Lettieri, Regimes of surface roughness measurable with light scattering, *Applied Optics*, Vol. 32, 1993, pp. 3401-3408.

48 D. Leger and J.C. Perrin, Real-time measurement of surface roughness by correlation of speckle patterns, *J. Opt. Soc. Am.*, Vol. 66, No. 11, 1976, pp. 1210-1217.

49 M. Zahidi, M. Assoul, J. Mignot and B. Bellaton, A fast 2D/3D optical profilometer for wide range topographical measurement, *Wear*, Vol. 165, 1993, pp. 197-203.

50 U. Persson, Real-time measurement of surface roughness on ground surfaces using speckle-contrast technique, *Opt. Lasers Eng.*, Vol. 17, No. 2, 1992, pp. 61-67.

51 D. Leger, E. Mathieu and J.C. Perrin, Optical surface roughness determination using speckle correlation technique, *Appl. Opt.*, Vol. 14, No. 4, 1975, pp. 872-877.

52 U. Persson, Measurement of surface roughness on rough machined surfaces using spectral speckle correlation and image analysis, *Wear*, Vol. 160, 1993, pp. 221-225.

53 G. Bitz, Verfahren zur Bestimmung von Rauheitskenngrossen durch Specklekorrelation, *Ph.D. thesis*, University of Karlsruhe, Karlsruhe, Germany, 1982.

54 F.T. Arecchi, D. Bertani and S. Ciliberto, A fast versatile optical profilometer, *Opt. Comm.*, Vol. 31(3), 1979, pp. 263-266.

55 Y. Fairman, E. Lenz and J. Shamir, Optical profilometer: A new method for high sensitivity and wide dynamic range, *Appl. Opt.*, Vol. 21(7), 1982, pp. 3200-3208.

56 M. Dobosz, Optical profilometer: A practical approximate method of analysis, *Appl. Opt.*, Vol. 22(24), 1983, pp. 3983-3987.

57 M. Dobosz, Accuracy of profile measurements by means of a focused laser beam, *Wear*, Vol. 98, 1984, pp. 117-126.

58 M.A. Schmidt and R.D. Compton, Confocal microscopy, ASM Handbook, Vol. 18, Friction, Wear and Lubrication, ASM International, 1992, pp. 357-361.

59 R.V. Anamalay, T.B. Kirk, D. Panzera and Z.L. Xu, Numerical descriptors for the analysis of surfaces using laser scanning confocal microscopy, *Wear*, Vol. 181-183, 1995, pp. 771-776.

60 A. Boyde, Quantitative photogrammetric analysis and quantitative stereoscopic analysis of SEM images, *Journal of Microscopy*, Vol. 98A, 1973, pp. 452-471.

61 P. Podsiadlo and G.W. Stachowiak, 3-D Imaging of Wear Particles Found in Synovial Joints, *Wear*, Vol. 230, 1999, pp. 184-193.

62 G.H. Shehata, A.M.A. Moussa and P.A. Molian, Nd:YAG laser alloying of high-speed steel tools with BN and Ti/BN and the effects on turning performance, *Wear*, Vol. 170, 1993, pp. 199-210.

63 B. Zhu, X. Zheng, T. Tao, S. Yang and K. Cui, Coarse cemented WC particle ceramic-metal composite coatings produced by laser cladding, *Wear*, Vol. 170, 1993, pp. 161-166.

64 L. Xu, S. Clough, P. Howard and D. StJohn, Laboratory assessment of the effect of white layers on wear resistance for digger teeth, *Wear*, Vol. 181-183, 1995, pp. 112-117.

65 J. Singh and A.T. Alpas, Dry sliding wear mechanisms in a $Ti_{50}Ni_{47}Fe_3$ intermetallic alloy, *Wear*, Vol. 181-183, 1995, pp. 302-311.

66 J-E. Sundren and H.T.G. Hentzell, A review of the present state of the art in hard coatings grown from the vapour phase, *Journal of Vacuum Science and Technology*, Vol. 4A, 1986, pp. 2259-2279.

67 P.J. Burnett and D.S. Rickerby, Assessment of coating hardness, *Surface Engineering*, Vol. 3, 1987, pp. 69-76.

68 A. Thomas, Microhardness measurements as a quality control technique for thin, hard coatings, *Surface Engineering*, Vol. 3, 1987, pp. 117-122.

69 N.A. Stilwell and D. Tabor, Elastic recovery of conical indentations, *Proc. Physical Society, London*, Vol. 78, 1961, pp. 169-179.

70 T.W. Wu, C. Hwang, J. Lo and P. Alexopoulos, Microhardness and microstructure of ion-beam-sputtered, nitrogen doped NiFe films, *Thin Solid Films*, Vol. 166, 1988, pp. 299-308.

71 V.P. Alekhin, G.S. Berlin, A.V. Isaev, G.N. Kalei, V.A. Merkulov, V.N. Skvortsov, A.P. Ternovskii, M.M. Krushchov, G.D. Shnyrev and M.Kh. Shorshorov, Micromechanical testing of materials by microcompression, *Zavodskaya Laboratoriya*, Vol. 38, 1972, pp. 619-621.

72 A.P. Ternovskii, V.P. Alekhin, M.Kh. Shorshorov, M.M. Khrushchov and V.N. Skvortsov, Micromechanical testing of materials by depression, *Zavodskaya Laboratoriya*, Vol. 39, 1973, pp. 1620-1624.

73 S.I. Bulychev, V.P. Alekhin, M.Kh. Shorshorov, A.P. Ternovskii and G.D. Shnyrev, Determining Young's modulus from the indenter penetration diagram, *Industrial Laboratory* (USSR) (English translation of Zavodskaya Laboratoriya), Vol. 41. 1975, pp. 1409-1412.

74 J.B. Pethica, R. Hutchings and W.C. Oliver, Hardness measurements at penetration depths as small as 20 nm, *Philosophical Magazine*, Vol. A48, 1983, pp. 593-606.

75 P.J. Blau, Applications of microindentation methods in tribology research, Microindentation Techniques in Materials Science and Engineering, editors: P.J. Blau and B.R. Lawn, ASTM STP 889, American Society for Testing and Materials, Philadelphia, PA, 1986, pp. 209-226.

76 B. Bhushan, Tribology and Mechanics of Magnetic Storage Devices, Springer-Verlag, New York, 1990.

77 B. Bhushan, Nanomechanical properties of solid surfaces and thin films, Handbook of Micro/Nano Tribology, editor: B. Bhushan, CRC Series Mechanics and Materials Science, CRC Press Inc., 1995.

78 W.C. Oliver and C.J. McHargue, Characterizing the hardness and modulus of thin films using a mechanical properties microprobe, *Thin Solid Films*, Vol. 161, 1988, pp. 117-122.

79 M. Nastasi, J-P. Hirvonen, T.R. Jervis, G.M. Pharr and W.C. Oliver, Surface mechanical properties of C implanted Ni, *Journal of Materials Research*, Vol. 3, 1988, pp. 226-232.

80 P.J. Burnett and D.S. Rickerby, The mechanical properties of wear-resistant coatings. II. Experimental studies and interpretation of hardness, *Thin Solid Films*, Vol. 148, 1987, pp. 51-65.

81 E. Vancoille, J.P. Celis and J.R. Roos, Mechanical properties of heat treated and worn TiN, (Ti, Al)N, (Ti, Nb)N and Ti(C, N) coatings as measured by nanoindentation, *Thin Solid Films*, Vol. 224, 1993, pp. 168-176.

82 S. Alba, J.L. Loubet and L. Vovelle, Evaluation of mechanical properties and adhesion of polymer coatings by continous hardness measurements, *Journal of Adhesion Sci. Technol.*, Vol. 7, 1993, pp. 131-140.

83 K. Kinosita, Recent developments in the study of mechanical properties of thin films, *Thin Solid Films*, Vol. 12, 1972, pp. 17-28.

84 M. Nishibori and K. Kinosita, Ultra-microhardness of vacuum deposited films I: Ultra-microhardness tester, *Thin Solid Films*, Vol. 48, 1978, pp. 325-331.

85 M.J. Mayo, R.W. Siegel, A. Narayanasamy and W.D. Nix, Mechanical properties of nanophase TiO_2 as determined by nanoindentation, *Journal of Materials Research*, Vol. 5, No. 5, 1990, pp. 1073-1082.

86 B.N. Lucas and W.C. Oliver, The elastic, plastic, and time dependent properties of thin films as determined by ultra low indentation, Proceedings of the Material Research Symposium, Materials Research Society, Vol. 239, 1992, pp. 337-341.

87 M. Yanagisawa and Y. Motomura, An ultramicro indentation hardness tester and its application to thin films, *Lubrication Engineering*, Vol. 43, 1987, pp. 52-56.

88 Y. Tsukamoto, H. Kuroda, A. Sato and H. Yamaguchi, Microindentation adhesion tester and its applications to thin films, *Thin Solid Films*, Vol. 213, 1992, pp. 220-225.

89 T.W. Wu, M. Moshref and P.S. Alexopoulos, The effect of the interfacial strength on the mechanical properties of aluminum films, *Thin Solid Films*, Vol. 187, 1990, pp. 295-307.

90 Standard Test for Microhardness of Materials, ASTM Standard Test Method E-384, Annual Book of Standards 3.01, American Society for Testing and Materials, Philadelphia, PA, 2000.

91 G.M. Pharr and W.C. Oliver, Measurement of thin film mechanical properties using nanoindentation, *Materials Research Society (MRS) Bull.*, Vol. 17, No. 7, July, 1992, pp. 28-33.

92 M.F. Doerner, D.S. Gardner and W.D. Nix, Plastic properties of thin films on substrates as measured by submicron indentation hardness and substrate curvature techniques, *Journal of Materials Research*, Vol. 1, 1987, pp. 845-851.

93 H.M. Pollock, D. Maugis and M. Barquins, Characterization of submicrometre surface layers by indentation, Microindentation Techniques in Materials Science and Engineering, ASTM STP 889, editors: P.J. Blau and B.R. Lawn, American Society for Testing and Materials, Philadelphia, PA, 1986, pp. 47-71.

94 D. Lebouvier, P. Gilormini and E. Felder, A kinematic solution for plane-strain indentation of a bi-layer, *Journal of Phys. D.*, Vol. 18, 1985, pp. 199-210.

95 N.A. Burnham and R.J. Colton, Measuring the nanomechanical properties and surface forces of materials using an atomic force microscope, *Journal of Vacuum Science and Technology*, Series A, Vol. 7, 1989, pp. 2906-2913.

96 T. Yokohata and K. Kato, Mechanism of nanoscale deformation, *Wear*, Vol. 168, 1993, pp. 109-114.

97 Y. Tsukamoto, H. Yamaguchi and M. Yanagisawa, Mechanical properties of thin films: measurements of ultramicroindentation hardness, Young's modulus and internal stress, *Thin Solid Films*, Vol. 154, 1987, pp. 171-181.

98 M.F. Doerner and W.D. Nix, A method for interpreting the data from depth-sensing indentation instruments, *Journal of Materials Research*, Vol. 1, 1986, pp. 601-609.

99 W.C. Oliver and G.M. Pharr, An improved technique for determining hardness and elastic modulus using load and displacement sensing indentation experiments, *Journal of Materials Research*, Vol. 7, 1992, pp. 1564-1583.

100 W.D. Nix, Mechanical properties of thin films, *Metall. Trans. A*, Vol. 20A, 1989, pp. 2217-2245.

101 J.L. Loubet, J.M. Georges, O. Marchesini and G. Meille, Vickers indentation curves of magnesium oxide (MgO), *Journal of Tribology*, Vol. 106, 1984, pp. 43-48.

102 M.M. Khruschov and E.S. Berkovich, Methods of determining the hardness of very hard materials, *Industrial Diamond Review*, Vol. 11(122), January 1951, pp. 42-44.

103 E.S. Berkovich, Three-faceted diamond pyramid for micro-hardness testing, *Industrial Diamond Review*, Vol. 11, 1951, pp. 129-132.

104 T.J. Bell, J.S. Field and M.V. Swain, Elastic-plastic characterization of thin films with spherical indentation, *Thin Solid Films*, Vol. 220, 1992, pp. 289-294.

105 D. Tabor, A simple theory of static and dynamic hardness, *Proc. Royal Soc. London*, Series A, Vol. 192, 1948, pp. 247-252.

106 I.N. Sneddon, The relation between load and penetration in the axisymmetric Boussinesq problem for a punch of arbitrary profile, *Int. J. Eng. Sci.*, Vol. 3, 1965, pp. 47-57.

107 G. Simmons and H. Wang, Single Crystal Elastic Constants and Calculated Aggregate Properties: A Handbook, 2nd edition, The MIT Press, Cambridge, Massachusetts, 1971.

108 J.B. Pethica and W.C. Oliver, Tip surface interactions in STM and AFM, *Physica Scripta*, Vol. T19, 1987, pp. 61-66.

109 J.B. Pethica and W.C. Oliver, Mechanical properties of nanometre volumes of material: use of the elastic response of small area indentations, Thin Films: Stresses and Mechanical Properties, editors: J.C. Bravman, W.D. Nix, D.M. Barnett and D.A. Smith, Proceedings of Materials Research Society Symposium, Pittsburgh, PA, Vol. 130, 1989, pp. 13-23.

110 G.M. Pharr, W.C. Oliver and F.R. Brotzen, An improved technique for determining hardness and elastic modulus using load and displacement sensing indentation experiments, *Journal of Materials Research*, Vol. 7, 1992, pp. 1564-1583.

111 B. Bhushan, V.S. Williams and R.V. Shack, In-situ nanoindentation hardness apparatus for mechanical characterization of extremely thin films, *Transactions ASME, Journal of Tribology*, Vol. 110, 1988, pp. 563-571.

112 V.S. Williams, A.L. Landesman, R.V. Shack, D. Vukobratovich and B. Bhushan, In situ microviscoelasticity measurements by polarization-interferometric monitoring of indentation depth, *Applied Optics*, Vol. 27, 1988, pp. 541-546.

113 P.E. Wierenga and A.J.J. Franken, Ultramicroindentation apparatus for the mechanical characterization of thin films, *J. Appl. Phys.*, Vol. 55, 1984, pp. 4244-4247.

114 T.W. Wu, R.A. Burn, M.M. Chen and P.S. Alexopoulos, Micro-indentation and micro-scratch tests on sub-micron carbon films, Symp. Proc., Materials Research Society, Pittsburgh, Vol. 130, 1989, pp. 117-121.

115 G.M. Pharr and R.F. Cook, Instrumentation of a conventional hardness tester for load-displacement measurement during indentation, *Journal of Mater. Res.*, Vol. 5, 1990, pp. 847-851.

116 J.L. Loubet, M. Bauer, A. Tonck, S. Bee and B. Gauthier-Manuel, Nanoindentation with a surface force apparatus, Mechanical Properties and Deformation Behavior of Materials Having Ultra-Fine Microstructures (editors: M. Nastasi, D.M. Parkin and H. Gleiter), Kluwer Academic Pub., Dordrecht, 1993, pp. 429-447.

117 Anonymous, Triboindenter, Nanomechanical Test Instrument, Hysitron Incorporated, Minneapolis, USA, (www.hysitron.com), 2002.

118 N.E. Dowling, Mechanical Behaviour of Materials, Prentice-Hall, New Jersey, USA 1999.

119 ASM Handbook, Vol. 8, Mechanical Testing and Evaluation, ASM International, Materials Park, OH, 2000.

120 Composites Engineering Handbook, ed. P.K. Mallick, Marcel Dekker Inc., New York, 1997.

121 ASM Metals Reference Book, ed. M. Bauccio, 3rd edition, ASM International, USA, 1993.

122 Woldman's Engineering Alloys, 7th edition, ASM International, USA 1990.

123 O.N. Dogan, G. Laird II and J.A. Hawk, Abrasion resistance of the columnar zone in high Cr white cast irons, *Wear*, Vol. 181-183, 1995, 342-349.

124 K.H. Zum Gahr, Modelling of two-body abrasive wear, *Wear*, Vol. 124, 1988, pp. 87-103.

125 D.H. Buckley, Surface Effects in Adhesion, Friction, Wear and Lubrication, Elsevier, Amsterdam, 1981.

126 F. Honda, T. Kometani and K. Nakajima, Tribochemical characterization of the worn surface of the ternary system Cu-Ni-Zn, *Wear*, Vol. 161, 1993, pp. 187-191.

127 J.H. Dumbleton, C. Shen and E.H. Miller, A study of the wear of some materials in connection with total hip replacement, *Wear*, Vol. 29, 1974, pp. 163-171.

128 W. Rostoker and J.O. Galante, Some new studies of the wear behaviour of ultrahigh molecular weight polyethylene, *J. Biomed. Mater. Res.*, Vol. 10, 1976, pp. 303-310.

129 R.M. Rose, E.V. Goldfarb, E. Ellis and A.N. Crugnola, Radiation sterilization and the wear rate of polyethylene, *J. Orthopaedic Research*, Vol. 2, 1984, pp. 393-400.

130 Y.M. Chen, B. Rigaut and F. Armanet, Friction and wear of alumina ceramics at high sliding speed, *Lubrication Engineering*, Vol. 47, 1991, pp. 531-537.

131 T.E. Fischer, M.P. Anderson and S. Jahanmir, Influence of fracture toughness on the wear resistance of yttria-doped zirconium oxide, *J. Am. Ceram. Soc.*, Vol. 72, 1989, pp. 252-257.

132 S.J. Cho, B.J. Hockey, B.R. Lawn and S.J. Bennison, Grain-size and R-curve effects in the abrasive wear of alumina, *J. Am. Ceram. Soc.*, Vol. 72, 1989, pp. 1249-1252.

133 C.P. Dogan and J.A. Hawk, Effect of grain boundary glass composition and devitrification on the abrasive wear of Al_2O_3, *Wear*, Vol. 181-183, 1995, pp. 129-137.

134 S.R. Hah, T.E. Fisher, P. Gruffel and C. Carry, Effect of grain boundary dopants and mean grain size on tribomechanical behaviour of highly purified α-alumina in the mild wear regime, *Wear*, Vol. 181-183, 1995, pp. 165-177.

7 SAMPLE PREPARATION AND EXECUTION OF TESTS

7.1 INTRODUCTION

Only a small number of tribometers have been standardized and appropriate guidelines provided for the execution of tribological tests on these rigs. As such guidelines are not available for the majority of tribometers, the best and most reliable testing procedures have to be specifically developed. The execution of wear or friction tests appears to be an apparently simple task but, in reality, it contains many difficulties, which can only be circumvented by careful application of laboratory techniques. The basic problem associated with the execution of tribological experiments is that tribology relates to the properties of thin surface layers which are difficult, if not impossible, to observe and control. Subsequently, the experiments and experimental techniques have to be organized to preclude any external influences on surface characteristics for the duration of each test. If procedures are not carefully devised then the data may be compromised by mundane problems as, for example, hydrocarbon contamination. The characteristic feature of friction and wear, which necessitates diligence in experimentation, is that friction and wear usually occur within a contact which is hidden from view. In order to obtain all possible data on friction and wear processes it is important to record all the events occurring in a dynamic contact as, for example, changes in noise emitted during wear. Often a large body of indirect evidence is needed for deduction of the hidden processes of friction and wear. With a large amount of data, the chances of some data corruption are very high unless experiments are carefully performed. Inadequately prepared samples can easily spoil otherwise most carefully executed friction and wear tests.

7.2 SAMPLE PREPARATION

Careful preparation of test samples is an essential part of experimental methodology if meaningful data is to be obtained and the results from tests conducted under identical operating conditions compared. Specific geometry of test specimens, such as balls, pins, cylinders, etc., requires that the machining operations such as cutting, turning, milling or grinding are employed in sample preparation. Consistent surface texture is another requirement for samples used in duplicate or identical experiments, as explained in Chapter 6. This is achieved by grinding or polishing.

Test materials are almost never supplied in an ideal pure form and are usually delivered with extraneous material such as packing substances or protective fluids as well as contamination. Layers of organic substances as thin as one molecular monolayer can affect friction and wear and these should be removed before testing. On the other hand, in tests simulating a particular tribological application, surface contaminants characteristic of real conditions should be preserved as they could affect the experimental data. In this section, methods and procedures involved in sample preparation are described.

Cutting, Grinding and Polishing of Specimens

Most tribological experiments require cutting, grinding and polishing of samples. When cutting a sample, care should be taken to ensure that the heat of cutting does not affect the sample. Since the final test sample is usually fairly small it can easily be heated to a high temperature when cut from the larger piece and its properties changed. Hard ferrous materials (>45HRC) and ceramics are likely to generate a lot of heat during cutting. For straight-section cuts the best way is to use a low speed cutting saw with diamond-impregnated blades for ceramics and carbides and Al_2O_3 blades for hard ferrous materials. The application of cutting fluids is always necessary. A lot of hard samples with complicated shapes have to be finished by grinding and similar precautions with regard to sample heating should be observed. When cutting a test specimen, the orientation of the wear surface with respect to internal microstructure needs to be considered. This is very important for heterogeneous materials such as composites where tribological properties tend to change with the orientation of wear surface. Cutting lubricants will contaminate the sample so that cleaning after cutting is essential to remove residual lubricant and cutting debris left on the sample.

Typical metallographical grinding and polishing procedures, that are slightly different for metallic and ceramic samples, are usually used in the preparation of tribological samples. However, care must be taken to avoid embedding hard abrasive particles into the sample surface. Microscopic diamond particles from diamond paste easily penetrate the surface of most metals and form a composite layer that can radically change the measured wear characteristics of the near surface regions.

In general, grinding or polishing produces an altered surface layer, known as the 'Beilby layer', with residual stresses and usually an extremely small grain size. The experimenter should be aware of the difference between the material immediately beneath a polished surface and the bulk material. This difference needs to be considered in any experimental assessment of material wear characteristics. In some cases metallic samples made of steel or cast iron are annealed after grinding or polishing to remove the residual stresses.

Abrading with abrasive paper and, to a lesser degree, polishing often results in a surface roughness with a distinct directionality, i.e. a predominant direction of surface features, unless the direction of polishing or grinding is regularly changed. If the direction of rubbing is constant then a series of straight grooves is produced on the surface caused by the ploughing and cutting action of abrasive grits. Directionality in surface roughness can affect wear and friction characteristics as described in Chapter 6. On the other hand, samples with clearly defined surface directionality are sometimes deliberately produced when its influence on wear and friction is the object of study. Polishing or finishing to specified roughness is an exacting process as the entire surface has to be rendered to the same level of surface finish. There is little value in leaving even one scratch on an otherwise polished surface as the scratch would affect wear. The problem of isolated scratches is particularly acute in tests involving a polymer sliding against a smooth hard counterface since even a single scratch will affect the wear of the polymer [1]. For example, in an artificial prosthesis where the combination of ceramic ball and ultra high molecular weight polyethylene (UHMWPE) cup is used as articulating surfaces, great care is taken to make sure that the ball surface is smooth and free from scratches. Polishing and grinding processes and related problems are illustrated schematically in Figure 7.1.

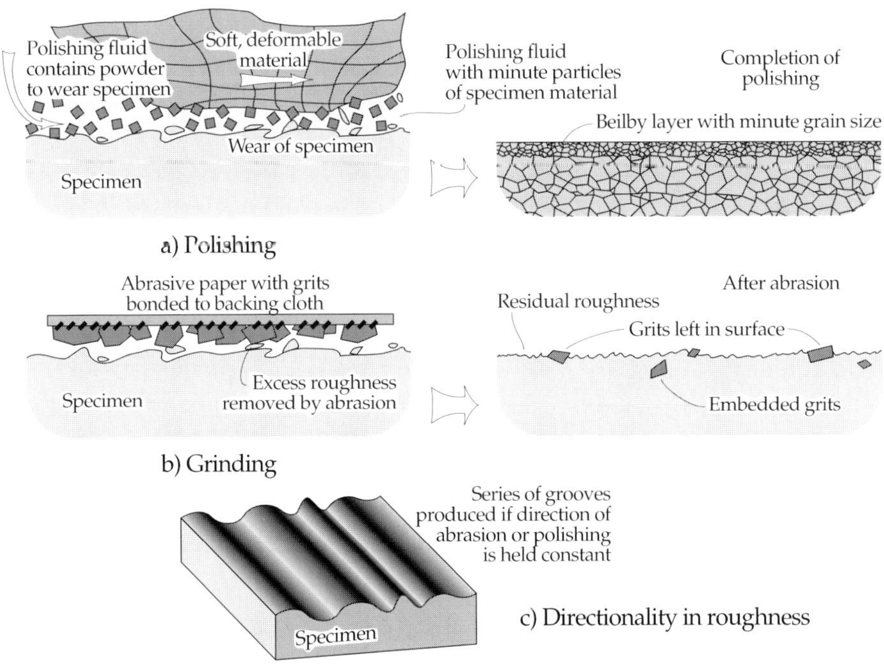

Figure 7.1 Schematic illustration of the grinding and polishing process of tribological samples.

Polishing and grinding require lubricants and cooling fluids, which should be removed from surfaces later by careful cleaning. In general, the polishing of samples should precede their final cleaning to ensure that friction and wear are not affected by residual polishing fluids or debris originating from the polished sample and abrasive particles.

Typical cutting, grinding and polishing methods for different materials are summarized in Table 7.1. The final choice depends on the surface finish (roughness) required for the tests and the best procedures for various materials are often recommended by suppliers of grinding and polishing materials.

Table 7.1 Cutting, grinding and polishing methods for different materials.

Preparation stage	**Metals** **Metallic alloys**	**Ceramics** **Carbides** **Very hard metals** **Composites**	**Polymers** **Polymer composites**
Machining	Turning, milling, cutting	Predominantly grinding (see below)	Dry machining
Grinding (abrasive wheels, papers or discs)	**Coarse grinding** SiC (non-ferrous and soft ferrous alloys) Al_2O_3 (ferrous alloys and soft metals) **Fine grinding** SiC papers (all materials)	Diamond wheels and discs (hard ceramics, carbides, very hard ferrous alloys) Metallic laps (ceramics) SiC papers (softer ceramics, hard metals)	Diamond and SiC (fine grinding)
Polishing (abrasive powders on cloth)	Diamond (steels) MgO (Al and Mg-based alloys) Cr_2O_3 (carbon and stainless steel) CeO (very soft metals) Al_2O_3 (final polishing of alloyed steels)	Diamond Al_2O_3 (super polishing of nickel alloys and ceramics)	Diamond CeO

Washing and Drying of Test Samples

For almost all tests, it is necessary to wash and dry samples before the experiments. If possible, as shown in Chapter 5, the period between sample cleaning and the test should be as short as possible to prevent re-contamination of the samples (see Figure 5.5). Little purpose is served if the samples are cleaned many days before the test as dust, moisture and pollutants from the air will deposit sufficient contaminants on the test samples during that time to affect the experimental data. Although it is relatively easy to remove most of the common contaminants such as grease and oil from the sample surfaces, it is very difficult to remove all of them to ensure ideal sample preparation. Often, remnants of oil can still be present on the sample even after very stringent cleaning [2]. The presence of surface contaminants, even as molecular monolayers, is responsible for the lower friction coefficient, which may occur during the running-in period of unlubricated wear testing. Usually, the less stringent the cleaning procedure, the lower is the initial value of friction coefficient. The effect of surface cleanness on friction was clearly demonstrated in self-mated sliding tests of PSZ ceramics [3]. Final cleaning with alcohol resulted in the coefficient of friction of 0.25 while blue gas flame cleaning increased the coefficient of friction to more than 0.45. Therefore, the sample cleaning protocol should be included in the description of tribological testing procedures.

For tests conducted in open air, where a period of initial wear and friction involving the removal of residual contaminants is inevitable, simple cleaning by immersion in an ultrasonic bath of hydrocarbon solvent followed by drying in hot air is adequate. The purpose of this type of cleaning is to remove most of the contaminants, apart from perhaps one monolayer of organic material attached to the surface, so that the period of removal of contaminants by wear is relatively short. In cases where cleaning is combined with polishing by abrasives, possible changes in the characteristics of polished surfaces, i.e. the tarnishing, should also be considered. For example, in one series of tests on stainless steel samples, polishing by emery paper was employed. It was found that the friction characteristics of a sample polished on Friday and then tested on Monday after the weekend was different from one prepared on either Monday, Tuesday, Wednesday or Thursday and tested the following day. The cause of this difference was that the period of three days between Friday and Monday was sufficient for corrosion to produce an oxide layer, on the stainless steel surface, of different friction characteristics than after one day during the working week [4].

Organic solvents such as acetone, toluene or ether are effective in cleaning most metallic samples and a domestic hairdryer gives rapid drying of both solvent and residual water. During the cleaning process, the vessel containing solvent with the immersed component is placed in an ultrasonic bath to ensure that all dirt particles are removed from the sample by vibration. A brief wash of the sample in clean solvent after it has been removed from the ultrasonic bath tends to be beneficial in removing any remaining particles.

Polymers are often difficult to clean with typical solvents because of the risk of chemical reactions. To avoid contamination, solvent-sensitive polymers can be machined without coolants using very clean cutting tools. Less reactive pure methyl or ethyl alcohols can also be used to clean polymers or polymer composite surfaces. Ceramics are usually cleaned using alcohol and detergent, followed by rinsing in distilled or de-ionized water. Thorough drying, often in an oven, is essential as water may chemically alter some ceramic surfaces. The recommended methods of sample cleaning are summarized in Table 7.2.

Table 7.2 Methods of cleaning wear samples for tribological tests.

Material	**Cleaning procedure**
Metals Metallic alloys Solvent-resistant polymers	• clean ultrasonically in solvent for several minutes • rinse with clean solvent or alcohol • dry in hot air
Polymers Polymer composites	• clean ultrasonically in ethyl or methyl alcohol • rinse with distilled or de-ionized water • dry in hot air
Ceramics Glasses	• clean in solvent first for highly contaminated samples • clean ultrasonically in alcohol and detergent or in commercial glass-cleaning solutions • rinse with distilled or de-ionized water • dry in hot air or oven • dry polish only for water-sensitive ceramics

For more specialized tests, such as measuring the frictional characteristics of prepared molecular monolayers of lubricants on polished surfaces, a more stringent cleaning process is necessary. A commonly used method is to apply the Soxlhet process where the sample is placed in a distillation column containing organic solvent. The organic solvent is then heated to boiling so that vapour condenses onto the component before dripping back into the boiling solvent. The solvent is already highly purified and the process of boiling produces still purer vapour to wash the sample. The raised temperature of distillation ensures that adsorbed contaminants on the sample are more easily desorbed than at room temperature. The Soxlhet process can be applied for several hours to completely remove any residual contaminant.

For all materials the closest possible approximation to a perfectly clean surface, without even oxide layers present, can be achieved by subjecting a sample to ion bombardment in a vacuum chamber. Ion bombardment involves an electric discharge in an inert gas, e.g. argon, which produces a stream of ionized gas that is directed by electric bias towards the cleaned sample. A disadvantage of this process, however, is uncontrolled heating during ion bombardment. Cleaning processes, including ion bombardment, of tribological samples are illustrated schematically in Figure 7.2.

Removal of oxide layers on metallic surfaces is also possible by heating the sample in hydrogen to initiate chemical reduction of the surface oxide and produce a pure metallic surface. This method, however, requires a high temperature, which may degrade the microstructure of the sample and allow hydrogen absorption by the metal.

Often tribological tests are specifically designed to model a particular application or industrial system. In these tests it is important to duplicate the real conditions, down to specific surface films, condensates or contaminants, as closely as possible. Testing very clean specimens in the laboratory may not provide valid data for real conditions where the surface contaminants are present.

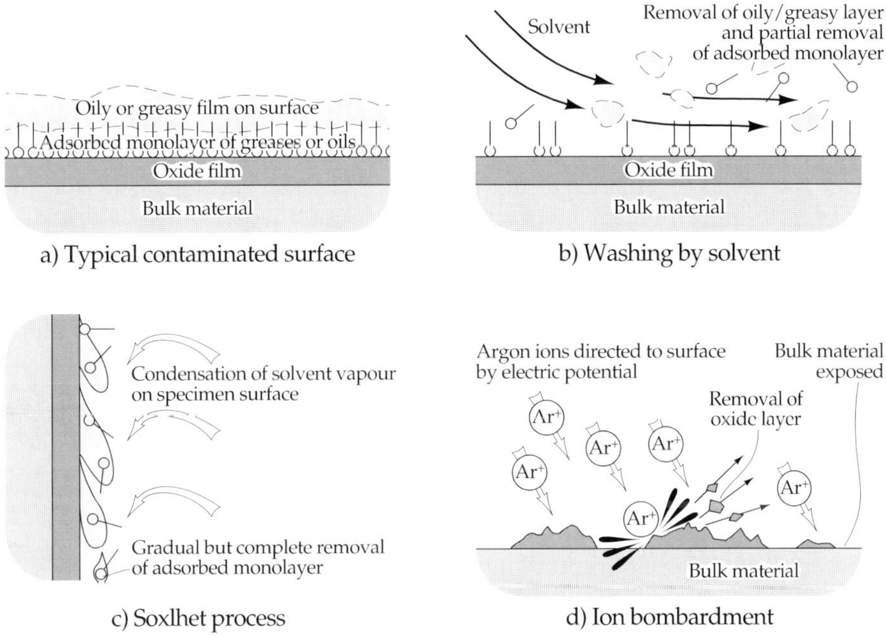

Figure 7.2 Schematic illustration of cleaning processes of tribological samples.

All cleaning measures are of no use unless the sample is carefully handled after cleaning. Contact with hands must be avoided as these are covered with oils and moisture. Tongs and tweezers are usually applied to handle the specimens. During very 'clean experiments' the experimenter might be required to wear a face-mask. During the 'ultra-clean experiments', particularly on oxide-free or perfectly dry samples, the cleaned sample usually remains in the same vacuum chamber used for cleaning. In practice, this means that the cleaning and wear and friction experiments are conducted in the same vacuum chamber [5].

7.3 EXECUTION OF TRIBOLOGICAL EXPERIMENTS

A classification of tribometers, compiled by the Society of Tribologists and Lubrication Engineers (STLE), includes 234 different test rigs. Since only a small number of them are standardized, e.g. ASTM G65, ASTM G99, there is not a single standardized experimental procedure suitable for all wear and friction tests. The execution of tribological experiments requires considerable care and the procedures developed must be strictly adhered to. A major difficulty associated with tribological experiments is to isolate fundamental influences on wear and friction, e.g. the role of materials or lubricants, from experimental artefacts. Great care is required to ensure that data relates to the fundamental factors of interest rather than to trivial influences such as daily temperature variations, external vibrations, humidity or minor changes in the experimental method. In this section, methods and procedures associated with the execution of tribological experiments are described.

Basic Principles of Operating a Tribometer

In most cases, the operation of a tribometer does not present excessive difficulty. Clean specimens are placed in the tribometer, checked for alignment and then loaded with the required contact force. The value of the contact force depends on the required contact pressure. Heating systems or controlled environments are then activated together with the instrumentation to measure friction and wear. Sufficient time must be allowed to stabilize the temperature of the environment before commencing the test. Sliding, rolling, impact or any other form of motion under study is then commenced and is maintained, usually at a constant rate, until the test is completed. Upon completion of the test, the steps described are applied in reverse. There is considerable flexibility in the experimental procedure applied with the exception that imposing the relative motion between samples is usually the last task to be performed. This is to prevent any anomalous wear occurring when the operating conditions are not yet set.

The basic procedural rules to follow when operating a tribometer are to run each experiment without interruption until it is completed and to maintain the same method of operating the tribometer. If an experiment is interrupted for some reason, e.g. a power failure, it is unlikely that the experiment can be resumed without some alteration in the wear and friction levels occurring. The alterations could be, for example, due to the fluctuation in contact temperature and bulk temperature of the apparatus, entrapment of a small quantity of lubricant between the hot opposing surfaces, which can cause uncontrolled chemical reactions. During the experiments, a steady state is established within a dynamic contact and when motion between the opposing contact surfaces is suddenly stopped, this steady state collapses to allow other, unintended, processes to occur, e.g. recontamination of the wearing surfaces by atmospheric pollutants. The problems caused by interruptions in the experiments are schematically illustrated in Figure 7.3.

However, it is not always possible to conduct the experiments without any interruptions. Often, wear characteristics such as wear volume or wear mass have to be measured at discrete intervals of time. Usually sample cleaning, to remove wear debris or lubricant, and drying must precede each weight loss measurement. On restarting the test, care has to be taken to position the sample in the same specific orientation to prevent any effects of surface texture on subsequent wear. In some test rigs, e.g. wet sand rubber wheel, incorrect repositioning of the test sample will cause a shift in the wear scar and subsequent errors in weight loss. In the tribological experiments, where components are subjected to only a moderate stress, e.g. studies of hydrodynamic lubrication, test interruptions are not particularly significant. However, temperature changes during the waiting periods may still affect the results.

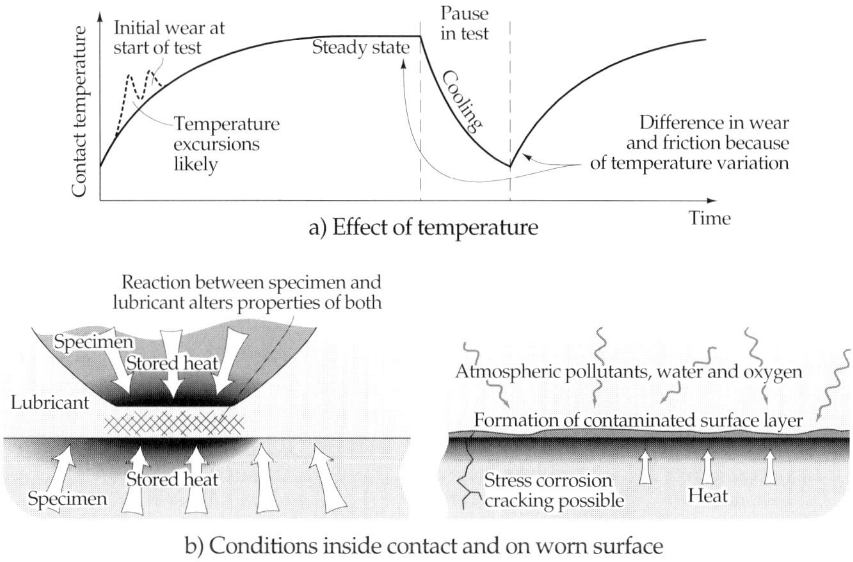

Figure 7.3 Schematic illustration of the undesirability of interruptions to tribological experiments.

Friction and wear are very sensitive to alignment between the test samples. Any misalignment between the samples can cause extremely high, localised contact stresses and uneven wear. The problem of misalignment is often encountered when a flat pin-on-flat disc configuration is used. When the two surfaces in contact are not parallel a cutting action by one of the edges may contribute to the abnormal wear loss. A method sometimes employed to produce a perfectly flat pin parallel to the disc surface is to lap the pin in-situ on the disc covered with a thin plastic sheet smeared with diamond paste. To eliminate cutting into the disc by sharp pin edges the edges are often given a narrow chamfer, especially when the pin contact area is large. In addition, a slight misorientation of the disc surface from normal to the direction of the pin axis may cause low frequency pin vibration and can also affect wear. Alignment between samples usually depends on the method of assembly of the tribometer and test samples. Naturally, the method of sample assembly needs to be kept unchanged during the entire program of tests.

Some test samples may also require a pre-loading period where the load is maintained in static contact between specimens for some time before the commencement of motion. The purpose of pre-loading is to allow the initial creep or plastic deformation that occurs upon loading to cease or at least reduce to the minimum level before wear measurements based on sample length change are conducted. This loading procedure is of particular importance in experiments with polymers [6]. Polymers are also affected by liquid absorption. When tested under lubricated conditions, e.g. water, control samples have to be used to allow for the change in weight of the sample due to the uptake of liquid.

Wear and friction data are easily influenced by external variations in experimental conditions. It is quite essential to record and note variations especially in humidity and room temperature, as already mentioned in Chapter 5. Humidity changes during testing can influence the results of unlubricated wear of ceramics and moisture-absorbing polymers, and are less important in testing metallic samples. The effect of external disturbances can be quite strong and any of these need to be recorded and checked for possible influence on the data. For example, locating a tribometer near a window may cause the test rig to be heated up by the sunlight during a few hours of the day. In one case a laboratory building was located next to a building site where a pile driver was used. The vibrations from the pile driver could be felt in the building for several weeks with inevitable effect on the data recorded. In many cases externally and some internally (e.g. the drive motor) induced vibrations in a test rig can be suppressed by using suitable damping mountings. The effects of vibrations can be important especially in unlubricated wear of ceramics where vibrations may promote wear by fatigue cracking [7]. In another case, engine performance data from an engine test facility showed a dependence on the weather. Investigations revealed that the exhaust outlet of the building in which the experiments were performed was located in a north-westerly direction from the air inlet. When the prevailing north-westerly wind blew, exhaust gas was directed towards the air inlet where it was sucked in contaminating the air supply. Locating various test rigs in a near proximity may also cause some problems since fumes or vibrations from one apparatus can affect the neighbouring rig.

During classroom experiments in tribology, which are attended by a large group of students, wear and especially friction characteristics measured in open test rigs may deviate from the expected pattern. The reason for this is probably that the combined exhaled breath of the audience conveys sufficient organic material to lubricate the wearing contact. Organic substances similar to many common lubricants lubricate the interior surfaces of the human lungs. The continuous passage of air through the lungs volatilizes the lubricants within [8].

Specific Problems in Abrasive and Erosive Tests

The specifics of some of the tribological experiments performed may pose entirely different challenges to an experimenter. For example, abrasive and erosive wear tests involving soil and rock materials present an experimenter with a new set of problems.

A characteristic feature of abrasive wear tests is the need to control abrasive particle characteristics. This is the reason why the standard abrasive wear tests such as ASTM G 65 and G 105 specify exactly the type of abrasive to be used (AFS 50/70 round quartz test sand). In non-standard tests the abrasive should be characterized precisely by either its source, e.g. river sand, silica sand, etc., or by the method of manufacture, e.g. crushed rock. Rock and sands are complex natural substances difficult to characterize. The geological description, which relates to the origin of the rock or sand, is generally available but may not be entirely suitable for tribological analysis [9]. Size and shape of the particles need also to be characterized since the geometrical factors such as the particles angularity or spikiness are known to control abrasive wear [10-13]. It is also necessary to ensure that the entire volume of abrasive grits required for tests is uniform.

In most abrasive materials, e.g. crushed rock or natural sand, there is a wide variation in grit size. Although the large, small and medium sized grits are randomly distributed at the beginning of a test, segregation of large and small particles caused by the motion of the sample and specifics of test rig design occurs. This process of abrasive grit segregation is illustrated schematically in Figure 7.4.

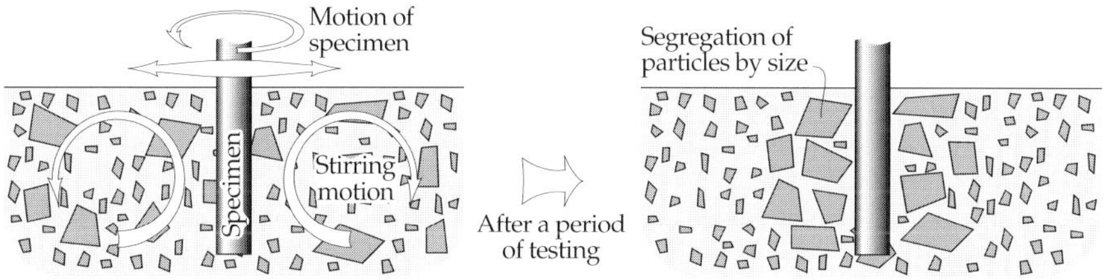

Figure 7.4 Schematic illustration of stratification process of abrasive grits during wear tests.

As a general rule, in order to avoid any scale effect, an abrasive is sieved before the tests. This allows the experimenter to use particles of a specific size range during the experiments.

In most cases segregation of grit by size causes a variation in wear rate with duration of the experiment. In short duration experiments, the abrasive can be replaced at the end of each experiment and the segregation effect is reduced to a minimum. In longer experiments, e.g. a full scale tractor test of abrasion resistant plough shares, there is a need for regular replenishment of abrasive and this can present some problems. Soil, which is often used as an abrasive, varies considerably over small areas and there may be insufficient area of a particular soil to allow a comprehensive test program if fresh unworked soil is required for each test [14]. It is therefore important to estimate the area of uniform soil required for tests before proceeding with the experiments. For example, if a tractor uses a guided test track and soil is repeatedly worked inside the test track, it is then necessary to monitor the soil segregation and replace it as soon as it changes its characteristics. The indications of these changes could be rounding of soil grits or a change of colour of the soil. In small-scale experiments, change of colour may imply that the soil grits have developed a transfer film of abraded material. Since the transfer film is capable of changing both the sharpness and the exterior hardness of the grits, e.g. when the grits are covered with a softer material, their abrasivity is almost certain to be affected. This, in turn, will affect measurements of abrasive wear resistance of materials. The problem of degradation of test abrasive and its effect on wear data is illustrated schematically in Figure 7.5.

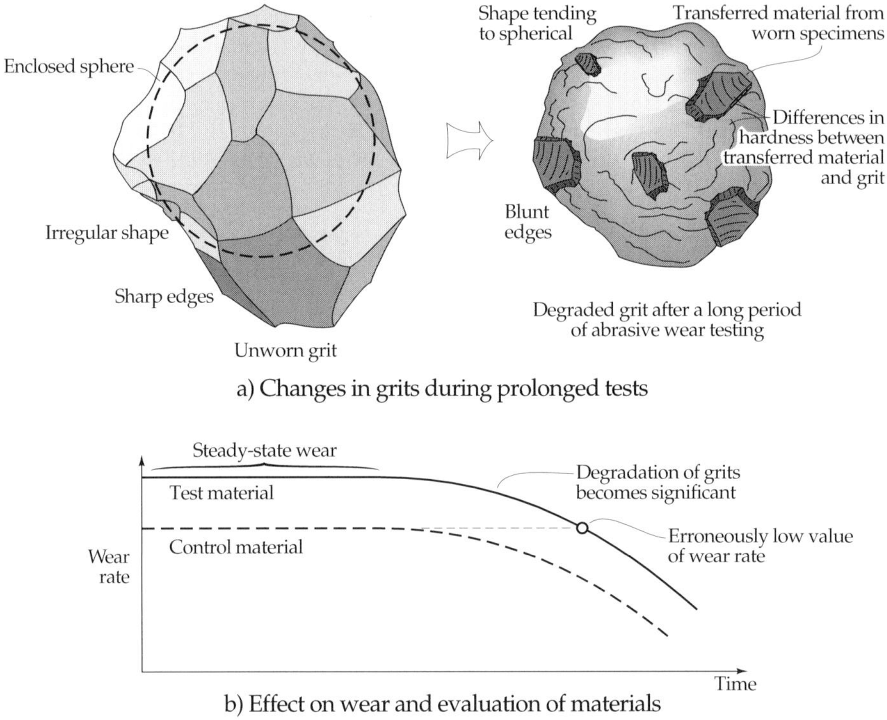

Figure 7.5 Schematic illustration of the degradation of abrasive grits during sustained testing and its effect on wear data.

A sample of control material, i.e. of known abrasive wear characteristics, might be used to monitor the grit degradation during the tests.

Similar problems are encountered during air erosion and erosion in slurries. In these experiments a large quantity of erodent particles is usually required and this necessitates recycling of particles. In some cases, the particles maintain their original shape after many impacts so the recycling is justified as a money saving measure. However, in other instances the particles suffer severe fracture during the first cycle of impact and need to be replaced [15]. Fracturing alters the shape of the particles which, in turn, can affect the wear rates measured. There appears to be a critical velocity below which particles are largely undamaged and above which most particles are fractured by the first impact cycle [15]. For example, when round particles are used as erodents, i.e. glass beads, fracture causes a formation of angular particles. The replacement of round particles by angular ones results in increased wear rates since angular particles are effective at cutting [15]. This also causes changes in wear mechanisms since cutting wear mechanisms replace wear by fracture or repeated plastic deformation (low cycle fatigue). This, in turn, affects experimental data on the relative wear resistance of materials. In the case of recycling of erodent particles, changes in their shape should carefully be monitored as it has been demonstrated that the shape of the particles is directly related to abrasive wear rates [11-13].

Phenomena Accompanying Friction and Wear Tests

A basic feature of tribological experiments is that unlike, for example, tensile strength obtained from a tensile test, friction and wear are not simple properties of a material. The particular values of friction and wear measured in any given experiment or test are the result of a complex interaction between several competing factors. Some of these factors may be of fundamental importance, e.g. changes of wear mechanism with sliding speed or load, while other factors, e.g. contamination of the worn surface by pollutants, may be an experimental artefact. It needs to be remembered that wear and friction are the result, rather than the cause, of interaction between solids in dynamic contact (i.e. sliding, rolling and impact). Thus, it is essential to collect all available information on possible influences on friction and wear. For instance, the noise from a dynamic contact can reveal whether stick-slip motion is occurring and whether structural vibrations of the tribometer are influencing conditions in the contact. A strange odour or fumes from a lubricated contact could indicate decomposition of the lubricant or wearing material. Fretting wear between steel surfaces can be detected from the red oxide powder that escapes from the contact. Mechanical activation of worn surfaces can exhibit a strong catalytic effect on chemical reactions and a reaction involving the lubricant or worn material may occur under conditions where it is not expected. Therefore careful and consistent observation is a basic feature of well executed tribological experiments.

Since wear and friction are affected by many different factors, the result of a single experiment can be misleading. Usually a series of duplicate tests under identical conditions are conducted and the mean values of the wear and friction data with their standard deviations are presented. Detailed guidelines for the presentation of tribological data are given in Chapter 11. It has been reported that intra-laboratory variation in wear and friction data from ball on disc tests can be as high as 13% and inter-laboratory variation can reach 20% [16]. Even a larger spread of results between the laboratories has been reported when the testing procedure was slightly altered [17].

7.4 SUMMARY

The task of performing tribological experiments and obtaining valid data is subject to many problems. In order to obtain meaningful results it is necessary to ensure that the basic conditions of test sample cleanness and control of test environment are met. Intensive efforts to clean or polish samples are of little use if the samples are later handled without gloves or tongs. Since only a few experimental rigs operate in a completely isolated environment it is important to check whether unexpected external factors, which are able to influence friction and wear, are excluded. Recycled test materials such as abrasive grits need to be carefully monitored for changes in their shape and size as this might affect the experimental data obtained. There is nearly always a paucity of data in friction and wear experiments to balance the complexity of phenomenon studied. Chance observations may provide the critical information to elucidate the basic mechanisms of friction and wear that occur. It is perhaps worth noting that many of the major discoveries in tribology occurred by chance, for example, the generation of pressure in hydrodynamic bearings, the development of toughened alumina ceramics, and many others.

REFERENCES

1 D. Dowson, S. Taheri and N.C. Wallbridge, The role of counterface imperfections in the wear of polyethylene, *Wear*, Vol. 119, 1987, pp. 277-293.

2 F. Pruvot, Ecole Polytechnique Federale De Lausanne, Laboratoire de Machines-Outils et Automates, Lausanne, Switzerland, Private communication, 1990.

3 N. Gane and R. Beardsley, Measurement of the friction and wear coefficients of PSZ and other engineering materials using a pin on disc machine, CSIRO, Report No. MTM/107, 1985, Melbourne, Australia.

4 K. Date, Imperial College of Science and Technology, London, Private communication, 1981.

5 K. Miyoshi and D.H. Buckley, Adhesion and friction of transition metals in contact with non-metallic hard materials, *Wear*, Vol. 77, 1982, pp. 253-264.

6 T.A. Blancet and F.E. Kennedy, The development of transfer films in ultra-high molecular weight polyethylene/stainless steel oscillatory sliding, *Tribology Transactions*, Vol. 32, 1989, pp. 371-379.

7 M.G. Gee and E.A. Almond, Effects from vibrations in wear testing of ceramics, *Mat. Sci. Technol.*, Vol. 4, 1988, pp. 655-662.

8 S. Hawgood and F.R. Poulain, Functions of surfactant proteins: A perspective, *Pediatric Pulmonology*, Vol. 19, 1994, pp. 99-104.

9 A.P. Mouritz and I.O. Smith, Present and future research in abrasive wear for the mining industry, Proc. Inst. Eng. Aust. Tribology Conf., Brisbane, 3-5 December 1990, IEA Nat. Conf. Publ. No. 90/14, 1990, pp. 28-32.

10 P.A. Swanson and A.F. Vetter, The measurement of abrasive particle shape and its effect on wear, *ASLE Transactions*, Vol. 28, 1985, pp. 225-230.

11 M.G. Hamblin and G.W. Stachowiak, A multi-scale measure of particle abrasivity, *Wear*, Vol. 185, 1995, pp. 225-233.

12 M.G. Hamblin and G.W. Stachowiak, A multi-scale measure of particle abrasivity and its relation to two body abrasive wear, *Wear*, Vol. 190, 1995, pp. 190-196.

13 D.V. De Pellegrin and G.W. Stachowiak, New technique of grit-profile analysis by cone-fitting, *Wear*, Vol. 247, 2001, pp. 109-119.

14 R.C.D. Richardson, The wear of metallic materials by soil - practical phenomena, *Journal of Agricultural Engineering Research*, Vol. 12, 1967, pp. 22-39.

15 A.J. Sparks and I.M. Hutchings, Effects of erodent recycling in solid particle erosion testing, *Wear*, Vol. 162-164, 1993, pp. 139-147.

16 H. Czichos, S. Becker and J. Lexow, Multilaboratory tribotesting: Results from the working party 'Wear Test Methods' of the Versailles Project on advanced materials and standards VAMAS, *Journal of Tribology*, Vol. 109, 1987, pp. 717-718.

17 E.A. Almond and M.G. Gee, Results from a UK interlaboratory project on dry sliding wear, *Wear*, Vol. 120, 1987, pp. 101-116.

8 SURFACE MICROGRAPHY AND ANALYSIS

8.1 INTRODUCTION

Surface micrography and analysis, conducted at completion of tribological testing, are vital sources of information on wear and friction processes. The measurements of friction and wear coefficients alone usually reveal little about the basic mechanisms and modes of friction and wear occurring. Microscopy examination of the complex morphology of wear scars helps in identification of wear mechanisms that produce the characteristic surface features such as film transfers, reaction products, grooves, pits, etc. For best results both interacting surfaces, as well as wear debris, should be examined.

All wear processes change the original topography of the surface and most of them change the chemical composition of the near-surface regions. During studies of worn surfaces the visual appearance of a surface is examined for signs of mechanical disturbance and chemical surface analysis is applied to find the composition of the worn surface, tribo-layers and the reaction products formed on the surface.

In tribological experiments the analysis of friction reducing films, usually forming on worn surfaces, is, because of their importance, of great interest. Composition analysis of films found on worn surfaces can reveal whether they are the product of reaction between the lubricant and worn material or atmospheric gases and water. Analysis of these films can provide information vital to a researcher. For example: Do the sulphur compounds present in the lubricant cause film formation? Is the tacky layer found on a worn surface the product of lubricant auto-decomposition at high surface temperatures?

Despite advances with in-situ observations of wear and friction, either by conducting experiments inside microscopes [1-3] or by using one transparent counterface such as glass [4,5], surface micrography and analysis are still the main sources of data or information on wear and friction mechanisms available. This method of analyzing worn surfaces has been likened to throwing a banana into a dark room. If the banana skin only is returned, we may conclude that there is a monkey in the room [5]. However, as is discussed below, much vital information may be gathered from careful examination of such 'banana-skins'. Micrographic examination of worn surfaces varies considerably according to the technique used, and there is an analogous variation between the relative utilities of different types of surface analysis. Efficient collection of experimental data depends on the ability of the instruments used to evaluate the changes in surface morphology and composition.

In this chapter the microscopic techniques used to characterize the topography, microstructure and chemical composition of tribological surfaces will be briefly described and their application illustrated by typical examples. More information on this topic can be found in several books or book chapters [6-9].

8.2 MICROSCOPIC OBSERVATION OF WORN SURFACES

In the majority of tribological experiments, the fine detail of worn specimens is examined. In any laboratory, an optical microscope is usually available and this can be used to provide some initial information about the worn specimen. The picture or image of the worn surface or wear particle obtained by optical microscopy is rarely sufficient for tribological analysis and it is nearly always necessary to proceed to a more sophisticated form of microscopy. There is a wide range of microscopes and microscopy techniques available and the more established techniques which have been used in tribology research are discussed in this section.

In simple terms, two scales of magnification are of interest. The lower scale of magnification, which can reveal detail as small as **100** [nm] is essential to the resolution of mechanical wear mechanisms occurring. The most suitable microscope for this scale of magnification is the Scanning Electron Microscope (SEM) which has proved to be extremely useful and is virtually indispensable in tribology research. A higher magnification allowing the resolution of atom-scale detail is used for specialised studies of the mechanisms of adhesion and subsurface material changes (e.g. of dislocation density). This level of microscopy represents the frontier of microscopy and there is a wide range of experimental or highly specialised equipment developed for the purpose. The most significant of these are: Transmission Electron Microscopy (TEM), Field Ion Microscopy (FIM), Low Energy Electron Diffraction (LEED), Atomic Force Microscopy (AFM) and Scanning Tunnelling Microscopy (STM). All of these techniques reveal different features of the surface or subsurface so that in many cases a combination of two or more types of micrography is necessary in order to obtain a comprehensive view of the worn surface.

Optical Microscopy

Optical microscopy is often used for initial observation of a worn surface, or when a more powerful microscope is not available. Compared to electron microscopes, optical microscopes have lower resolution and lower depth of field.

The resolution limit of a microscope is the minimum distance between two points on the specimen surface that can be distinguished as separate points. The theoretical resolution of an optical microscope, '**r**', is limited by a diffraction effect and is described by the following relationship (Rayleigh criterion) [10]:

$$\mathbf{r} = 0.61\lambda/\mu\sin\alpha$$

where:

λ	is the wavelength of the light [m];
μ	is the refractive index of the medium between the object and the objective lens;
α	is the semi-angle of the aperture.

The product '$\mu\sin\alpha$' is usually called the numerical aperture (NA). The resolution limit of an optical microscope is **0.2** [μm] (the absolute limit is **0.15** [μm]) [10]. Since the human eye can distinguish points separated by about **0.2** [mm] the highest useful optical magnification is about **1000**×. In practice the resolution limit is degraded by lens aberrations.

The depth of field, '**h**', is that distance around the plane of optimum focus where all the points of the object appear to be of the same sharpness. It can be described by the following relationship [10]:

$$\mathbf{h} = 0.61\lambda/\mu\sin\alpha\times\tan\alpha$$

It can be seen that to improve the depth of field the convergence angle α should be decreased. However, by doing it the resolution is made worse. In common optical microscopes the depth of field is similar in value to the resolution. With a very small angle 'α' of **5**° the depth of field would be about **40** [μm] but at the cost of the resolution of **3** [μm] [10].

Despite its limitations, optical microscopy of worn or failed surfaces can provide observations very useful in the analysis of wear behaviour [11]. There is no need for special preparation of surfaces and both conducting and non-conducting materials can be examined. The surfaces do not have to be very clean and traces of lubricants or other contaminants left on the worn surface can be examined at low magnification. Selected examples of the application of light microscopy to the analysis of failed and worn surfaces are described below.

Very low magnification (2 to 10×) light macrographs are very useful in failure analysis of, for example, bearings and gears [e.g. 11-14]. Those macrographs combined with stereoscopic light microscopy can help in preliminary diagnosis of failure causes. Optical macrographs of the surfaces of the failed machine components are shown in Figure 8.1.

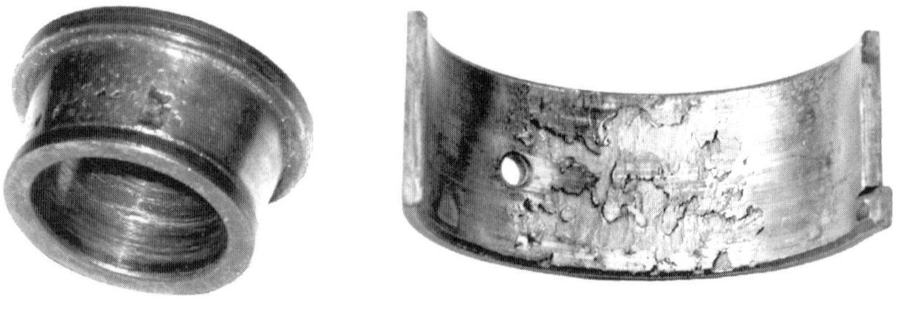

a) b)

Figure 8.1 Examples of surfaces of the failed machine components; a) fatigue spall on a rolling bearing raceway, b) fatigue of a white metal layer on a journal bearing raceway.

Light microscopy, combined with image analysis, has found application in measuring the dimensions of wear scars on pins or discs, as described in Chapter 4. The measurements are then used for determining the wear rates or wear coefficients. The magnification of 10 or 50× is often sufficient for the purpose and at this magnification the light images are reasonably sharp.

Most worn surfaces are rough and cause difficulty in imaging by a light microscope beyond a low magnification because of the limited depth of field available. Low depth of field of light microscopy results in fuzzy images of rough or worn surfaces when they are viewed at a higher magnification (above 200×) that is often necessary for distinguishing special wear features. The development of confocal light microscopy, described already in Chapter 6, has been an attempt to improve the depth of field of light images. Confocal microscopy allows for obtaining sharper images of rough or worn surfaces by eliminating the out-of-focus light from the image plane. Multiple optical sections, taken along the z-axis, are combined to produce a projected image of the surface [15-16]. Use of short wavelength laser light in confocal microscopy allows the resolution close to the theoretical limit to be achievable. Also, the depth of field can be extended to 200 [μm] [16]. Currently available commercial confocal systems incorporate powerful image processing software for construction of 3D images, topographical maps, stereo images, etc. [15-16].

Since the first commercial confocal systems were developed in the 1980s, this type of light microscopy is not yet widely used in tribological studies. However, over the last two decades laser scanning confocal microscopy (LSCM) has become an established technique in biological and medical fields of research. The published literature on the application of LSCM to the study of worn surfaces is quite limited and some examples are discussed in [15-16]. LSCM can also be used for the quantification of surface roughness, and this application is described in Chapter 6. Example of the confocal image of the fracture surface is shown in Figure 8.2. The varying gray scale levels represent varying feature heights. The black areas are the surface regions where no reliable signal was recorded.

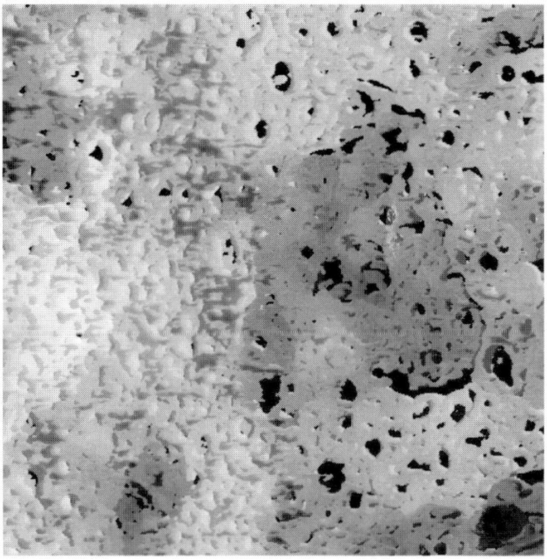

Figure 8.2 Example of a confocal microscopy height encoded image of the fracture surface [17].

A great advantage of light microscopy over electron microscopy is the ability to observe colours. Colour of worn surfaces or wear debris indicates whether free or oxidized metallic surfaces dominate, what level of frictional heating occurred or if lubricant decomposition or corrosion products are present. Colour is used to distinguish between different materials, such as for example brass and steel, and between different types of iron oxides. Polarized light is used to determine the presence of polymers and helps in distinguishing between crystalline and amorphous materials. An important application of light microscopy in tribological studies, where the possibility to observe colours is important, is the use of an optical system in wear particle analysis which is discussed in detail in Chapter 10.

A special attachment to a light microscope, called 'Nomarski differential interference contrast illumination', provides details of surface topography at low magnification [18,19]. A beam of polarized light is divided by a Wollaston prism into two beams that are then focused on two close points on the surface. The reflected beams interfere changing brightness of surface slopes. Nomarski contrast illumination greatly enhances contrast of surface relief compared to ordinary reflected light illumination. It has to be mentioned, however, that with this attachment the surface features on the image (peaks and valleys) are somewhat exaggerated compared to the real surface.

Prior to tribological testing light microscopy is often applied to examine the microstructures of specimens, as indicated in Chapter 6. One of the basic aims is to facilitate the analysis of the worn surface morphology. This is illustrated in Figure 8.3 where the surface of high chromium white cast iron worn in 3-body abrasive tests is compared to the original material microstructure. From this comparison it can be deduced that the matrix has been worn away leaving the hard protruding carbides.

Regular metallographic cross-sections are used to observe changes occurring in the subsurface zone due to wear [20,21]. The techniques used to prepare metallographic surfaces, cross-sections and taper sections are described in [19]. Great care has to be exercised while preparing the sections in order to avoid introducing artefacts such as cracks, surface smearing, chipping, edge rounding, etc. [19,20].

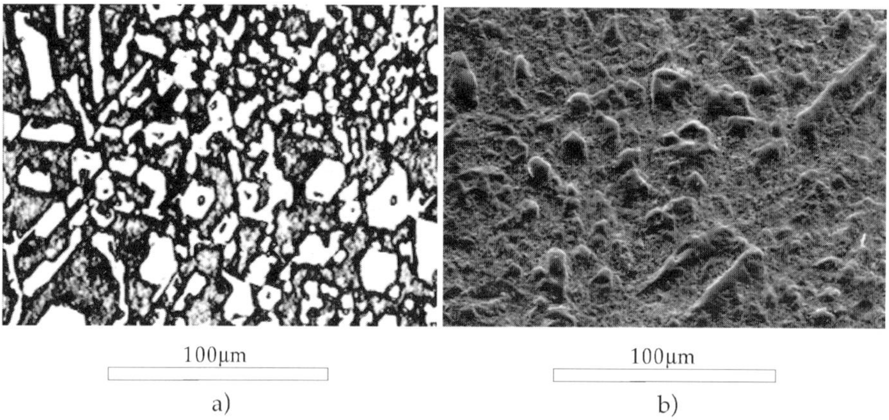

Figure 8.3 Optical micrograph of the microstructure of high Cr white cast iron consisting of hard carbides in a martensitic matrix (a); SEM micrograph of the same material after 3-body abrasive tests (b).

To protect the wear-surface features during cutting and polishing a layer of plated nickel is often applied [20,21]. When the cross-section of interest is very thin, as in cases of, for example, thin coatings, a taper section is used to optically enlarge the area under examination [11,20]. Subsurface changes due to wear observed on metallographic sections include plastic deformations, grain size reduction, white layer formation, cracking, delamination of coatings, etc. Special etchants and Nomarski contrast illumination can be used to enhance the topography of deformed layers [11,21]. Nowadays, because of limitations of optical microscopy, metallographic sections across the wear scars are often observed using either SEM or TEM microscopes [22]. Light microscopy example of 11.5° taper section across a worn surface containing white layers is shown in Figure 8.4 [23]. A microhardness tester has been used to determine the hardness of the deformed subsurface.

Scanning Electron Microscopy

Scanning Electron Microscopy (SEM) is a technique of imaging surfaces and near-surface regions of bulk specimens that utilizes various interactions between an electron beam and a specimen surface (elastic and inelastic scattering). In SEM a high-energy electron beam is directed onto and scanned (rastered) over a thick specimen surface generating signals such as secondary electrons, Auger electrons, backscattered electrons, X-rays and cathodoluminescence [10]. Each signal, captured by a special detector, can provide different information about the surface examined. The various SEM applications of the electron beam-specimen signals are listed in Table 8.1.

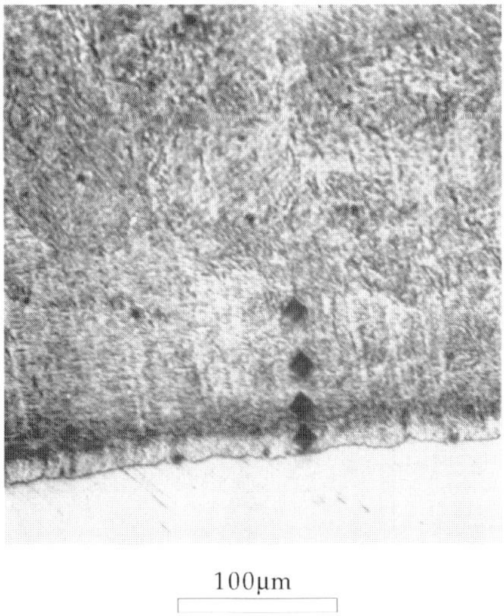

Figure 8.4 Example of 11.5° taper section across the worn surface containing white layers [23].

Table 8.1 SEM applications of electron beam-specimen interactions.

SEM Signal	Applications
Secondary Electron (SE)	• Surface topography, transfer films • Size and shape of wear debris and abrasive particles
Backscattered Electron (BSE)	• Variation in chemical composition over the surface • Variation in chemical composition amongst the wear debris
Auger Electron	• Chemical composition of top surface layers • Ability to detect light elements
X-Ray	• Chemical composition of deeper layers
Cathodoluminescence (CL)	• Chemical composition

The initial development of the Scanning Electron Microscope (SEM) was intended as an improvement on the Transmission Electron Microscope [24]. The earliest design of the SEM was really a modified form of the Transmission Electron Microscope (TEM) where a thin specimen was scanned by a rastering electron beam and an image formed by transmission electrons was collected [25,26]. A rudimentary form of the SEM was first used by Zworykin, Hillier and Snyder in 1942 [27] to provide fine resolution images of thick specimens (which could not be imaged by the TEM). A refined version of the SEM, incorporating many safeguards to prevent damage by unskilled use, was first released in 1964 by the then Cambridge Instrument Company [28,29]. Magnified topography images of a surface are obtained in an SEM by detecting the emitted secondary electrons when an electron beam is rastered over the surface studied. Secondary electrons are the electrons that are dislodged from atoms in the specimen surface during bombardment by high-energy electrons from the electron source (gun) of the microscope. The operating principle of an SEM is illustrated schematically in Figure 8.5.

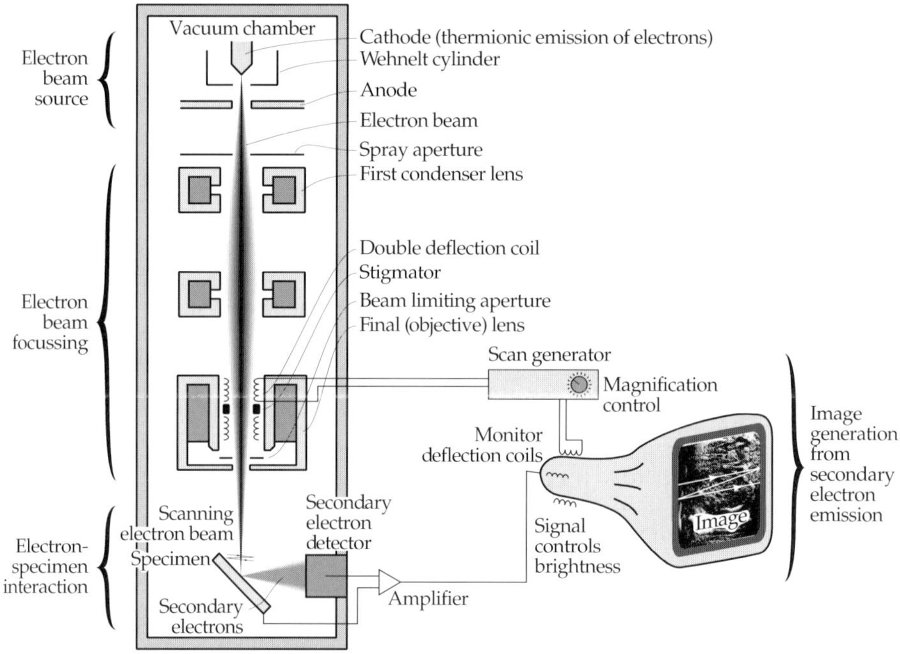

Figure 8.5 Schematic illustration of the operating principle of a scanning electron microscope (adapted from [24]).

As can be seen from Figure 8.5 the SEM consists of various subsystems such as: a vacuum chamber allowing free passage of electrons without ionization of gas molecules, a thermionic electron source filament (electron gun), a system of electromagnetic lenses to control the electron beam and a secondary electron detector to collect the secondary electrons emitted after interaction of the beam with a specimen. The final image is displayed on a television monitor (CRT), where the spot's intensity is modified or modulated by the amplified signals from the secondary electron detector. The scanning of the electron beam is synchronised with that of the CRT spot. The magnification depends on the ratio of the size of the CRT screen to the size of surface area scanned by the electron beam, i.e. to obtain a high magnification a very small area on the specimen is scanned. The Wehnelt cylinder provides the initial control of the electrons once they leave their thermionic source. Without this cylinder the electrons would tend to be emitted in all directions instead of being concentrated in a useful beam [24]. Recent developments of the SEM include introduction of computer control – new SEMs are now operated from a keyboard or mouse instead of from dedicated knobs. In modern SEMs the beam position on the specimen is controlled digitally and the image is displayed on a computer screen [10]. Image analysis software can then be used to analyse various features on the collected images. Apart from the secondary electron detector most SEMs have additional detectors that collect other signals such as BSE, CL and X-rays (see Table 8.1). Since an ultra-high vacuum is necessary to collect low-energy Auger electrons, a separate instrument, Scanning Auger Microscope, is devoted to this purpose.

The critical advantage of an SEM compared to a light microscope is the combination of finer resolution and larger depth of field [24]. In an SEM the resolution depends on the diameter of the incident electron beam, or more exactly the sampling volume. Apart from the filament type, in practice the diameter of the electron beam is controlled by the strength of the condenser lens and the working distance (WD) [10]. WD is the distance between the objective lens and the specimen surface. The beam diameter is decreased by increasing the condenser lens strength and by decreasing the WD. The minimum attainable diameter is limited by spherical and diffraction aberrations of the objective lens. It must be remembered that by increasing the condenser lens strength the beam current is also decreased. The beam current determines the contrast level between two points on the surface. Since both the diameter of the beam and the beam current affect the image quality the optimum conditions have to be found for each examination. When setting the optimum conditions the magnification used should also be taken into consideration [10]. The magnification determines the size of each point scanned by the beam on the specimen surface. Ideally, for best resolution, this point should have an area equal to that of the beam.

In the SEM the minimum observable dimension of detail on an object (or resolution) is proportional to the diameter of the electron beam. The minimum diameter of an electron beam is typically 4 [nm] for a heated tungsten filament [30]. In some more advanced instruments with much brighter LaB_6 or field emission filaments this diameter might be as small as 3 and 1 [nm] respectively. The limiting resolution of a typical SEM is about twice the beam diameter, i.e. about **7** [nm] [7]. In contrast, the resolution of typical white light microscopes is about **0.5** [μm] (500 [nm]). The best potential optical microscope resolution is about **0.2** [μm] when using blue light and numerical aperture (NA) of 1.5 (oil-immersion lens).

The electron beam of the SEM is extremely long and thin compared to the light beam of optical microscopy. This ensures that even when this beam is focused at a specific distance from the electron source, spreading of the electron beam is very limited for a considerable distance from the focal point. This lack of electron beam spreading allows focus to be maintained on objects with pronounced topographical relief. The depth of field for SEM is typically 10 to 100 times larger than that for the optical microscope [24]. Even when a worn surface has an amplitude of roughness of grooves reaching 1 [mm] or more, it is still possible to observe fine detail of fracture or cracking on both the peaks and troughs of grooves in the same focused image. It is this characteristic of SEM which makes it so important a tool for tribological investigations. A further advantage of SEM is that materials which are transparent to light, e.g. glass, are nearly always opaque to electrons [30]. Transparent substances are very difficult to observe by light microscopy and SEM facilitates the study of, for example, wear processes on glass or polymethacrylate (a transparent hydrocarbon polymer).

High resolution and large depth of field of SEM are the main reasons that this technique is almost exclusively used nowadays for determining wear mechanisms on the worn or failed tribological surfaces. Almost all currently published papers related to wear contain SEM micrographs. For this purpose a secondary electrons signal is used. The number of secondary electrons emitted depends strongly on the accelerating voltage but less strongly on the atomic number of the specimen [10]. Secondary electrons originate from a region a bit larger than the diameter of the incident beam and have a smaller sampling volume than the other SEM signals. Therefore the spatial resolution of secondary electrons is better than that of other signals and they are most suited for topography imaging. The number of secondary electrons depends on the tilt between the incident beam and the surface and more electrons are emitted from tilted regions. Therefore for SE topography images the specimen should be tilted towards the detector by 20-40°. The examples of SEM secondary electron images illustrating common wear mechanisms, i.e. adhesive, 2-body abrasive, oxidative, erosive, fatigue and tribochemical wear, are shown in Figure 8.6.

As shown in Table 8.1, there are several modes of electron emission and radiation released during operation of the SEM and these other forms of response to the impinging electron beam can provide image detail which is not found with secondary electron emission. This other commonly used mode is backscattered electron emission (BSE) which is sensitive to both surface topography and atomic number of the specimen. Backscattered electrons are electrons elastically scattered from the specimens. By subtracting or adding the electronic signal from two back-scattered electron detectors fitted in most of the SEM microscope chambers it is possible to exclude either information on surface topography or its composition. Imaging by backscattered electron emission is most effective in the examination of metallic specimens where the details of surface composition are not obscured by a conductive coating deposited on the specimen.

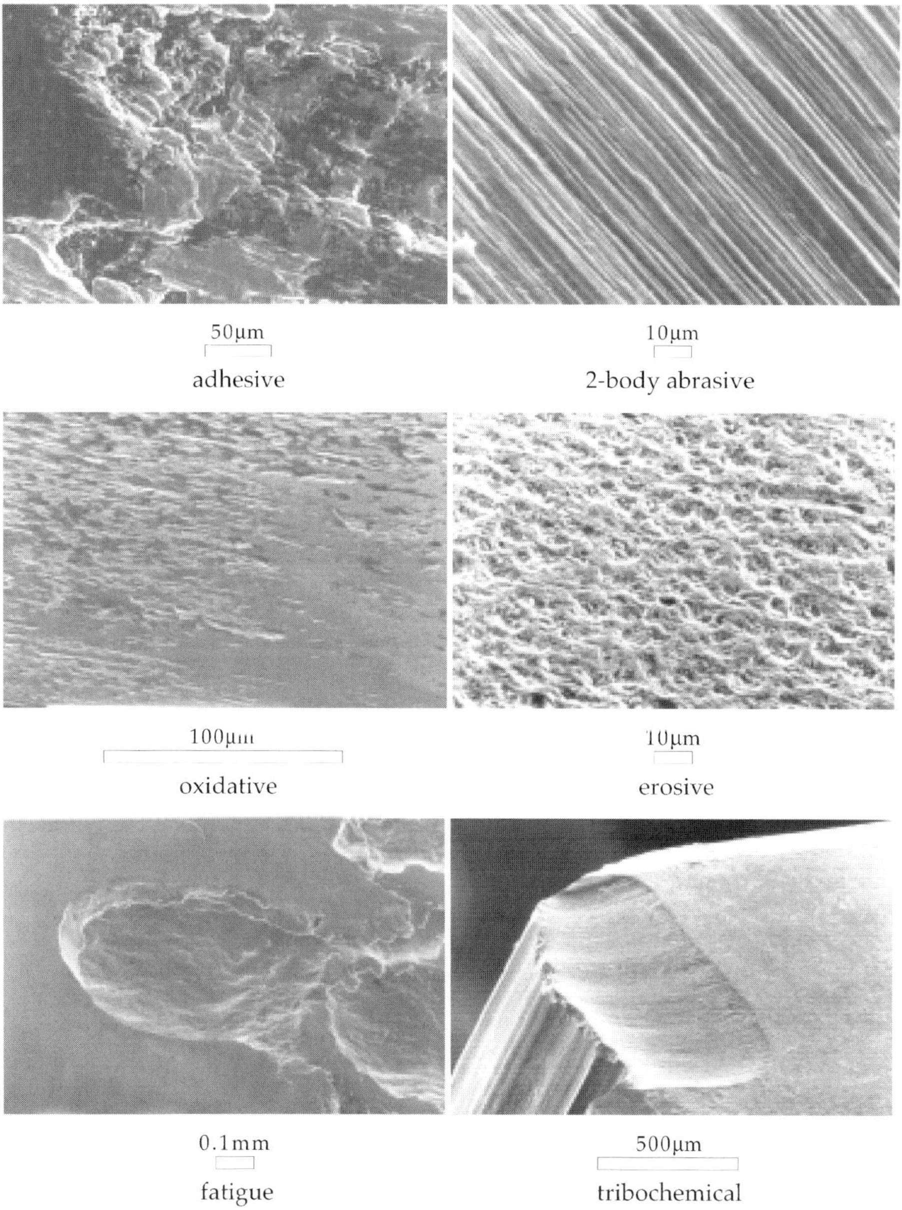

Figure 8.6 Examples of SEM secondary electron images of common wear mechanisms.

Specimen preparation for SEM is quite simple. Specimens up to 100 [mm] in largest dimension can be loaded into the modern SEMs and there is no requirement for elaborate cleaning of the specimen apart from washing in solvent to remove grease contaminants and thorough drying. Contaminated porous specimens should be cleaned by a low-temperature vacuum bakeout. The surfaces should be free from dust or other particle contaminants that may cause charging during imaging. Insulating specimens usually require a sputtered carbon or gold coating to allow electron conduction away from the impinging electron beam during observation but this procedure does not usually pose difficulty. Carbon coating is preferred if compositional analysis, such as energy dispersive X-ray spectroscopy, is to be conducted as well. Gold or gold-palladium coatings produce better topography images using secondary electrons. The use of metal coatings is especially important when viewing surfaces containing low atomic number elements. However, in some instances, e.g. during the microscopic observation of nylon specimens, conductivity of the polymer was found to be sufficient to allow microscopy without a conducting coating [31].

In contrast to optical microscopy, real colours are not visible on the SEM images. For example, red iron oxide, a typical product of steel fretting, is easily seen under an optical microscope but looks like any other oxidized debris in the SEM. Also, various oxidised layers on steel cannot be differentiated on the SEM images. Sometimes the SEM images are given false colours as an improvement on the original monochrome image.

SEM not only offers a high quality visual image of the surface but the impinging electron beam can also provide much information about the chemical composition of the surface. The SEM microscopes are usually equipped with the X-ray spectrometer, i.e. X-ray Energy Dispersive Spectroscopy (EDS), facilitating this type of analysis. The combined visual and chemical analysis of a worn surface is extremely important to tribology and will be discussed later in this chapter.

The disadvantage of a conventional SEM is that examination occurs only under a high vacuum which precludes the observation of liquids or wetted objects, e.g. oily layers on worn surfaces. A recently developed related technique, the Environmental Scanning Electron Microscope (ESEM), allows the observation of wet objects, e.g. biological samples [32,33]. ESEM has been used to investigate wear processes on the cartilage of knee joints [34] where dehydration under a vacuum by conventional SEM would destroy the original cartilage surface structure.

ESEM has several key features that differentiate it from conventional SEM. Firstly, the use of multiple pressure limiting apertures in the path of the electron beam allows elevated pressures to be maintained in the specimen chamber. Chamber pressures can be as high as **2.7** [kPa] (about 20 [torr]), though a range between **0.13** and **1.07** [kPa] (between 1 and 8 [torr]) is more common. Secondly, it utilizes a different secondary electron detector than the traditional Everhart-Thornley detector used in SEM [35]. The gaseous, or environmental, secondary electron detector (GSE) utilizes ionisation of chamber gas molecules to amplify the secondary electron (SE) signal, allowing normal SE imaging to be conducted under the elevated pressures used in ESEM. In addition, the GSE detector can be used in the imaging of uncoated samples, including samples that exhibit insulating properties. Charging at the specimen surface is neutralised by positive ions generated after collisions between secondary electrons and gas molecules present in the specimen chamber. Thus very little, if any, sample preparation is required for ESEM analysis. One of the most important advantages of ESEM is its ability to image a sample in 'wet' environment, i.e. with water present in the sample chamber. By utilizing an operating pressure of at least **0.61** [kPa] (about 4.6 [torr]) (the vapour pressure of water at 0°C), it is possible to sustain liquid water in the specimen chamber. The temperature of the cooling stage can then be manipulated to cause either the hydration or dehydration of a sample, e.g. articular cartilage.

· *Examples of Use of Scanning Electron Microscopy in Tribological Studies*

During the SEM examination, in most cases, secondary electron emission is used as this gives good surface detail on almost any object. Secondary electron SEM images of worn surfaces have provided critical data, which helped to formulate the basic theories of wear and friction mechanisms, as shown in Figure 8.6. The classical example is adhesive wear, where detailed micrographs of worn surfaces allow one to resolve the wear mechanisms operating. The earliest studies of adhesive wear related to metals but with the introduction of engineering ceramics the question arose whether adhesive wear was a significant cause of wear between metals and ceramics [36]. Tests of Partially Stabilized Zirconia (PSZ) sliding against ferrous and non-ferrous metals revealed that the sliding combination of PSZ and brass had a comparatively low friction coefficient but high wear rate of brass. There was considerable practical as well as theoretical interest in the reasons for this unusual characteristic of low friction and high wear. Micrographic examination of the worn surfaces revealed that the brass formed a thick sacrificial transfer layer on the PSZ surface [36]. An example of this transfer film is shown in Figure 8.7a.

Signs of adhesion, such as shearing of the transferred brass layer in the direction of sliding and close contact with the substrate are clearly visible. The SEM micrographs revealed that after an initial short period of direct sliding between brass and PSZ a transfer layer developed on the PSZ and this transfer film controlled the long term wear and friction of the PSZ/brass sliding contact. Similar behaviour is found in other ceramic/metal combinations but the morphology of the transfer film varies. An example of grey cast iron film transfer on PSZ is shown in Figure 8.7b.

Backscattered electron (BSE) imaging has been found to be extremely useful in the studies of worn metallic surfaces as this mode of imaging is less easily disrupted by poor surface quality than secondary electron imaging. Worn metallic surfaces are often covered with non-conducting deposits (oxides) and particles, which tend to gather electric charge during observation and disrupt photographic recording of the secondary electron image. Fretting wear often presents severe imaging difficulties because of the accumulations of non-conducting iron oxide or oxidized debris on the surface of the wear scar [37]. In this instance, BSE imaging is particularly useful since it provides a reliable means of obtaining a high quality photographic record of the worn surface.

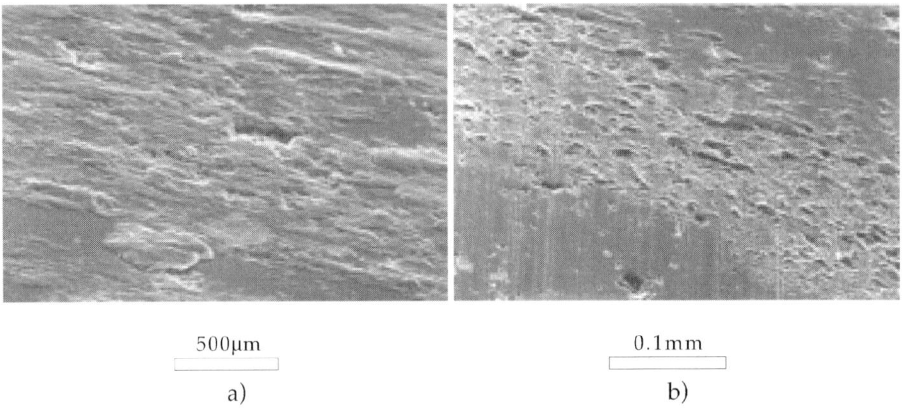

Figure 8.7 SEM micrograph of PSZ surface after sliding against; a) brass and b) grey cast iron.

Since the intensity of BSE image depends on the atomic number this technique can also provide some fundamental information about the wear processes which is not revealed by secondary electron imaging. For example, a BSE image of a fretting wear scar on steel wire is shown in Figure 8.8a. Various levels of greyness in the image can be seen corresponding to different degrees of oxygen content in the wear debris. In this instance, greyness indicates the oxidized iron while whiteness indicates largely unoxidized iron (or with only a very thin oxide layer). Based on the information obtained from this image it is possible to distinguish between areas which were at the stage of forming oxidized wear debris (grey) and areas which contained compacted accumulations of highly oxidized wear debris (black). Areas of the metal devoid of an oxidized debris layer can also be seen in the wear scar (light).

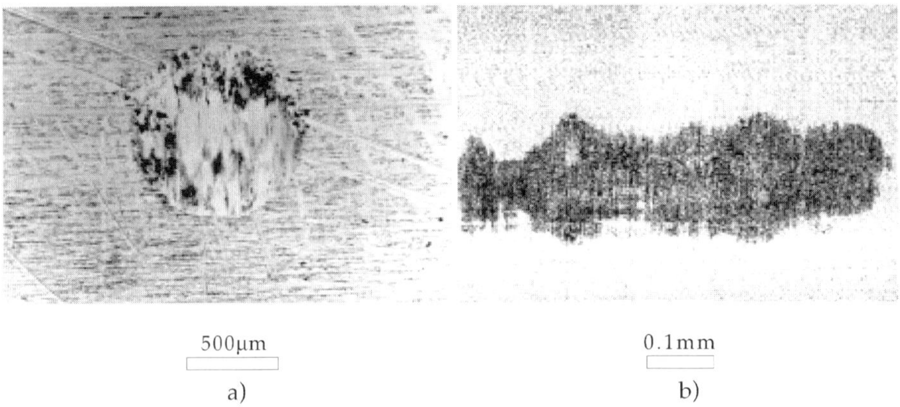

Figure 8.8 Examples of the ability of BSE imaging to reveal; a) the oxide debris layers in a fretting wear scar on steel wire and b) film transfer of oxidized steel on zirconia ceramics.

Wear scars often contain pits and these can present particular difficulties with imaging. The edges of the pits, often found in polymers and other non-conducting materials, rapidly accumulate electric charge and prevent imaging by secondary electron emission [31]. Since BSE imaging is usually difficult to perform with polymers, it is often necessary to accumulate and integrate a large number of images by applying a rapid scan of the electron beam. The digital images acquired are then integrated to obtain a final image.

- *Examples of Use of Environmental Scanning Electron Microscopy in Tribological Studies*

Environmental Scanning Electron Microscopy has been used in bio-tribological studies, in particular in studies of wear of articular cartilage. Since ESEM operates under much higher pressures than conventional SEM, dehydration artefacts caused by low pressures can be avoided. This is a great advantage since the biological surfaces in their natural state can be examined. Specimens can be kept hydrated during the examination in the ESEM through use of a Peltier-cooled specimen stage in the microscope chamber. In this manner the specimen can be kept wet and its dehydration prevented. In one work during the examination of articular cartilage worn surfaces its temperature was maintained at approximately 4-6°C while the chamber pressure was maintained at about 0.8 [kPa] (about 6 [torr]). It needs to be mentioned that biological surface can also be examined in the SEM. However, in contrast to ESEM examination the articular cartilage samples for SEM analysis must be dry and conductive. Before drying the samples are fixed in an osmium tetroxide solution and ethanol to prevent the dehydration artefacts. Whether the fixation process introduces any artefacts still needs to be investigated. Examples of the ESEM images of the worn surface of articular cartilage are shown in Figure 8.9.

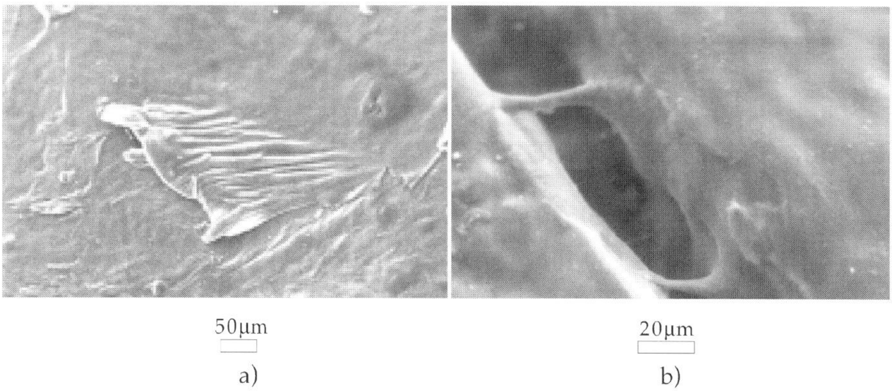

Figure 8.9 Examples of ESEM images of the surface of articular cartilage; a) surface damage after 15 [min] of wear, b) splitting of articular cartilage after 240 [min] of wear [38].

Transmission Electron Microscopy

Transmission Electron Microscopy (TEM) is a technique of imaging the internal structure of solids using a beam of high-energy electrons transmitted through the solid. This arrangement can be compared to the basic optical microscope with transmission illumination (sometimes called a biological microscope). The TEM images are usually more difficult to interpret than the SEM images of the external morphology of solids. While the principles of TEM imaging have not changed, many new facilities, mainly analytical, have been added over the years to a basic TEM microscope. Apart from simple imaging, modern TEMs contain facilities for X-ray and electron spectrometry, electron diffraction and others [10].

The wave-like properties of electrons were first observed by the physicist de Broglie [39]. By devising electric lenses to focus the electrons, the first Transmission Electron Microscope (TEM) was developed by Ernst Ruska from Germany in 1933 [40]. The basic principle of the TEM is that a photographic image is recorded from the electron flux after it has passed through a thin sample of the specimen under study. The operating principle of TEM and image formation is illustrated schematically in Figure 8.10.

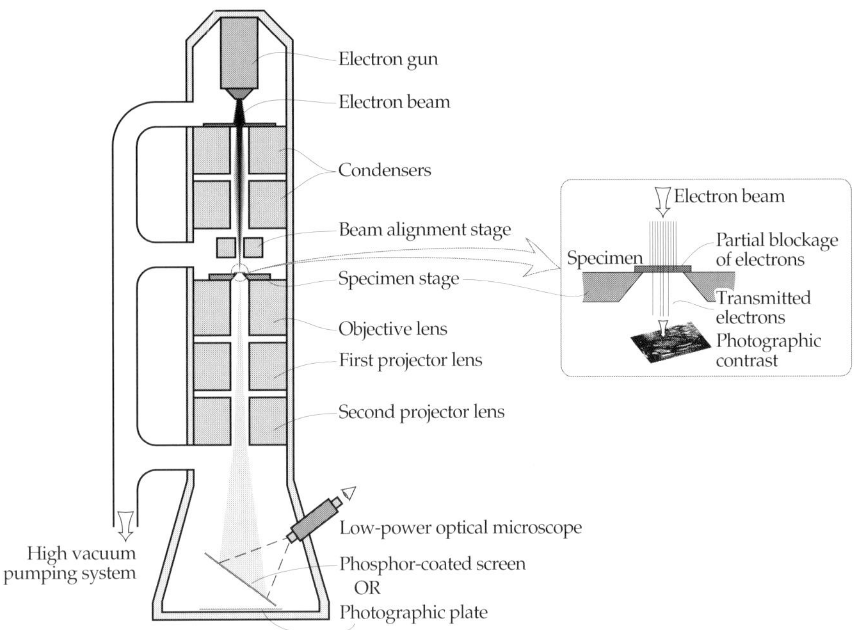

Figure 8.10 Schematic diagram of TEM and principle of image formation.

Resolving power or resolution of a TEM is determined by the wavelength of an electron flux 'λ', convergence angle 'α' and spherical aberration. The theoretical resolution of a TEM can be described by a simplified equation [10]:

$$r = 0.61\lambda/\alpha$$

However, this theoretical resolution cannot be obtained due to lens aberrations, mainly spherical aberration. The wavelength of electron flux is inversely proportional to the square root of the electron voltage [30]. By applying a high electron voltage of, for example, 50 [kV], a wavelength of 5 [nm] can be obtained and with 100 [kV] a wavelength of 2 [nm] is possible. In principle, it is possible to achieve a resolving power close to those wavelengths but various technical problems prevent this. With TEM, extremely high magnifications are possible and the microscope has found wide application, particularly in the micrography of biological cells and viruses. In tribology, the most important role of TEM has been the imaging of dislocation tangles in deformed surface layers beneath wear scars [41,42]. Images of cellular dislocation tangles in a layer of metal within 1 [μm] from the worn surface provided the evidence of wear particle formation controlled by the formation of a surface layer with a cellular microstructure completely different from the bulk material.

The process of preparing metallic specimens for TEM can be tedious and time consuming as it is necessary to cut a thin foil of metal from the wear specimen and then subject this foil to further thinning. The thinning process is based on ion-bombardment, which is usually slow. Care is required to ensure that the preparation processes do not alter the microstructure of the wear specimen. The thickness of a specimen suitable for viewing in a TEM depends mainly on the accelerating voltage (AC) and the average atomic number (AN) of the sample. The higher the AC and the lower AN the thicker the sample can be. The size of the sample is limited by the diameter of metallic grids, usually 3 [mm], that support the sample. TEM has also been used to observe crystalline and amorphous films formed on a worn metal surface by a lubricant additive, Zinc Dialkyl Dithiophosphate (ZnDDP) [43,44]. It is necessary, however, to use a specialised film replication technique in order to obtain suitable specimens of wear surface films for TEM [44].

8.3 ATOMIC-SCALE MICROSCOPY TECHNIQUES

Field Ion Microscopy, Low Energy Electron Diffraction, Atomic Force Microscopy and Scanning Tunnelling Microscopy are advanced techniques of observation that are applied to fundamental studies of the mechanisms of friction and wear on a nanoscale level. In most cases, these forms of microscopy are not routinely applied to the examination of worn surfaces, as is the case for SEM. The information gained, however, by these more complicated microscopy methods is still of great importance to the fundamental studies of tribological phenomena.

Field Ion Microscopy

The Field Ion Microscope (FIM) was developed in a rudimentary form by Erwin Wilhem Muller in 1937 and later refined by Muller and Tsong [45]. The FIM has been widely used to study the atom-scale detail of metallic surfaces. Although this microscope is capable of extremely high magnifications, e.g. even individual atoms can be resolved, it can only be used for carefully prepared metallic specimens in the shape of a sharp needle. A schematic diagram of the FIM is shown in Figure 8.11.

The operating principle of FIM is based on placing a sharp metallic needle in an atmosphere of low pressure inert gas and applying a positive high electric potential to it. The high electric potential causes the formation of positive gas ions on the needle surface, the gas ions subsequently escape from the pin surface to discharge the potential. An enlarged image of a specimen with detail corresponding to individual atom sites can be obtained by inserting a fluorescent screen in the path of the escaping positive gas ions. Since the positive gas ions fly along the electric field lines the atoms on the metallic tip are radially projected onto the screen. In order to produce high electric fields and clear high magnification images the needle tip must have a radius of curvature between a few and a few tens of a nanometre.

Restrictions on observable specimens have severely limited the applicability of FIM to tribology. Adhesion processes between metallic surfaces [46] and the effect of oxygen adsorption on intermetallic adhesion have been observed by FIM [47] where the needle was loaded against a metal counterface and then separated from the counterface for collection of images. Other examples of the application of FIM in tribology studies can be found in [48]. FIM images from adhesion and friction experiments show deformations and film transfers, already familiar from SEM images, with atomic resolution [48]. The application of FIM to tribology presents considerable difficulties. Adhesion and friction experiments have to be conducted using a specially designed, delicate rigs placed in the FIM chamber evacuated to an ultra-high vacuum. Apart from restrictions on the material type and the difficulties in preparing fine needle-tips, the analysis of the FIM images, especially surface defects or deformations, requires considerable experience. Computer-simulated FIM patterns can provide great help in the interpretation of FIM images [48].

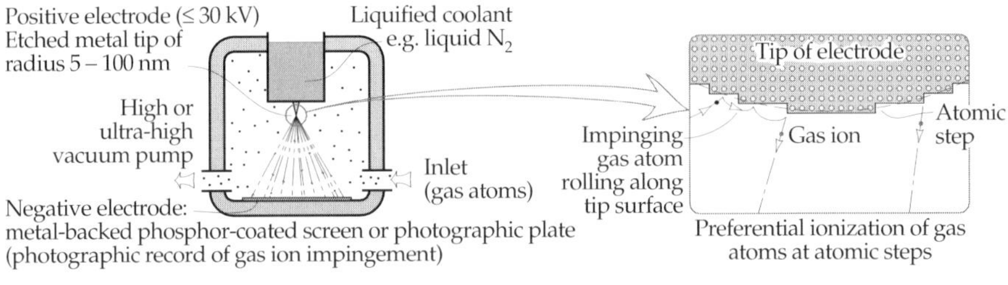

Figure 8.11 Schematic diagram of the operating principles of Field Ion Microscope.

Low Energy Electron Diffraction

In contrast to FIM, the Low Energy Electron Diffraction (LEED) microscope has been extensively used to study the mechanism of adhesive transfer between clean metallic surfaces and the interaction of a clean metallic surface with oxygen and sulphur [46]. LEED is related to electron diffraction imaging but is different because the low electron voltage used restricts electron penetration of the specimen to the outermost atomic layers. LEED is typically used in surface science experiments to check the cleanliness of a specimen surface. The sensitivity of LEED to surface contaminants lends this microscopy technique to investigations of the effect of minute amounts of surface contaminants on adhesion and adhesive wear. Imaging of atom-scale detail from the surface of the specimen is based on observing the diffraction patterns established by the electron flux reflected from the specimen surface. A schematic diagram of the LEED is illustrated in Figure 8.12 together with a diagram of its application to the study of adhesive transfer.

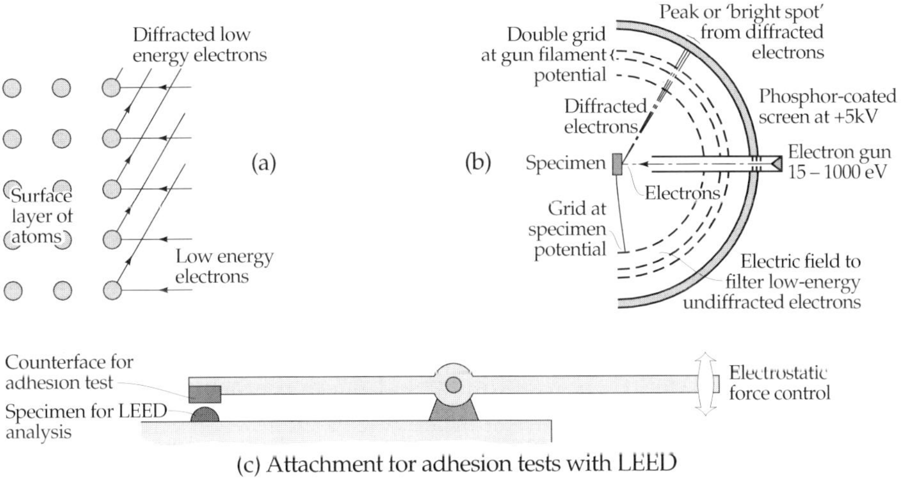

Figure 8.12 Schematic diagram of the operating principles of LEED and its application to the study of adhesive transfer.

The operation of LEED and the preparation of specimens is not simple so that this microscopy technique is only suitable for specialized experiments where other forms of imaging are inadequate. Another imaging technique of a surface at atomic resolutions that is related to LEED is Reflective High Energy Electron Diffraction (RHEED), where a high energy flux of electrons is directed at a high angle of incidence to the surface. The high angle of incidence or small angle of approach ensures that a significant fraction of the electrons is reflected by surface atoms. A surface image at resolution corresponding to atomic details is obtained by locating a fluorescent screen in the path of the reflected electrons. However, this technique does not appear to have been used for tribological studies so far. Both LEED and RHEED produce only diffraction images of electron flux from which various surface features can be deduced. Neither of these techniques produce a visual image of the surface analogous to the images produced by SEM.

Scanning Tunnelling Microscope and Atomic Force Microscope

The Scanning Tunnelling Microscope (STM) and the Atomic Force Microscope (AFM) are two much newer forms of microscopes than the microscopes discussed in the previous sections. The Scanning Tunnelling Microscope was invented by two researchers at IBM in Zurich, Gerd Binnig and Heinrich Rohrer, in 1981. The STM was one of the first instruments allowing the acquisition of images of surfaces with atomic resolution and earned Binnig and Rohrer the Nobel price for Physics in 1996. Atomic Force Microscope (AFM) was developed five years later by Binnig, Quate and Gerber, in 1986 [49], to counterbalance one of the major limitations of STM which is its inability to analyse insulators unless they are present as ultra thin films on conducting substrates. These two types of microscope provide extremely high resolution at high magnifications of surfaces which allows the observation of atom-scale details.

The STM also provides some limited chemical information, which allows atoms of different atomic numbers on a surface to be distinguished. For example, adsorbed oxygen atoms can be distinguished from nickel atoms on a nickel surface [50]. This information is limited due to the fact that the STM probe only detects the electron energy levels of conduction and valence as opposed to core electrons of an atom which control the fine structure of chemical molecules [51]. However, this type of information and other data of a similar kind are extremely pertinent to the understanding of wear and friction mechanisms. AFM and STM are developing rapidly, both in technical sophistication and because of the increasing number of new microscopes derived from the basic principles embodied in the AFM and STM [52]. Nowadays, a working knowledge of AFM and STM is becoming almost essential to experimental work in tribology. As friction and wear depend on the chemical and physical properties of atoms and molecules at the interface between interacting surfaces [53] understanding the phenomena occurring on the surfaces at atomic level has become critical to understanding the tribology at macroscopic level. The great advantage of AFM is that it allows to examine surface characteristics at an atomic level in gas (air) or liquid environment without the need for high vacuum. A whole new branch of tribology, nano-tribology, has been created utilizing these two instruments and many other devices which subsequently evolved [e.g. 54].

In the majority of cases the applications of AFM have focused on studies of dry surfaces, coatings, polymer films, etc. However, AFM has also been applied in the studies of thin liquid films, films formed by liquid lubricants, reaction films formed by lubricant additives, etc. [e.g. 55-61].

The basic operating principle of both STM and AFM shows some similarities to a surface profilometer. Conventional microscopes in current use obtain images from reflected radiation or electrons. The basic limitation of these methods is the wavelength of radiation or electron flux. When a radiation or electron flux of very short wavelength is used, the radiation or electrons may have too high energy to interact with the specimen and produce a useful image. In order to overcome these limitations of resolution, STM and AFM compile an image by traversing a sharpened tip, usually **2** [μm] long and less than **100** [Å] in diameter, across a surface in a manner similar to a surface profilometer.

Scanning Tunnelling Microscope uses a conducting tip with a bias voltage applied to the tip and the sample. The operating principles of an STM are schematically illustrated in Figure 8.13. The tip of the STM probe is maintained at a very small distance above the surface, sufficiently close to sample the electron cloud of the specimen. The density (i.e. number of electrons per unit volume) of the electron cloud declines exponentially with distance from the surface so that an extremely close co-ordination between the probe and the surface is required before the electron cloud can be detected. In STM the sharp tip is brought to within **10** [Å] distance of the specimen surface. At this close distance the electrons from the sample or from the tip begin to 'tunnel' to the tip or to the sample depending on the sign of bias voltage. The tunnelling current is measured and used as an imaging mechanism of STM. The exponential relationship between the tunnelling current and the distance between the tip and the sample gives these instruments extraordinary sensitivity. For example, if the distance between the tip and sample changes by 1 [Å], i.e. 10%, then the tunnelling current changes by an order of magnitude. This allows one to image surfaces with sub-angstrom vertical resolution and atomic lateral resolution [62].

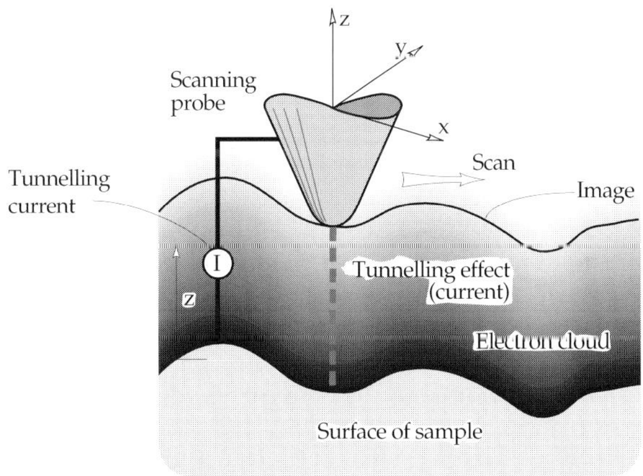

Figure 8.13 Schematic illustration of operating principles of STM.

The STM can operate in two modes, i.e. in a constant-height mode and in a constant-current mode. In the constant-height mode the tip is traversing the surface at fixed height and the tunnelling current varies while in the constant-current mode the tunnelling current is kept constant by adjusting the height of the tip by a feedback system. In the constant-current mode the changes in tunnelling current cause the adjustments to the voltage supplied to the piezoelectric scanner which in turn adjust the position of the tip above the surface under examination. These two modes of STM are schematically illustrated in Figure 8.14.

The constant-height mode is essentially limited to relatively smooth surfaces while constant-current mode works quite well with rougher surfaces. However, because the system does not have to adjust the position of the scanner at each point of measurement the constant-height mode is faster than the constant-current mode. One of the drawbacks of STM is that both the tip and the sample must be made of conducting materials. This may cause some problems if the area of the sample is oxidized. In the constant-height mode the current will drop when the tip encounters the oxidized area while in constant-current mode the tip will move unrealistically close to the surface in order to maintain the constant tunnelling current. Subsequently the STM cannot be used to image insulating materials. To overcome this restriction an Atomic Force Microscope has been developed.

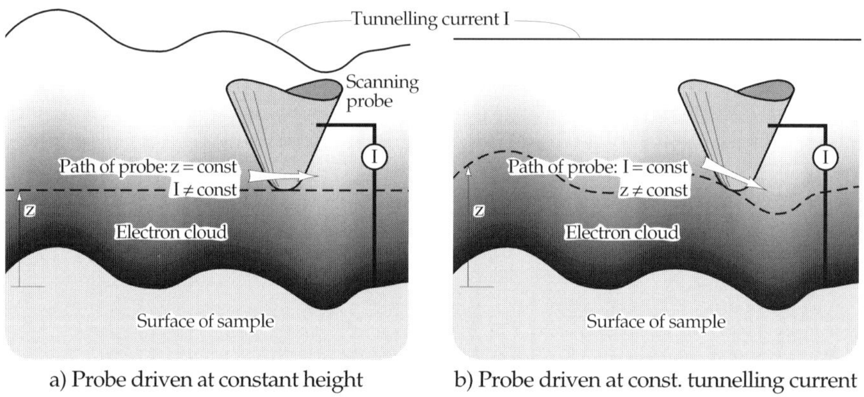

Figure 8.14 Schematic illustration of two modes of operation of STM.

Atomic Force Microscope (also known as the Scanning Force Microscope SFM) uses a sharp tip with a radius of **10-30** [nm] located on a cantilever about **100** to **200** [μm] long with some spring constant '**k**'. The spring constant '**k**' is lower than the typical elasticity constants between atoms. The tip is then moved across the specimens surface by a piezo system similar to that developed for STM. The displacement of the cantilever by the specimen is due to van der Waals forces and Born repulsion [53]. During the measurements the position of the tip is monitored and recorded as the tip is scanned over the sample or the sample is scanned under the tip [52]. In most AFMs optical techniques are employed to trace the position of the cantilever. Laser beams are used to bounce off the cantilever onto the position sensitive photodetector, i.e. as the cantilever moves the position of the beam spot on the detector shifts. Attempts are being made to manufacture the cantilever from a piezoresistive material so its deflections can be directly monitored. The displacements of the tip can be plotted as a function of '**x**' and '**y**' co-ordinates resulting in the image of the sample surface. Sometimes this image is called a 'force map' since the displacements of the tip can be, according to the Hooke's law, related to force, i.e. **F = -kz**, where '**k**' is the spring constant and 'z' is the displacement [64,65]. The operating principles of AFM are illustrated schematically in Figure 8.15.

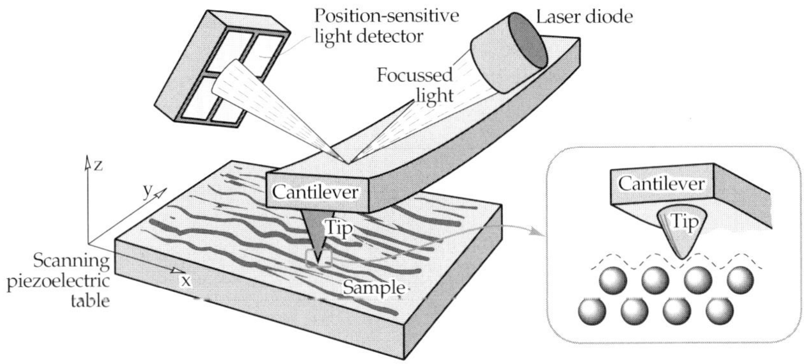

Figure 8.15 Schematic illustration of the operating principles of AFM.

The lateral resolution of AFMs is excellent, i.e. usually better that 10 [nm]. This resolution depends on imaging forces (contact pressures). With very low imaging forces, i.e. < 1 [nN] true atomic resolution can be obtained [63,66]. When the imaging forces are in the nano-Newton range and above, the atomic scale defects are no longer visible although the atomic structure may still appear to be present [63].

AFM can also be used to measure the vertical force that is applied to the tip during image acquisition. This is achieved simply by switching off the 'x' and 'y' piezos and measuring the deflection of the cantilever as a function of sample displacements, i.e. from large separations down to contact with the specimen and back out to large separations. As the sample moves towards the tip there are several forces acting on it, i.e.: cantilever spring force (**F = -kz**), van der Waals forces (generally attractive) and the Born repulsion forces acting at small separations [63]. When the sample and the tip are close enough, at the critical point, the tip jumps into the surface in a similar way as two magnets brought sufficiently close to each other. As the sample and the tip continue to move together the cantilever deflects away from the surface approximately in a linear fashion [62] until the direction of sample movement is reversed. During the retraction, in the absence of scanner hysteresis, cantilever deflection follows the same curve as the tip is pulled away from the surface as illustrated in Figure 8.16.

In air there is usually a monolayer or a few monolayers of water present on the surface. As the tip approaches the surface of the sample, meniscus forms between the tip and surface, resulting in a capillary force which is very strong and attractive. During the retraction of the scanner the water will hold the tip in the contact with the surface until the tip springs free from the contact as illustrated in Figure 8.16. This is known as a snap-back point. With more complex layers of fluid, e.g. water and lubricant, multiple snap-back points can occur [62]. This effect has been utilized in analysing and characterizing thin liquid lubricating surface films [67], contaminants' viscosity, thickness of lubricant layers, local variations in elastic properties of the surface, etc. [62]. The slope of the curve shown in Figure 8.16 is related to the elastic modulus of the system. When the cantilever is softer than the sample surface then the slope mostly reflects its spring constant. On the other hand, when the cantilever is stiffer than the surface the slope provides an information about the elastic properties of the sample.

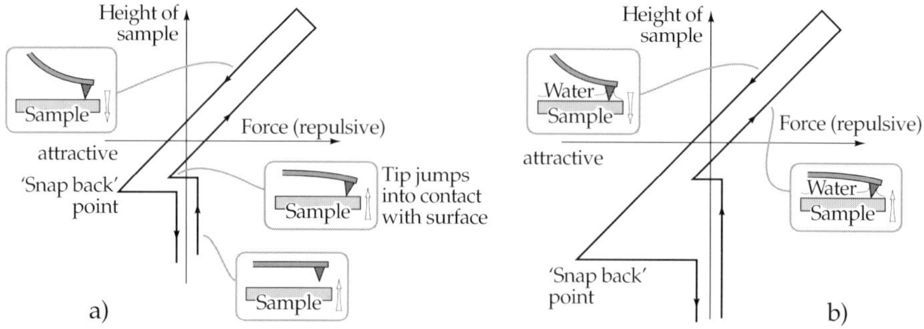

Figure 8.16 Relationship between the cantilever displacement and tip-sample separation in; a) vacuum and b) air.

Depending on the application, the AFM has been designed to operate in three different modes:

- contact,
- non-contact and
- tapping.

In contact mode the tip makes the direct contact with the surface examined. As the tip scans the surface the cantilever bends to accommodate the changes in surface topography due to the contact forces acting upon it. When the tip is brought close to the surface the atoms from the tip and surface will initially weakly attract each other. This attraction will increase until the electron clouds of the tip and the sample atoms start to repel each other electrostatically at the same time. These forces will balance each other when the distance between the atoms is about 0.2 [nm]. When atoms are in contact the total van der Waals force becomes positive, i.e. repulsive. Since the slope of the van der Waals force, in the contact region, is very steep, as illustrated in Figure 8.17, any attempt to push the atoms closer together will be balanced by an increase in the van der Waals force. The cantilever will bend or the sample will deform in order to maintain a constant distance between the sample and the tip [62]. During the measurements in air, the sample is always covered with a layer of water, which exerts strong attractive capillary force. This force depends on a distance between the tip and the sample and is virtually constant when the tip is in contact with the sample. The cantilever also exerts a force, which depends on the cantilever deflection and its spring constant. The cantilever and capillary forces are balanced by the van der Waals force and they vary from 10^{-7} to 10^{-6} [N].

In non-contact mode, the tip is separated from the sample by a distance between **50** to **100** [Å]. At this distance, electron clouds covering the surfaces of the tip and the sample begin to interfere and the electron orbits within the atoms of the tip and sample begin to synchronize resulting in a weak attraction. In non-contact mode a stiff cantilever must be used since the soft cantilever could flop onto the surface of a sample because of these attractive forces. Since a stiff cantilever is less responsive to small force variation than a soft cantilever its sensitivity is increased by vibrating it near its resonant frequency (200-300 [kHz]) with an amplitude equal to a fraction of an angstrom [62]. Attractive van der Waals forces vary with the distance between the tip and the sample and cause the tip's resonant frequency to change. As the tip scans the surface, the changes in the resonant frequency or amplitude are kept constant by moving the scanner up and down. From the scanner movement the set of surface data is obtained and sub-angstrom vertical resolution comparable to contact AFM achieved. In non-contact mode a very small force of about 10^{-12} [N] is generated. This mode allows one to study soft or sticky samples, where the tip could become stuck in the surface, or powdered materials, where the tip moves the particles around [63]. Use of this mode avoids the contamination of the tip. The non-destructive character of this method is at the cost of lower spatial resolution compared to a contact mode. Also in the case of samples covered by thin layers of liquids, the results obtained might be misleading. AFM operating in contact mode will penetrate these layers while AFM operating in non-contact mode will not, as illustrated schematically in Figure 8.18.

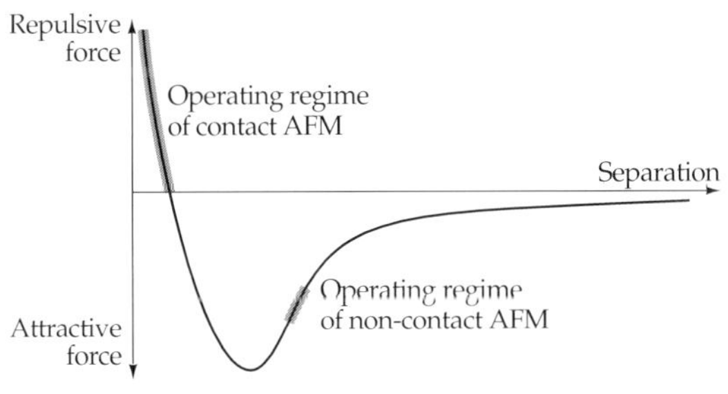

Figure 8.17 Relationship between the van der Waals force and the separation distance between the tip and sample [62].

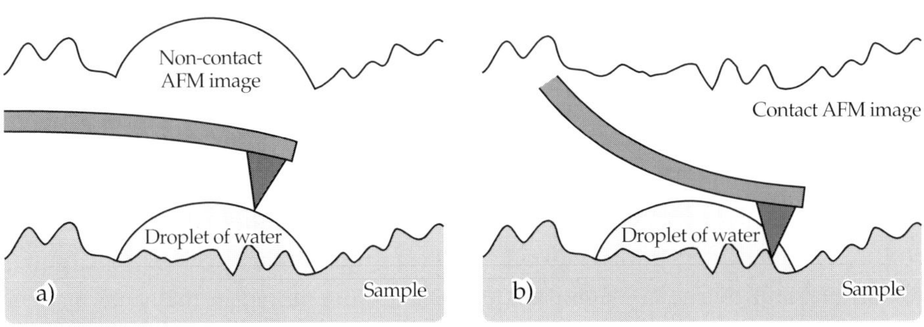

Figure 8.18 Differences between contact and non-contact AFM imaging of surfaces covered with layers of fluid (adapted from [62]).

In tapping mode the cantilever oscillates at its resonant frequency at a very high amplitude of **100** [nm] and with each oscillation the tip touches the sample. Since in this mode the lateral forces, i.e. frictional drag, are eliminated the sample is less likely to get damaged. This is particularly important when imaging the biological materials. Tapping mode AFM can measure liquid covered surfaces which further enhances its applications [68]. Forces acting on the sample are essentially normal to the surface hence this mode is far less destructive than contact mode with its inherently large lateral forces. However, normal forces in this mode are significantly higher than the capillary forces, i.e. 10^{-8} [N] since the tip must be able to penetrate a layer of liquid and emerge from it as, for example, in the case of imaging surfaces covered with layers of liquids [62]. In this mode the AFM tip is 'tapped' on the surface at regular intervals of 500 [nm] [69]. This allows to obtain a large-scale image of the surface from which smaller areas of interest can be selected for detailed investigation by the AFM contact or non-contact modes.

Lateral Force Microscope (LFM), also known as Friction Force Microscope (FFM), is a recently developed variation of AFM [e.g. 70-72]. These instruments allow one to measure local variations in surface friction, which can arise from inhomogenity in surface material and changes in the slope as illustrated in Figure 8.19.

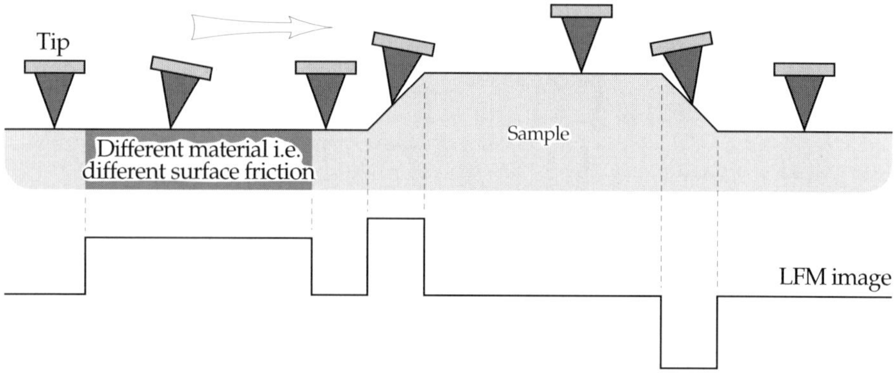

Figure 8.19 Schematic illustration of tip and cantilever deflections resulting from surface friction changes and slope changes (adapted from [62]).

As the tip traverses the surface it is subjected to lateral forces, which are related to local coefficients of friction, i.e. ratio of the lateral force to the normal force acting on a tip is a coefficient of friction. This view of friction is naturally a simplification as it avoids considerations such as contact stress, which would normally be included in a more rigorous analysis of friction. These forces are transmitted to a cantilever, which twists along its length. The cantilever deflections and twists, i.e. lateral deflections, are monitored in the same manner as in a traditional AFM using position sensitive photodetectors. The operating principles of LFM are illustrated in Figure 8.20. As a result of this refinement the LFM can provide a frictional map of the surface with nanometre resolution together with the surface topography data in one scan. The application of various force modulation techniques [e.g. 73,74] to intermittent contact between the tip and a sample allowed to determine other material properties such as adhesion, elastic and viscoelastic properties of the sample [75]. Chemical differences between the regions of similar morphology can be revealed almost down to atomic scale [63]. However, the detailed information about the chemical nature of surfaces molecular interactions is provided by Chemical Force Microscopy (CFM) [76]. In these instruments a chemically modified cantilever/tip assemblies allow imaging with chemical sensitivity.

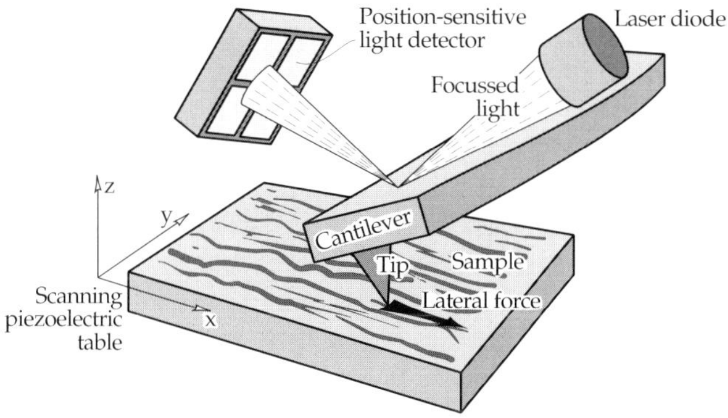

Figure 8.20 Schematic illustration of operating principles of LFM.

An example of a surface frictional map obtained by LFM from an area of 650 nm^2 covered with boundary films is shown in Figure 8.21 [75]. Those films were generated during sliding of a steel ball on a polished steel plate in a molybdenum diethylhexyl-dithiocarbamate and mineral base oil solution at 100°C [75]. It can be seen from Figure 8.21 that darker (lower lateral force) and brighter areas (higher lateral force) are oriented with the sliding direction. The lateral force histogram showed two clear peaks (Figure 8.21c). It was found that the areas with lower lateral force were corresponding to the grains while the areas with higher lateral force were corresponding to the area outside the grains. It can also be seen from Figure 8.21 that low lateral force patches are constituted of small grains agglomerated in long paths oriented along the sliding direction [75]. It is believed that the grains correspond to MoS_2 sheets and MoS_2 single crystals [75,77-79]. As these grains are present only on the high surface spots they may significantly affect the overall friction.

It can be seen from Figure 8.21 that there is a close correspondence between the local presence of a film and low friction although some secondary variation between film thickness and friction coefficient is also visible. Data of this kind provides further confirmation of the basic tribological principle that friction is largely controlled by the presence of surface films. The microscopic scale of variation in friction coefficient shows that frictional force in any contact between asperities of opposing surfaces will vary significantly from the average value deduced from a bulk coefficient of friction. Some asperity contacts will sustain high frictional stress leading possibly to the formation of a wear particle while others remain at a much lower level of frictional stress.

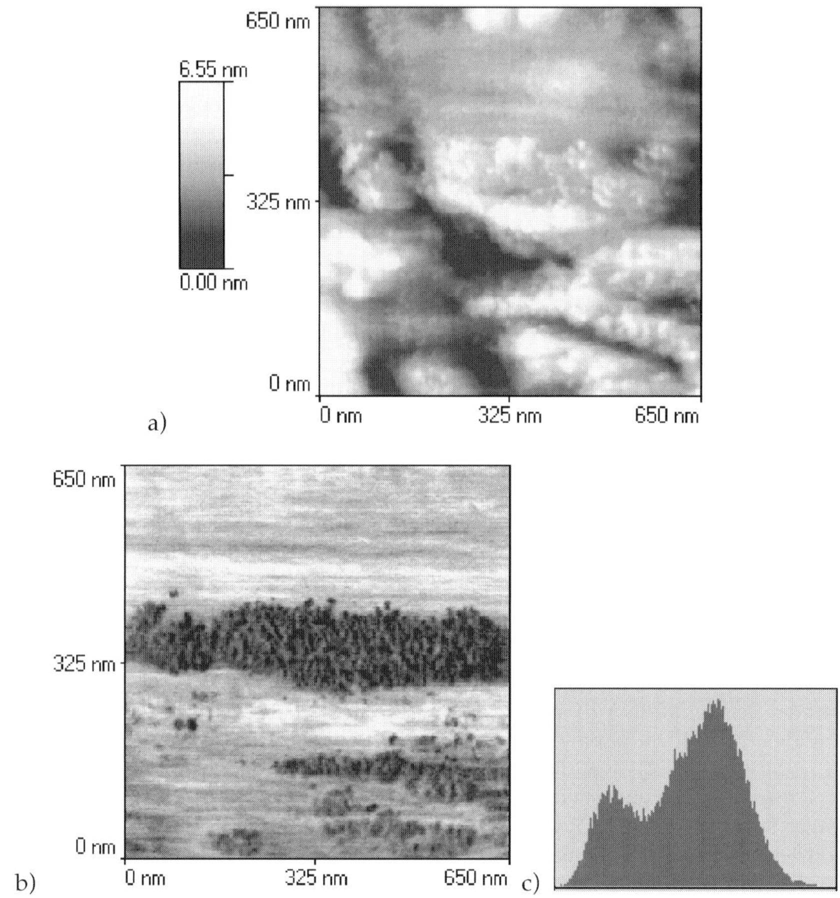

Figure 8.21 An example of an AFM surface topography a), LFM lateral force image b) and corresponding lateral force histogram c) over 650 [nm^2] area covered with boundary films [75].

Magnetic Force Microscopy (MFM) is a variation of AFM, which utilizes magnetic forces for imaging samples with a magnetic structure. The MFM uses a magnetic tip, e.g. the tip is coated with a ferromagnetic film, and operates in a non-contact mode. The changes in the resonant frequency of the cantilever caused by changes in the magnetic field are detected. These changes depend on a tip-to-sample separation. This microscope allows one to obtain information about both the topography and the magnetic properties of the surfaces. This type of measurement is essential in studies of magnetic structure of the surfaces, e.g. in magnetic data storage devices [80].

Scanning Near-Field Optical Microscope

Scanning Near-Field Optical Microscope (SNOM) is a member of the family of scanning probe microscopes. For imaging it uses the visible light as an ordinary optical microscope but has the resolution improved by an order of magnitude compared to a conventional optical microscope, i.e. beyond the diffraction limit [81]. The first idea how to overcome this diffraction limit was proposed about seventy years ago by Synge in 1928 [82]. The diffraction limit is overcome by applying a very small aperture, i.e. about **200-500** [Å] in diameter very close to the sample surface, i.e. at the distance much less than the wavelength of the light used. In this case the image formation depends only on the aperture size and the aperture-to-sample separation. The practical application of Synge's original idea has become possible with the introduction of STM and especially with its probe positioning system allowing for an accurate control of the probe position above the sample surface at the distances in nanometres. The first practical SNOM with spatial resolution of 25 [nm] was introduced in 1984 by Pohl, Denk and Lanz [83]. The advantage of SNOM over AFM is that the traditional contrast methods used in optical microscopy can be applied. This allows for a chemical identification of the films present on the surfaces, which is almost impossible to obtain with AFM. In general, SNOM allows for a simultaneous acquisition of optical contrast and topographic information with almost AFM resolution [81]. The operating principles of SNOM are schematically illustrated in Figure 8.22.

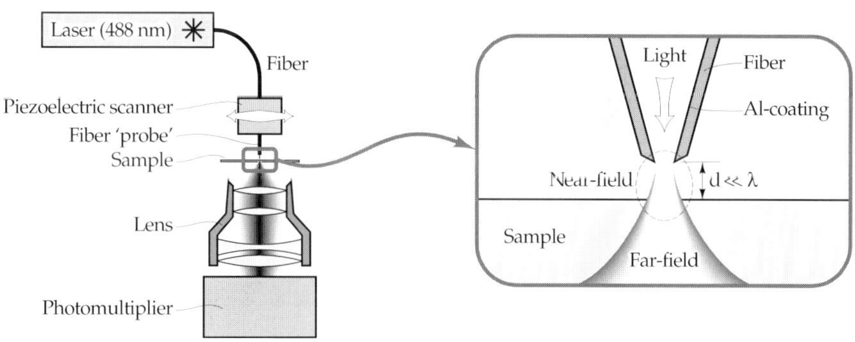

Figure 8.22 Schematic illustration of operating principles of SNOM.

As can be seen from Figure 8.22 laser light is coupled into a near-field probe made of optical fibre with very small aperture of about 50 [nm]. This probe is kept in the near-field distance by a system similar to that used for non-contact AFM (shear-force detection system). The optical near-field that protrudes from the aperture is disturbed by the presence of a sample. Light emitted from the location opposite to the aperture, either in transmitted or reflected modes, is collected by a lens and detected by a photomultiplier. The sample or the probe are scanned and the signals are collected from each point. Simultaneously, topographical (shear-force signal) and optical contrast images can be obtained.

In principle an AFM can be modified to function as a SNOM. An aperture can be achieved by drilling a tiny hole in the tip and shining the light from the above. The requirement that the SNOM has to be at a constant distance from the sample surface is also easily accommodated by a typical AFM. With SNOM, it is also possible to perform optical spectroscopy using a laser emitting different wavelengths. Currently, the developments are being continued into application of fluorescence, cathodoluminescence and Raman spectroscopy with SNOM systems [81].

Sampling Problems With High Resolution Microscopes

The basic problem with increasing magnification in microscopic analysis is the vast quantity of visual data that is produced and hence some means of generalization of the visual data obtained are becoming essential. For instance a 10-fold increase in magnification creates, in theory, a 100-fold decrease in the area of image obtained. The AFM allows to analyse minute surface areas in detail whilst the greater part of the surface remains unobserved. AFM and STM in their present form appear to be most suitable for examining the microscopic detail of minute objects. A basic problem in the microscopy of worn surfaces is that minute detail of a large surface must be observed and analysed. In tribological studies, analysis of large worn surfaces is often required and for such applications AFM and STM are not suitable. A possible solution to combining micro-analysis of large worn surfaces with nano-analysis of minute details would be to use both SEM and AFM to examine the sample. The AFM would be directed at small areas of the worn surface, which are representative of the bulk of the worn surface or else contain critical evidence of wear processes while overviews of larger areas would be conducted using SEM. In other words, it is important to be able to determine whether, for example, the area observed by AFM is located on a surface elevation or in a depression which is sheltered from asperity interaction.

It is evident that for the practical observation of worn surfaces, data collection (i.e. observation) by AFM and STM should be at least partly automated to allow the researcher to analyze the collective properties of innumerable surface features on a typically complex worn surface. The application of SEM stereoscopy and image analysis techniques might help to alleviate this problem. Some basic ideas on image analysis are presented in Chapter 10 and 11 but this aspect of tribological research remains essentially untouched.

The comparison between the range and resolution limits for various imaging techniques is illustrated schematically in Figure 8.23.

8.4 SURFACE ANALYSIS

The presence and nature of surface films or wear debris on a worn surface is a controlling factor of friction and wear. Analysis of the content and distribution of film material or residual debris on a worn surface is an essential part of any tribological investigation. Chemical degradation of the worn surface, e.g. degradation of polymers by the heat and stress of dynamic contact, may also need to be assessed in some instances. The scope and capability of surface analysis has expanded considerably in recent decades and is continuing to develop rapidly. Each surface analysis technique offers a different range of data, e.g. some analytical techniques are suitable for area distribution of elements on a surface while others provide elemental distribution versus depth into the surface. Planning and execution of a surface analysis of worn specimens should be optimised according to the requirements of the research conducted and the capabilities of the available surface analysis technology.

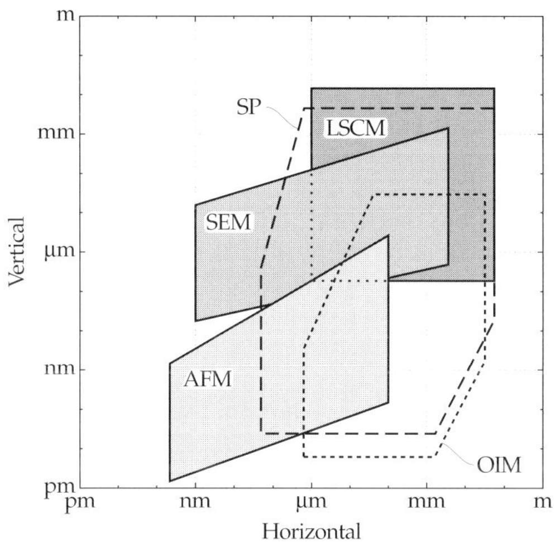

Figure 8.23 Comparison of range and resolution limits for various imaging techniques (SP-Stylus Profilometry, LSCM-Laser Scanning Confocal Microscope, SEM-Scanning Electron Microscope, OIM-Optical Interference Microscope) (adapted from [84]).

Surface Analysis Simultaneous With Scanning Electron Microscopy

The electron beam used to generate an image in Scanning Electron Microscopy (SEM) also causes the emission of X-rays from the specimen. A large part of this radiation is released with a continuous spectrum of photon energy and is known as Bremstralung radiation. However, this radiation is not often used in surface analysis and is not discussed further. In the remainder of the emitted X-ray radiation the wavelengths or photon energy of the X-rays are characteristic of particular elements. In this form of X-ray emission, the photon energy is determined by quantum energy levels of electron jumps within an atom during electron bombardment. Since the quantum energy levels are constant and independent of SEM operating conditions this results in a very useful form of element identification. The relative intensity of the emitted X-rays at various wavelengths depends on the proportions of chemical elements present in the specimen examined. These unique features of X-ray emission during electron bombardment were first applied to surface analysis by Castaing in 1951 [85]. The technique commonly used for analysis by X-ray emission in SEM is known as Energy Dispersive Spectroscopy (EDS). EDS is sometimes referred to as Energy Dispersive Analysis by X-ray (EDAX), which is the trademark of a commercial device. Although there are other types of equipment and technology used to analyse X-ray emission EDS, because of its simplicity to use in conjunction with SEM, is routinely applied in the majority of tribological studies.

EDS is widely used in tribological studies for qualitative analysis of a surface or in other words detection of specific chemical elements present in the surface. The X-ray detector of the EDS system usually contains a semiconductor which emits an electrical signal proportional to incident photon energy. This semiconductor system is much simpler to operate than older X-ray analysis systems based on diffraction where tuning of the detector to each chemical element of interest was required. The diffraction based system, usually referred to as Wavelength Dispersive Spectroscopy (WDS), is used for quantitative metallurgical analysis requiring fine resolution of the X-ray energy. The WDS is slower than EDS because much time is spent tuning the diffractometer and also because only a small fraction (which is centred on the tuned wavelength) of the entire X-ray spectrum is collected at any given stage in the analysis. EDS has the advantage of collecting all or most of the emitted X-rays that reach the X-ray detector and then sorting the X-rays according to energy. The disadvantage of EDS is the poor resolution of X-ray energy so that some of the characteristic X-ray energies emitted by differing elements tend to overlap. A common example of X-ray overlap is sulphur and molybdenum and in this case it is necessary to check for a higher energy X-ray of molybdenum [24]. Also when a low electron voltage is used in SEM, a signal from this higher energy X-ray may be absent or very weak. A schematic diagram of an EDS system is shown in Figure 8.24.

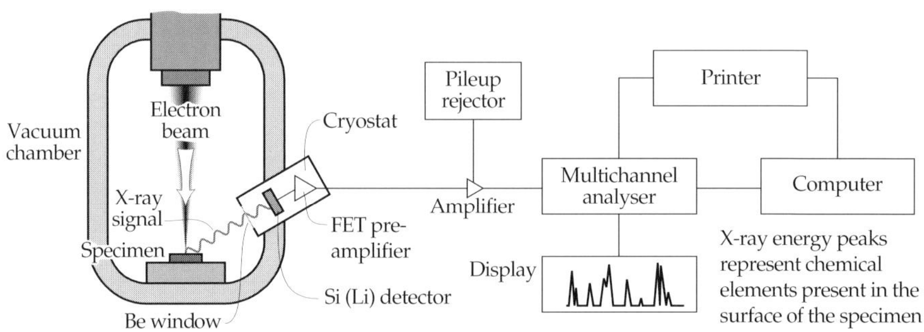

Figure 8.24 Schematic illustration of the operating principles of Energy Dispersive Spectroscopy (adapted from [24]).

EDS can be used in quantitative analysis of surfaces. However, this requires a smooth surface not a rough worn surface and hence the quantitative analysis by EDS usually works the best with metallographical or geological specimens. Chemical elements with atomic number equal to sodium or higher can be detected by EDS without any difficulty. In tribological applications, this means that sulphide or chloride film formation by Extreme Pressure additives can be detected. Also, transfer films between different metallic surfaces, e.g. bronze against steel, can readily be detected from the EDS signal. There are numerous examples of the application of EDS to determine film formation on worn surfaces reported in the literature.

The limitation of EDS analysis to elements with atomic number higher than sodium is often unsatisfactory since basic elements controlling wear processes such as oxygen are not detected during the routine applications of EDS. The only way of detecting oxygen by EDS is the application of the, so called, 'EDS windowless' or 'thin window' mode. In conventional mode the X-ray detector of the EDS system is separated from the specimen by a window made of beryllium which is largely impervious to the low energy X-rays emitted by elements with smaller atomic numbers than sodium. When the observation chamber of the SEM system is brought to a high degree of vacuum by prolonged pumping, the beryllium window can be opened allowing the detection of the weaker X-rays emitted by lighter elements. Elements with atomic number equal to boron or higher can be detected. In some cases, it is necessary to tolerate the inconvenience of windowless EDS in order to determine the role of light elements in wear processes. For example, fretting wear is partly controlled by the accumulation of oxidized debris in the wear scar. In a study of the fretting between steel wires, windowless EDS was used to confirm that within the wear scar, oxygen was present in much higher quantities than outside the wear scar [37].

In some experiments, the composition of very thin films on the surface of the wear scar needs to be determined. For the observation of metallic surfaces, an electron beam voltage of 15 [kV] is typical and this gives an electron penetration into the specimen of about 1 [μm] [24]. If the composition of a layer 100 [nm] or less in thickness is required then a more shallow electron penetration is required to ensure that a sufficient fraction of the emitted X-rays comes from the surface layer and not beneath this layer. In this case, it may be helpful to reduce the electron voltage to, for example, 5 [kV], which reduces electron penetration to approximately 0.1 [μm] in most metals of practical interest [24]. Low voltage EDS was found to be essential to the detection and analysis of thin sulphur films produced by E.P. additives on steel surfaces [86]. The disadvantage of voltage reduction is that high atomic weight elements, e.g. metals such as copper or chromium, may not emit characteristic X-rays at these low voltages so the low voltage technique is suitable only for non-metals or light metals.

In some cases the EDS can be used for quantitative measure of thickness of surface films provided that an external calibration of film thickness is available [24]. For example, in a study of sulphur reaction with steel surfaces, calibration of EDS sulphur data was obtained by gravimetric measurements of sulphur reaction with thin steel foil [86,87]. A calibration constant between the EDS sulphur X-ray signal and the weight of sulphur on the surface was found by EDS analysis of the reacted steel foil.

Another problem associated with the analysis of thin surface films is oxidation and hydration occurring during the interval between testing (i.e. film formation in a wear scar) and surface analysis. In cases of thin films, less than 100 [nm] thickness, film oxidation and hydration pose a serious experimental problem [86].

It is generally accepted that the period of time between testing and surface analysis should be as short as possible but no longer than 24 hours. More subtle experimental artefacts than complete modification of a thin surface film are the changes to a complex wear scar film consisting of thin and thicker surface films. In this case, the thin films which are likely to be found on asperity peaks and are therefore critical to lubricant function [88], tend to be chemically modified while the thicker films remain virtually unchanged. Although the thicker films will receive a similar amount of oxidation and hydration as the thin films the proportional change in film composition is much less. This concept is illustrated in Figure 8.25.

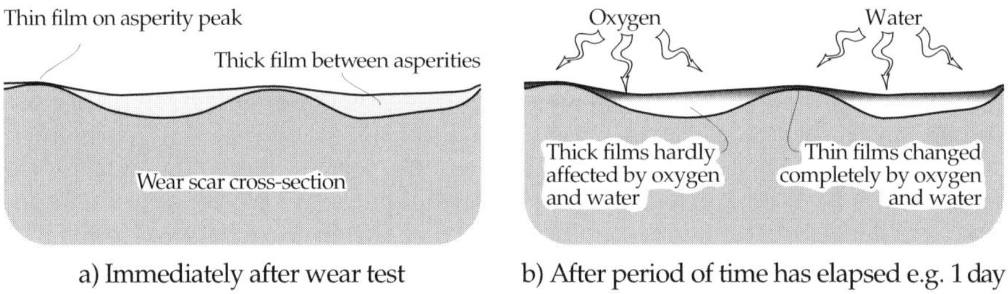

Figure 8.25 Formation of experimental artefact in thin surface films caused by oxidation and hydration after completion of wear test.

When performing surface analysis on wear scars, it is essential to remember that the data collected only relates to the wear scar at the time of analysis. At present, it is still very difficult to monitor and analyse wear scar surface films as they are formed in a wear process. Some changes or even loss of surface material is inevitable before a wear scar can be analyzed even with the most rapid and well organized experimental procedure. However, experiments were recently successfully conducted to directly observe and study wear scar formation, debris formation, transportation and oxidation during fretting between a steel ball and sapphire [89].

Infra-Red Spectroscopy and Raman Spectroscopy

Infra-red (IR) Spectroscopy and Raman Spectroscopy in tribological studies provide much information about the molecular structure of surface films. These techniques can be used to analyze liquid or semi-liquid surface films in the open air which overcomes the limitations of e.g. ESCA (Electron Spectroscopy for Chemical Analysis) which requires a vacuum and is only suitable for solid films. A further advantage of Infra-red Spectroscopy is that it can be used to study lubricant films, i.e. chemical composition and mechanisms involved in their formation, inside a sliding contact [90]. Raman spectroscopy has been found particularly useful in studies of chemical films formed on the surface during wear processes.

· *Infra-Red Spectroscopy*

Infra-red spectroscopy is based on the measurement of changes to infra-red radiation during its transmission through a specimen, e.g. a layer of lubricant. In almost all cases, infra-red radiation interacts with a specimen material by causing the atoms within the specimen to vibrate. In the case of solid specimens, e.g. metallic surfaces, heat is retained within a solid mostly by vibrations of atoms so that the infra-red radiation is in fact heating the specimen. Vibrations of atoms within a molecule or crystal lattice can be accurately modelled by a sphere and spring model, with the exception that the frequencies of vibration are determined by quantum mechanics not by Newtonian mechanics. Atoms can be modelled as rigid spheres while the bonding between atoms is modelled as a spring joining these spheres. This is not a bad analogy since atomic vibration can be reasonably compared to the vibrations of hard spheres joined by springs. When a molecule is exposed to infra-red radiation with a frequency that matches the resonant frequency of a vibration mode in the molecule, enhanced absorption of the infra-red radiation occurs. This process is analogous to the accumulation of mechanical energy in the model spheres and springs which are excited at their resonant frequency. The resonant absorption of infra-red radiation causes a reduction in a transmitted amount of this radiation over a narrow frequency range. When this absorbed radiation is plotted against the transmitted radiation wavelength, then a series of 'absorption peaks' is revealed. The absorption peaks are specific to each type of chemical bond, e.g. carbon-carbon or carbon-hydrogen, and from these bondings the individual molecules can be identified. For example, carbon-hydrogen bonding without any carbon-carbon bonding is indicative of methane. The basic principles of infra-red spectroscopy are illustrated schematically in Figure 8.26.

Infra-red spectroscopy found applications in the analysis of surface films formed during wear. In simple cases these film can be analysed by bouncing the infra-red beam directly off the surface. However, these films must be thin enough to allow for the penetration of the infra-red beam and the surface must be sufficiently smooth to limit the beam scattering, also the spectrum obtained by this technique is weak.

In fact most deposits on wear scars or lubricant films are too thin to cause a detectable alteration of the infra-red radiation so that it was necessary to develop a system where infra-red radiation repeatedly passed through the film in order to accumulate sufficient transmitted distance. A technique which delivers a much stronger signal is the internal reflection spectroscopy (IRS), also known as attenuated total reflectance spectroscopy (ATR) [91]. In ATR, the infra-red beam is multiply reflected through a crystal of a high refractive index, e.g. GeSe, silicon, germanium, etc. In this technique, the sample is placed on the crystal and the infra-red beam interacts with the sample on the surface. Lubricating oils can be spread on the crystal surface and analysed. With this technique it is possible to study the interaction between the lubricant and metallic surface, i.e. formation of surface films. It has been shown that this could be achieved by depositing a thin layer, 70 [Å] thick, of iron [92]. This layer can then be covered with the lubricant and the reaction between the lubricant compounds and the deposited iron film takes place. An infra-red beam penetrates through the iron layer into the reacted layer producing a spectrum. Since the amplitude of the evanescent beam decays exponentially with the distance through the metal, the metallic interface must be very thin [92].

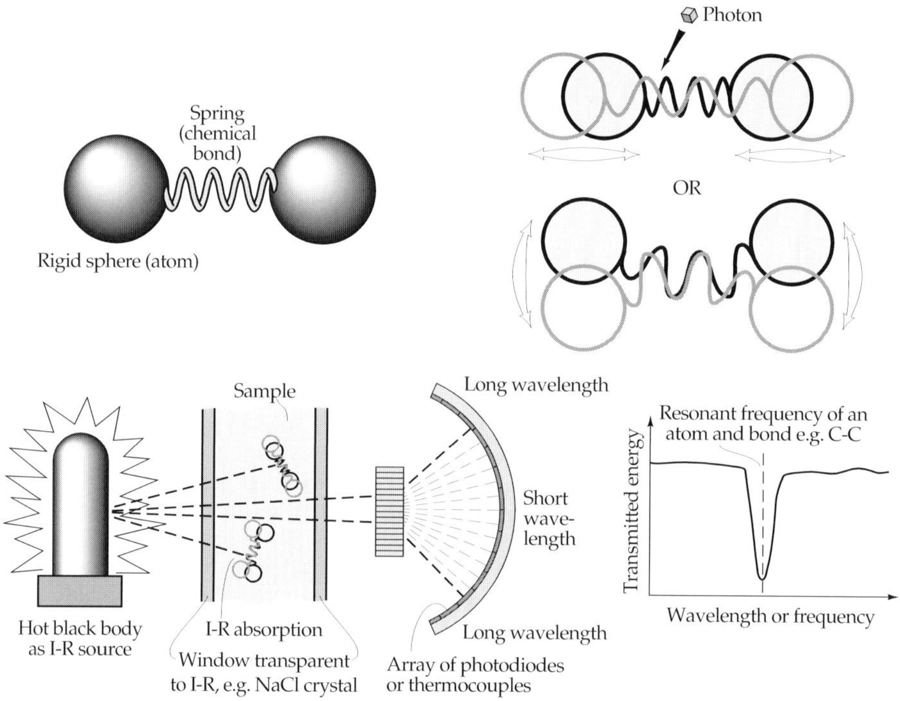

Figure 8.26 Schematic illustration of the operating principles of Infra-Red Spectroscopy.

Another variation of this technique, i.e. external reflection spectroscopy, involves bouncing the infra-red beam between a surface containing the film examined and either an inert backing plate or another sample of the surface investigated. However, this system suffers from the alignment problems since it is necessary to accurately position the two samples. Also, the surfaces of the samples may not be perfectly flat and this creates further problems of alignment. The advantage of this system is that it allows films present on worn surfaces to be studied. The principles of internal and external reflection spectroscopy are illustrated schematically in Figure 8.27 [92].

Infra-red spectroscopy has been used in various tribological studies. For example, it has been shown that fatty acids do not decompose significantly upon adsorption to steel or other metal surfaces [92,93]. A comparison of an infra-red spectrum from worn surfaces and a similar spectrum from unworn surfaces heated to various temperatures revealed that under conditions of moderate wear and friction, adsorption of fatty acid lubricants occurs at relatively low temperatures with the original molecular structure of the fatty acid molecules remaining largely unaffected [93]. The infra-red data obtained also showed that reactions on unworn surfaces are different from reactions on worn surfaces at equivalent temperatures [93]. It should be noted that in the original studies of Bowden and Tabor into lubrication by fatty acids, much of the evidence of adsorption and chemi-adsorption on the surfaces was indirect. Direct confirmation of the structure of molecular films forming on the worn surfaces had to await the development of reflectance infra-red spectroscopy.

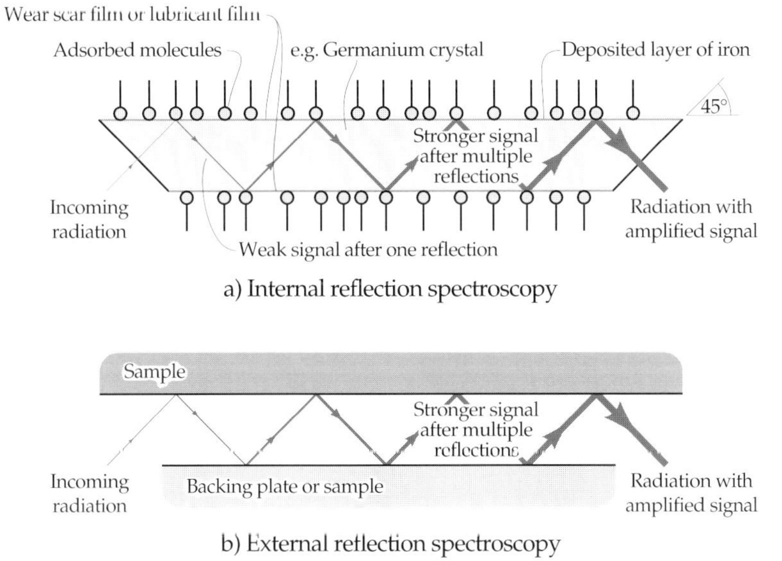

Figure 8.27 Schematic illustration of the principles of the internal (a) and external (b) reflection spectroscopy.

Single reflectance spectroscopy is possible for a surface film as thick as an EHL film [90] which is in the range of 0.1 to 1 [μm]. A specially modified EHL apparatus, where the contact occurs between a steel ball and a transparent window made of diamond or sapphire, was used in direct studies of changes occurring in a lubricating film within a heavily loaded EHL contact. This apparatus also allowed details of the surface film formation and deposition by lubricant additives to be studied. The schematic illustration of this apparatus is shown schematically in Figure 8.28 [90].

The use of infra-red spectroscopy in these studies provided a unique combination of physical and chemical data of the EHL contact. The measured change in frequency of the C-H (carbon - hydrogen) bond from the absorption peak provided an accurate index of lubricant pressure after calibration by an external pressure standard. When a phosphonate additive was added to the lubricant the deposition of an inorganic phosphate film on the sliding surfaces was detected from data based on the P=O-M bond (phosphorus - oxygen - metal). There was also evidence of shear alignment within the contact of molecules generated from soap, oil and polymer viscosity index improver.

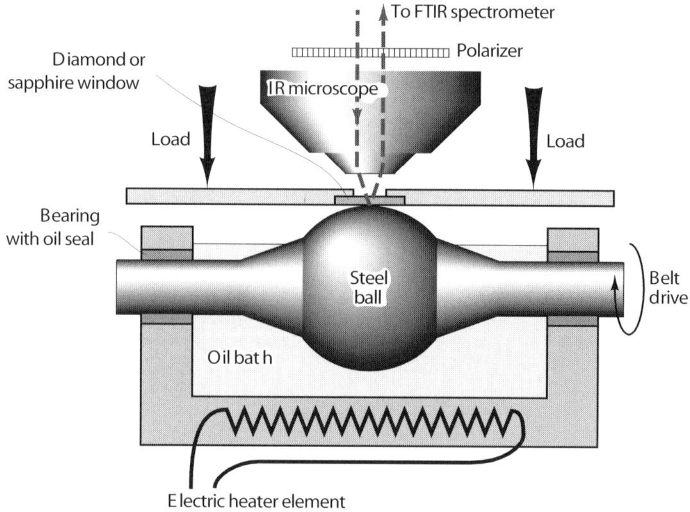

Figure 8.28 Schematic illustration of the apparatus utilizing the direct infra-red spectroscopy in studies of surface film formation in EHL contacts (adapted form [90]).

Infra-red spectroscopy has also been applied in studies of the deposition mechanisms of Polytetrafluoroethylene (PTFE) transfer films on steel surfaces [94]. The strong infra-red absorption of fluorine-carbon bonds allowed the detection of PTFE films as thin as 1 [nm] present on the surface. Molecular alignment with the sliding surface was also detected. Summarizing, the research data reported in the literature indicate that infra-red spectroscopy is a versatile experimental technique with a wide range of applications in tribology.

· *Raman Spectroscopy*

Raman Spectroscopy utilizes the specific properties of scattered light. When a beam of light is directed at a surface some of the light is scattered. Although most of the scattered light remains unchanged, i.e. with unchanged frequency or wavelength, a small proportion of the light is scattered with a different wavelength. In most cases, the wavelength of this altered radiation is longer than that of the original light because of the dissipation of photon energy. The scattered light with unchanged wavelength is known as 'Rayleigh scattering' while the altered radiation is known as 'Raman radiation'. The change in wavelength is caused by partial absorption of photon energy by an atom.

For the vast majority of interactions between atoms and photons, the photon energy is only retained for a very small period of time of about 10^{-12} [s] and the photon is released in its original form. A small fraction of atoms, about 10^{-6}, release only part of the photon energy and a fraction of the original energy corresponding to a quantum energy level change is retained. The retained energy is characteristic of the molecule or atom according to the principles of quantum physics and is not affected by other factors such as incident light intensity or wavelength. The retained energy is later released as photons of lower energy, which are not usually analyzed or else the energy is progressively dissipated. When the spectrum of scattered radiation (light) is collected and analyzed, it shows a very strong signal corresponding to original radiation surrounded by a series of minor peaks originating from the Raman radiation. Analytical data is obtained from a graph or spectrum of the shift frequencies, i.e. the difference in frequencies between the original light and the various Raman radiations emitted from the sample. A schematic illustration of the principles of Raman spectroscopy and the quantum mechanics of scattered light is shown in Figure 8.29.

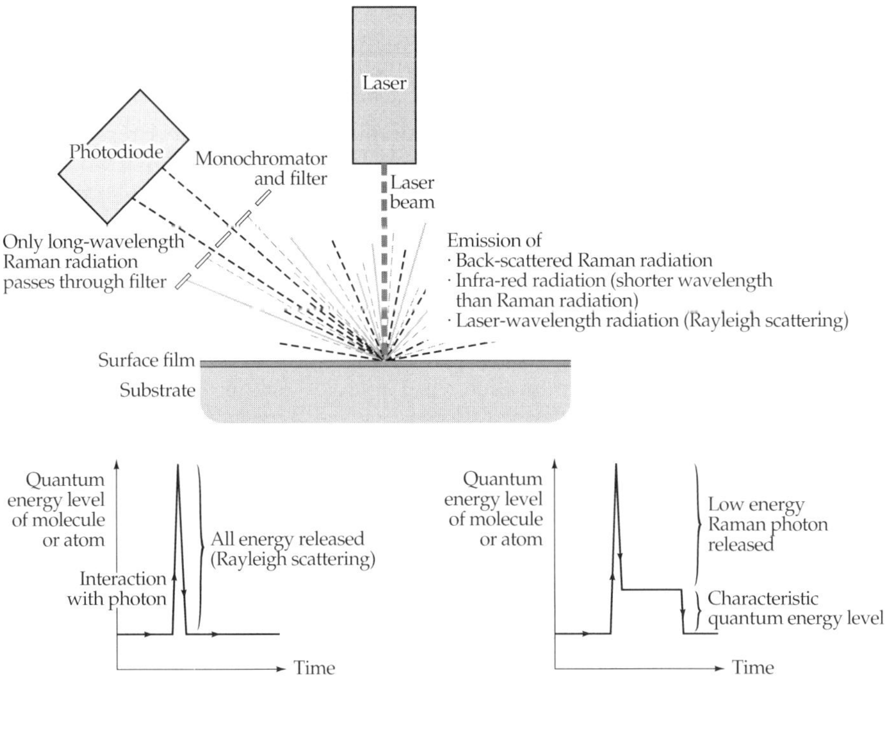

Figure 8.29 Schematic illustration of the principles of Raman spectroscopy and the quantum mechanics of scattered light.

A laser is usually applied in order to provide a sufficiently concentrated source of radiation for a detectable Raman radiation and the data is obtained from sensitive photodiodes placed in the path of the scattered radiation.

Raman spectroscopy is a complementary technique to infra-red spectroscopy since a lower frequency range of the same radiation as infra-red is selected as the data source during the measurements. Raman spectroscopy is particularly sensitive to the surface complexes forming during the adsorption of organic molecules on metal surfaces [95]. Raman spectroscopy provides information about the overall molecular structure of a particular chemical species while infra-red spectroscopy supplies information about chemical bonds which in turn can be used to deduce a molecular formula. When the type of molecule is completely unknown, as is the case for the majority of wear scar films, the process of deduction by infra-red spectroscopy is unlikely to be successful. The low frequency of absorption peaks, which are characteristic of large organic molecules, present additional difficulty to analysis by infra-red spectroscopy [95].

Raman spectroscopy is not as widely used as infra-red spectroscopy in the analysis of lubricating and wear scar films. Raman spectroscopy is suitable for the analysis of thin surface films. However, fluorescence occurring under laser irradiation could be a problem and also very rough worn surfaces may present an additional difficulty. In Raman spectroscopy it is required that worn surfaces are free of dust as this could cause extraneous light scattering called the Tyndall effect.

Raman spectroscopy has been applied in fundamental tribological research. For example, Raman spectroscopy has been used in a study of the reaction stages between a fatty acid and oxidized metal surface [95]. The details of adsorption of the carboxyl group of the fatty acid onto the oxidized metal surface were analyzed by this technique. Models of lubrication by fatty acids which were formulated many years earlier without conclusive experimental evidence, were confirmed by surface analysis using Raman spectroscopy. Raman spectroscopy can also be used in the analysis of chemical changes occurring within a lubricant. For example, the oxidation of samples of ZnDDP dissolved in cyclohexane was observed by placing the ZnDDP solution in capillary tubes and monitoring the degradation of the molecular structure of ZnDDP from the change in Raman shift peak intensities [96].

Surface Analysis With Emitted Electrons, Short Wavelength Radiation or Low Energy Ion Beams

In tribological studies, the properties or composition of the 'interacting' surface are usually the focus of interest. While the analyses provided by EDS and EPMA (Electron Probe Micro-Analysis) reveal much information about the content of material within 0.1 to 1 [μm] of the surface, these techniques are unsuitable to assess the material characteristics in the uppermost layer of the thickness between 1 and 10 [nm] beneath the surface. Very often, it is important to know either the composition of a film or a surface layer or the variation in material composition with depth from 1 to 100 [nm] below the surface, with a spatial resolution of a few nanometres. For example, this information is important in studies of the composition and structure of films formed on metal surfaces by lubricant additives [97].

Most commonly used near surface analysis techniques in tribological studies include:

- Auger Electron Spectroscopy (AES) which depends on bombarding the surface by an electron beam, similar to that used in SEM, to cause an emission of Auger electrons from the top surface layers,
- X-Ray Photoelectron Spectroscopy (XPS) which depends on irradiating the surface with X-rays and utilizing the resulting photoelectron emission from the surface as a data source,
- Secondary Ion Mass Spectroscopy (SIMS), where a stream of impinging ions dislodge atoms or ions of the surface material. The ejected particles are then analyzed by an adjacent detector.

All three systems, although related in technology and purpose, are quite distinct and are described in separate sections. The techniques of AES and XPS are well developed in the engineering sense and therefore are the most widely used for surface analysis in tribological investigations. More information on AES, XPS and SIMS, with particular relevance to tribology, can be found in [9]. There are many other types of surface analysis technologies in use or under development, but those that are listed here have gained acceptance as analytical tools for tribological studies. In future, other techniques may become superior so that it is not implied here that only the techniques described below are appropriate in tribological studies.

· *Auger Electron Spectroscopy*

Auger Electron Spectroscopy (AES) was developed, in a form suitable for regular use as an analytical tool, in the 1960s [98]. Surface analysis data is acquired from the energy of Auger electrons released from the surface under electron bombardment. An Auger electron is an electron released from an atom after the electron bombardment of the atom has already dislodged one electron from it. Since the energy of the Auger electron is specific to each chemical element, the AES provides information about the elemental composition of a surface under examination. In a typical Auger spectrometer a focused electron beam is applied to provide data from a small volume of material usually **1** to **5** [nm] in depth [30], the slenderness of the electron beam renders this technique suitable for a localised analysis of worn surfaces. The technique allows for the investigation of the variation in surface film composition, e.g. the difference between films generated at the bottoms of wear scar grooves or at the tops of grooves or on plateaux.

A further advantage of AES is that a stream of argon ions can be used to slowly 'drill', i.e. erode, a hole into the surface in order to reveal changes in the film composition versus depth. The electron beam and the argon ion stream are applied simultaneously to the surface and the size of the hole drilled by the argon ions is much larger than the cross section of the electron beam. Often several electron beams are directed together into the drilled hole for more rapid data acquisition. The rate of hole deepening is extremely slow, about one or two hours is required to reach 100 [nm] below the original surface so that the Auger spectrometer is most suitable for the analysis of film composition in a depth range from **1** to **100** [nm]. Calibration of depth versus time requires special measurements so that in most studies only an approximate measure of depth is obtained.

A commonly used practice involves a deposition of a film of known thickness on a substrate and measurement of the time necessary for the hole eroded by argon ions to reach the substrate. Upon reaching the substrate the concentration of elements recorded by Auger spectrometer abruptly changes providing a clear indication of film penetration. The rate of hole deepening is then found from the quotient of film thickness divided by drilling time. Since the wear scar films usually have unknown composition it is necessary to assume that the rate of drilling by argon ions of the calibration film is similar to the drilling rate of the wear scar film. AES provides information about the elemental concentration with very little data on chemical bonds. All elements except hydrogen and helium can release Auger electrons and be identified. Hydrogen and helium cannot release Auger electrons because they only have one electron shell. The schematic illustration of the origin of Auger electrons and the operating principles of Auger electron spectroscopy and its application to the analysis of a surface film is illustrated schematically in Figures 8.30 and 8.31 respectively.

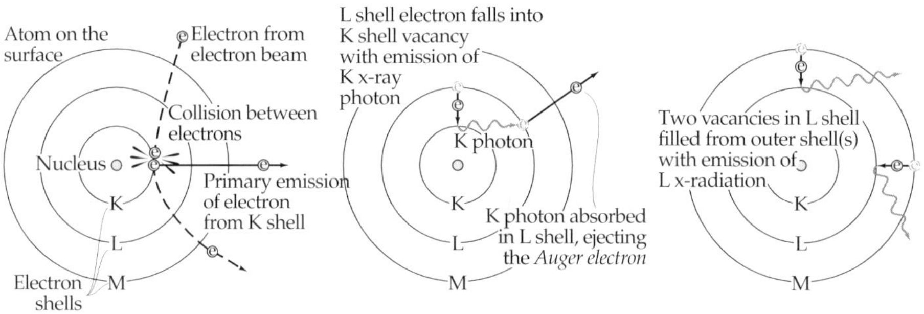

Figure 8.30 Schematic illustration of the origin of Auger electrons.

Auger electron detectors are sometimes fitted to SEM as an accessory but this practice is not very popular as Auger spectroscopy requires a higher quality of vacuum and more intense electron source than is usually used in the SEM. In most instances, Auger spectroscopy is installed as a separate apparatus.

· *X-Ray Photoelectron Spectroscopy*

X-ray photoelectron spectroscopy (XPS) was developed in the 1960s mostly by the efforts of Siegbahn et al. [99,100]. A surface irradiated by X-rays releases photoelectrons that are characteristic of both the element each electron was emitted from and of the bonding between that atom and neighbouring atoms. Photoelectrons emitted from the uppermost 1 to 5 [nm] of the surface are not modified by collision with intervening atoms and are suitable for analysis. The energy of photoelectrons released from deeper below the surface is severely diminished by atomic collision. These photoelectrons either do not reach the detector or else have such a distinctively lower energy that they can be easily discriminated and discounted. The XPS technique provides information on both elemental composition and chemical bonding which is an advantage. For example, it allows one to detect whether sulphur on a surface is present as either elemental sulphur, sulphide or sulphate [97]. The operating principles of XPS are illustrated schematically in Figure 8.32.

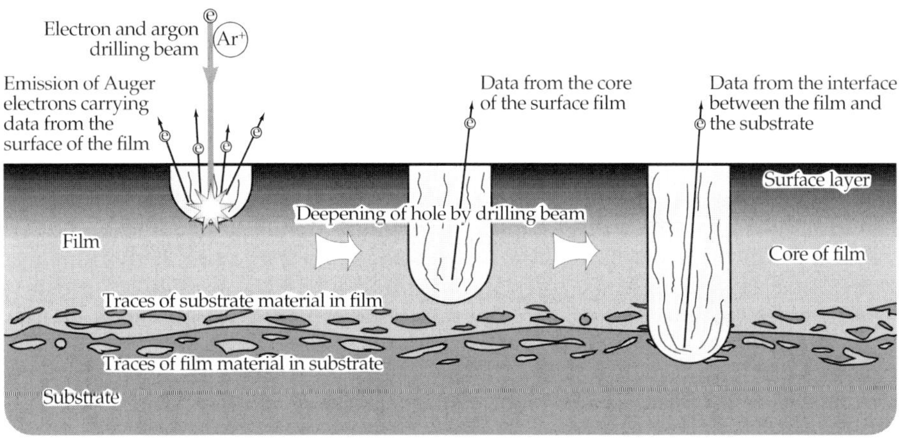

Figure 8.31 Schematic illustration of operating principles of Auger electron spectroscopy and its application to the analysis of a surface film structure.

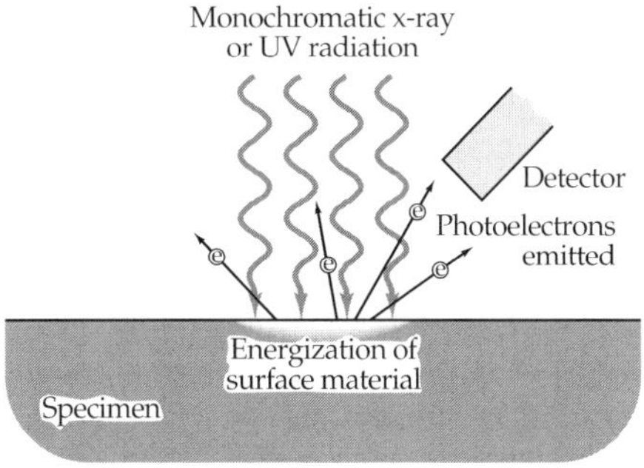

Figure 8.32 Operating principles of X-ray photoelectron spectroscopy (XPS).

The experimental technique, XPS, is also often referred to by the corresponding surface analysis technology which is known as Electron Spectrometer for Chemical Analysis (ESCA). XPS provides a most suitable tool for the investigation of thin films (reacted surface layers) generated during wear processes. For example, severity of adhesive wear is largely determined by the presence and nature of surface films **1** to **5** [nm] thick so that XPS is a most suitable tool for the investigation of this form of wear. Although XPS is mostly used for the qualitative detection of thin surface films, quantitative analysis of wear scar films by XPS is also possible [97].

· *Examples of Surface Analysis Using XPS and AES*

A widely used test to determine the suitability of lubricating oils for metalworking purposes is the tapping test. The test involves tapping holes with different taps in a flat bar of 1020 steel or 304 stainless steel [101]. In one study, the tapping test was used to evaluate over-based metal sulphonates as additives for metalworking oils intended for the machining of stainless steel [101]. Tapping efficiency was measured and a combination of XPS and Auger electron analysis was used to investigate the chemical composition of films and surface deposits on the tapped surfaces which revealed signs of severe wear. XPS analysis enabled the detection of iron sulphide, ferric oxide, chromic oxide and calcium carbonate on the tapped surfaces of stainless steel specimens. The structure of the film was further investigated by Auger electron spectroscopy and it was found that instead of a nearly homogeneous mixture of the various chemical compounds already listed, a highly structured film formed. An example of the data obtained is shown in Figure 8.33 [101].

It can be seen from Figure 8.33 that as the lubricant additives are changed from calcium sulphonate to sulphurized olefin and then to a combination of the two additives, the composition of elements on the surface changes accordingly. In all cases, the subsurface iron signal declines to a low level at the surface where there is a strong signal from either sulphur, calcium or a combination of both depending on the type of additives used. Oxygen, presumably atmospheric, is also present in the surface films. Chromium remains in comparatively high concentrations at the surface unlike iron, so in relative terms, there is a chromium enrichment occurring at the surface. The significance of this compositional change in terms of wear and friction is still unknown and awaits further research. The subsequent structure of the film on the worn surface as deduced from the Auger data is shown in Figure 8.34.

It can be seen from Figure 8.34 that the composition of the film changes with the chemical composition of the additive present in the lubricating oil. For the calcium sulphonate enriched oil, a layer of calcium carbonate forms above a complex layer containing iron sulphide, iron sulphate, ferric and chromic oxides. When the sulphurized olefin is substituted for calcium sulphonate, the calcium carbonate layer disappears and is replaced by a layer of iron sulphide and ferric oxide. A layer of iron sulphide mixed with chromic and ferric oxide forms beneath this outer layer. When both the calcium sulphonate and sulphurized olefin are present then a more complex film structure is found which consists of a basal oxide layer superimposed by a mixed iron sulphide, ferric and chromic oxides layer. A superficial calcium carbonate layer is also present on the outer surface. However, in this work the relationship between film structure and wear was not determined. It was speculated that the increased film thickness and presence of a basal oxide layer improved wear properties when both additives were used.

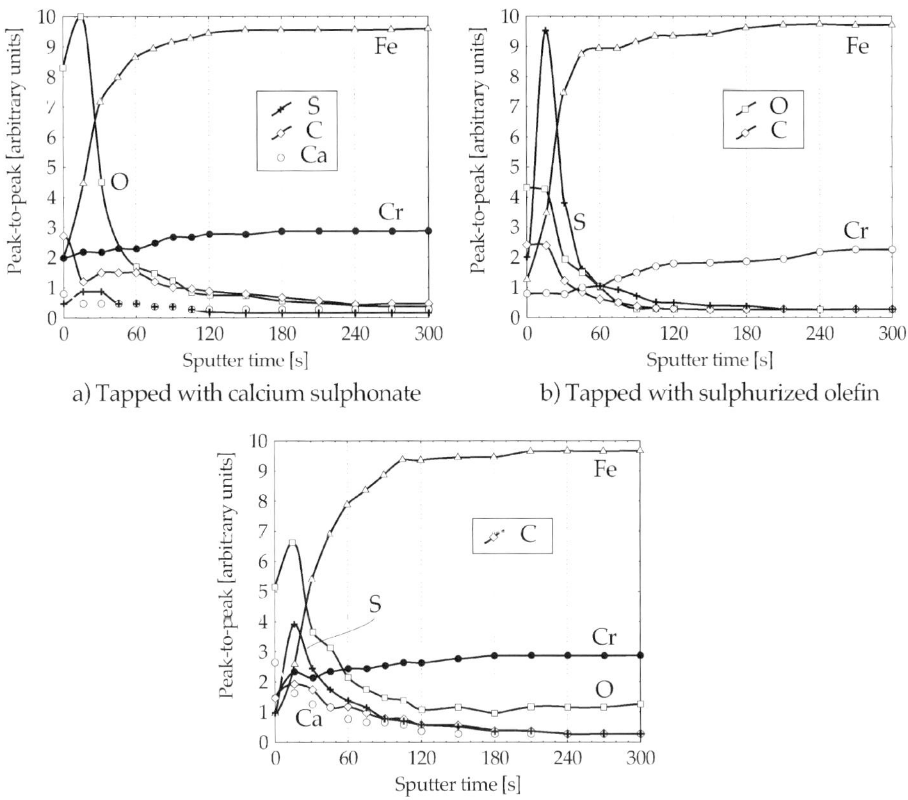

c) Tapped with a combination of calcium sulphonate and sulphurized olefin

Figure 8.33 Example of Auger data used to determine the structure of films formed on worn surfaces. The data is relative signal intensity of sulphur, iron, chromium, oxygen, carbon and calcium from a worn stainless steel specimen after tapping test (adapted from [101]).

In another study a tribometer was fitted into Auger/XPS system to study adsorption effects on friction. Oxygen is a very reactive gas, which rapidly chemisorbs on any material capable of producing oxide films. Metal surfaces, when exposed to air, are always covered by the mixture of oxide films and the contaminant layers, i.e. water, hydrocarbon, etc. In fact any solid surface which has been in contact with air is covered with adventitious carbon which can be detected by XPS. Because of this carbon contamination, our knowledge about the tribological properties of 'native' oxide films, i.e. when isolated from the contaminant layers, is very limited. Tests conducted under high vacuum revealed that nanometre thick 'native' oxide films forming on steel surfaces exhibited exceptional wear resistance and high friction [102]. Even in high vacuum there is always a small amount of molecular oxygen present. It was found that these 'native' films tend to fail if more oxygen is introduced by increasing the oxygen partial pressure [102].

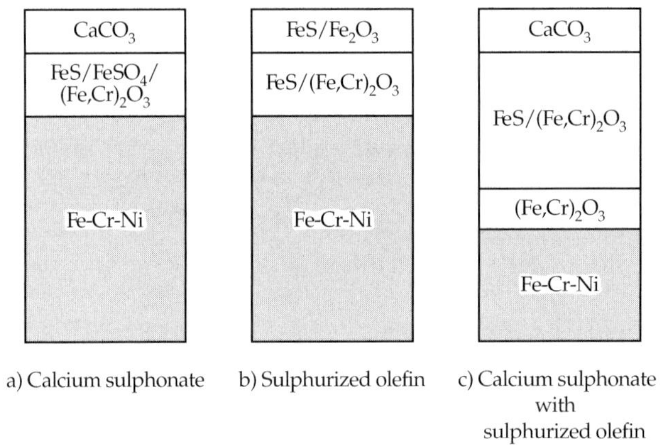

Figure 8.34 Deduced structure of films formed on tapped surfaces by calcium sulphonate and sulphurized olefin enriched lubricating oils [101].

In common with other forms of surface analysis, the data from Auger and XPS is subject to experimental artefact. For example, detection of oxygen and carbon close to the outer surface of the wear scar film may be due to contamination from greasy fingers or by oxidation on exposure to the atmosphere.

The strength of tribological research based on surface analysis is that the effectiveness of additives in terms of changes occurring on worn surfaces can be measured accurately. The limitation of the data obtained is that it reveals nothing about the causal relationship between deposits on a worn surface and wear or friction characteristics. Almost all tribological studies involving surface analysis invoke an association between the formation of films and a reduction in friction or wear. The mechanisms of how a film controls friction or wear remain largely uninvestigated. Maybe the increasing use of Atomic Force Microscopes (AFM) will remove this limitation.

· *Surface Analysis by Ion Streams*

The information or data that can be obtained from any specimen surface is always dependent on the type of surface analysis technique used. For example, if the impinging electron beam in EDS is substituted by an ion stream then a correspondingly different picture of the surface results. Secondary Ion Mass Spectroscopy (SIMS) is perhaps the most widely used analytical technique that is based on the principle of an ion stream impinging on the surface investigated [30]. Impinging ions dislodge atoms or ions from the surface examined, which are then picked up and analyzed by the detectors.

A related but different technique is called Ion Microprobe Analysis (IMA). A gas such as oxygen, nitrogen or argon is used as the source of ions, which may be either positively or negatively charged. The impinging ions cause ions from the specimen surface to be emitted. Analytical data of the specimen surface is obtained by ensuring that a detector or spectrometer is located to collect the emitted ions. A schematic diagram of the operating principles of SIMS and IMA is shown in Figure 8.35.

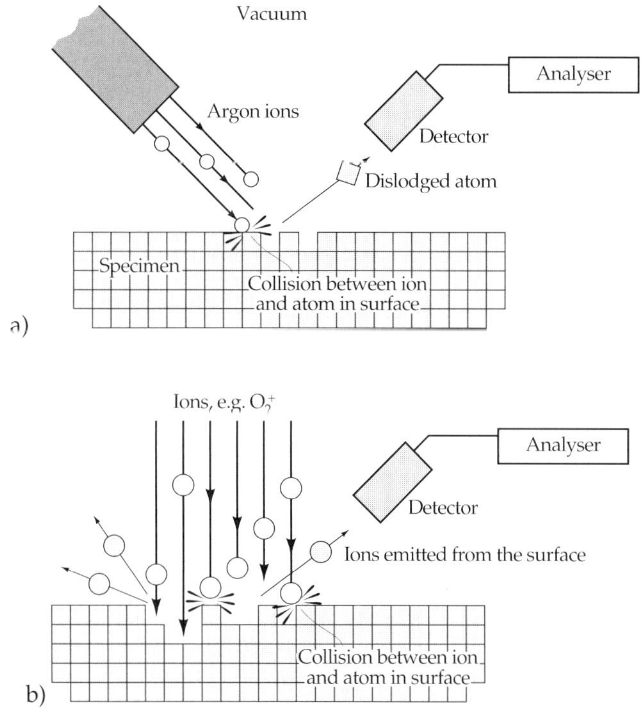

Figure 8.35 Schematic diagram of operating principles of; a) Secondary Ion Mass Spectroscopy (SIMS) and b) Ion Microprobe Analysis (IMA).

Both IMA and SIMS allow the detection of any chemical element without the limitations of EDS. SIMS is more sensitive to trace elements than EDS and concentrations of elements as low as **1** in **1,000,000** can be detected compared to **1** in **1000** for EDS [30]. SIMS utilizes a diffused ion beam, which covers an area more than 1 [mm] in diameter (or minimum dimension), to obtain data about the surface atomic layers of a specimen. On the other hand, an IMA uses an ion beam which is focused on a much smaller area of surface, i.e. with a size ranging from **3** to **400** [μm] [30]. Surface penetration by the ion beam with IMA, typically **1** [μm], is much deeper than with SIMS. This means that many thin surface films found in wear scars cannot be analyzed by IMA without allowing for subsurface interference. The concentrated ion beam of IMA creates a progressively deepening hole in the specimen allowing for a depth profiling of surface films in a way similar to the Auger technique. This problem is avoided with SIMS. However, the disadvantage of SIMS is that although it provides good data about very thin surface films, there is also a large area averaging effect. Film composition in a typical wear scar varies significantly over even a few micrometers so that data based on an area of more than 1 [mm] in diameter will not reveal the inherent variability in wear scar film composition, which is so important from the tribological view point.

Despite its shortcomings SIMS has found wide applications in tribological studies [9]. For example, in one study SIMS was used to analyze wear scar surface films of steel implanted with oxygen and aluminium ions [103]. SIMS was used in this instance to detect oxygen in a concentration range with a maximum of $10^{29}/m^3$ and a minimum of $10^{25}/m^3$ which is a relative variation of 10,000. This large range in detectable oxygen concentration cannot be obtained by EDS which can only detect a concentration variation of 100 before background random variation in X-ray signal corrupts any weak signal from a characteristic X-ray. This shows that the application of SIMS can provide unequivocal evidence of whether oxygen, for example, is present or effectively absent in a wear scar as opposed to presence in quantities below or above the limits of detection.

Other Methods of Surface Analysis

Most of the techniques discussed so far are unsuitable for the investigation of the crystalline state of the surface material in a wear scar. X-ray diffraction is the technique commonly used in materials research to analyze the crystalline structure of materials. Although this technique is suitable for the analysis of powders including wear debris and thin sections of materials, its applications to the analysis of wear scar surface material are extremely limited. Wear scar material, in most cases, exhibits a very complex structure and pronounced surface topography, which often prevents an accurate data from being collected. A technique, which allows for the analysis of structures of the solids is Extended X-ray Absorption Fine Structure (EXAFS). EXAFS is based on the principle of irradiating the sample material by short wavelength X-rays and measuring the energy spectrum of transmitted X-rays. The energy of the X-rays used is typically about 1 - 2 [GeV] so that a synchrotron is required as the source of X-rays. Analytical data is obtained from changes in X-ray energy when the X-ray photon energy of the impinging radiation is similar to the characteristic X-ray emitted from the sample under examination. A phenomenon known as the 'absorption edge' occurs at this point, where imposed X-ray radiation induces a resonant emission of characteristic X-rays. This effect is analogous to shaking a mechanical structure at a frequency close to its resonant frequency. In addition to the primary

X-ray response of the absorption edge there are secondary emissions at higher X-ray energies. These higher energy emissions are the source of data for EXAFS. The operating principles of the EXAFS system are shown in Figure 8.36.

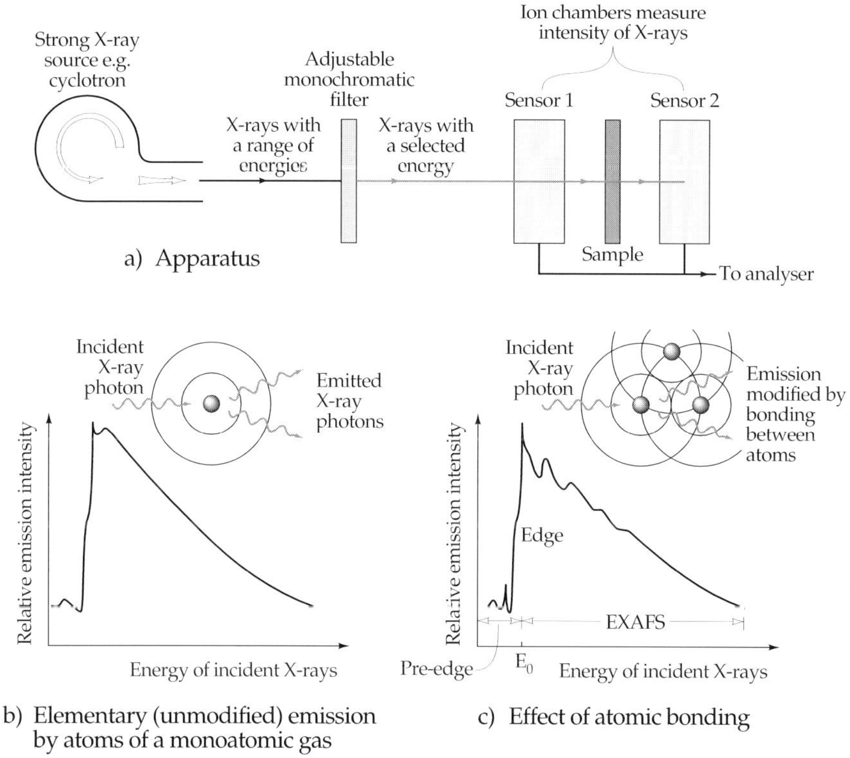

Figure 8.36 Operating principles of Extended X-ray Absorption Fine Structure (EXAFS) (adapted from [104]).

Secondary scattering of X-rays between adjacent atoms in a solid generate a superimposed variation in energy of the X-rays after passing through the specimen. Plotting the X-ray energy versus wavelength reveals various peaks. The peaks in X-ray energy spectrum beyond the absorption edge are sensitive to the distances between atoms within a crystalline structure or amorphous entity. This allows EXAFS to provide data about the solid state structure of a specimen. The X-ray spectrum is subjected to Fourier transformation and other analytical techniques, which are too specialised to describe here, before measurements of interatomic distance are obtained. When analysed by EXAFS, amorphous solids reveal one characteristic interatomic distance while crystalline solids show a series of interatomic distances [104].

EXAFS has been applied in the analysis of surface film formation in a sliding contact lubricated by Zinc Dialkyl Dithiophosphate [104]. Analysis of wear debris, collected from sliding steel surfaces lubricated by Zinc Dialkyl Dithiophosphate (ZnDDP), by EXAFS confirmed that ZnDDP reacts with a steel surface to form an amorphous surface film. An amorphous surface film lacks grain boundaries for rapid diffusion of reacting elements which in this case are zinc, oxygen, sulphur and iron. The amorphous film would therefore not grow as rapidly (by chemical reaction of zinc etc. to form more film material) which would limit corrosive wear. One of the drawbacks of EXAFS is that this is a highly specialised technique requiring access to a synchrotron.

It should be mentioned that in most cases, evidence of either amorphous solid state or ultra-fine crystalline structure can be obtained directly from Transmission Electron Microscope (TEM) analysis.

To provide information of surface films on wear scars, radio-active energy has also been used in some applications. Radio-isotope tracers provide an extremely sensitive means of detecting deposits of specific elements on a surface. For example, in order to detect sulphur deposited in surface films on a wear scar after lubrication by Extreme Pressure additives (E.P. additives), the sulphur isotope S_{35} can be used [105]. A basic difficulty however with the use of radio-isotope tracers are safety considerations in the laboratory during their use. Radio-isotope tracers can only provide a measure of the quantity of a particular element but unfortunately they do not provide data on the chemical state of an element. For example, the radio-isotope data on sulphur deposited in surface films on a wear scar after lubrication by E.P. additives does not indicate whether the sulphur is physically trapped on the surface as small granules or whether it has reacted to form a metal sulphide film. The limitations of data and severe safety problems have greatly curbed the use of radio-isotope tracers for tribological studies.

Surface films can also be detected by ellipsometry, which employs the measurement of the change in polarization state of light after reflection from a surface under examination. The refractive index and light absorption by any film on a surface can be determined by ellipsometry. Comparing the values of the refractive index and light absorption with external measurement of known material gives a simple form of surface film detection. However, the information available from ellipsometry is limited and has only been applied rarely in tribology [e.g. 94].

8.5 SUMMARY

An ever widening range of microscopy and surface analysis instruments and techniques are available for tribological studies. Some of the techniques offer convenience of use for the rapid acquisition of data while others are highly experimental and intended to provide fundamental information about worn surfaces. The most commonly used microscopy instrument in tribological studies is the Scanning Electron Microscope. The SEM can greatly be supplemented by the Atomic Force Microscope, which provides much more detailed information especially at higher magnification. In addition the AFM allows local variations in surface friction to be measured, which can arise from surface material inhomogenity, or reveal the chemical differences between the regions of similar morphology almost down to atomic scale.

Surface analysis ranges from a simple assessment of the concentration of elements present on a surface to the detailed analysis of molecular structure of film material. Three-dimensional surface analysis is possible with mapping of the distribution of elements with respect to the position on the worn surface and depth beneath the surface.

The rapid advances in microscopy and surface analysis present almost a surfeit of information available to the researcher. However, there still remains a lack of knowledge of how surface characteristics, e.g. the type of surface film present, control friction and wear characteristics.

REFERENCES

1 K. Hokkirigawa and K. Kato, An experimental and theoretical investigation of ploughing, cutting and wedge formation during abrasive wear, *Tribology International*, Vol. 21, 1988, pp. 151-157.

2 M. Chandrasekaran, A.W. Batchelor and N.L. Loh, Direct observation of frictional seizure of mild steel sliding on aluminium by X-ray imaging, Part I - Methods, *Journal of Materials Science*, Vol. 35, 2000, pp. 1589-1596.

3 P. Cong, J. Imai and S. Mori, Effect of gas pressure on tribological properties and tribochemical reactions of alumina sliding against zirconia in HFC-134a, *Wear*, Vol. 249, 2001, pp. 143-149.

4 M. Godet, The third body approach: A mechanical view of wear, *Wear*, Vol. 100, 1984, pp. 437-452.

5 M. Eriksson, J. Lord and S. Jacobson, Wear and contact conditions of brake pads: dynamical in situ studies of pad on glass, *Wear*, Vol. 249, 2001, pp. 272-278.

6 T.F.J. Quinn, Physical Analysis for Tribology, Cambridge University Press, 1991.

7 L.K. Ives, Electron microscopy, ASM Handbook, Friction, Lubrication and Wear Technology, Vol. 18, ASM International 1992, pp. 376-392.

8 Characterization of Tribological Materials, ed. W.A. Glaeser, Butterworth Heinemann, Boston, 1993.

9 Surface Diagnostics in Tribology: Fundamental Principles and Applications, eds. K. Miyoshi and Y.W. Chung, World Scientific, New Jersey, 1993.

10 P.J. Goodhew, J. Humphreys and R. Beanland, Electron Microscopy and Analysis, 3rd edition, Taylor & Francis, London 2001.

11 W.A. Glaeser, Light microscopy, ASM Handbook, Friction, Lubrication and Wear Technology, Vol. 18, ASM International 1992, pp. 370-375.

12 T.A. Harris, Friction and wear of rolling-element bearings, ASM Handbook, Friction, Lubrication and Wear Technology, Vol. 18, ASM International, 1992, pp. 499-514.

13 Lubrication Fundamentals, Gear Failure Atlas, *Lubrication Engineering*, Vol. 56(11), 2000, pp. 17-24.

14 Lubrication Fundamentals, Gear Failure Atlas: Part II, *Lubrication Engineering*, Vol. 57(2), 2000, pp. 38.

15 M.A. Schmidt and R.D. Compton, Confocal microscopy, ASM Handbook, Vol. 18, Friction, Wear and Lubrication, ASM International, 1992, pp. 357-361.

16 D.N. Hanlon, I. Todd, E. Peekstok, W.M. Rainforth and S. van der Zwaag, The application of laser scanning confocal microscopy to tribological research, *Wear*, Vol. 251, 2001, pp. 1159-1168.

17 T.K. Kirk, Department of Mechanical and Materials Engineering, University of Western Australia, Private communication, 2002.

18 G.F. Vander Voort, Metallographic Principles and Practices, McGraw-Hill, 1984.

19 ASM Handbook, Vol. 9, Metallography and Microstructures, ASM International 1992.

20 A.A. Torance, The metallography of worn surfaces and some theories of wear, *Wear*, Vol. 50, 1978, pp. 169-182.

21 P.J. Blau, Measurement and interpretations of sliding wear damage in metals, *J. Tribol.*, Vol. 107, 1985, pp. 483-490.

22 D.A. Rigney and J.E. Hammerberg, Unlubricated sliding behavior of metals, *MRS Bulletin*, Vol. 23, 1998, pp. 32-36.

23 L. Xu and D. StJohn, White layer formation on back milled surfaces and its effect on wear performance during three-body abrasive wear tests, Proc. Intern. Tribology Conf. Austrib'94, editor: G.W Stachowiak, 5-8 December 1994, Perth, Australia, pp. 205-210.

24 J.I. Goldstein, D.E. Newbury, P. Echlin, D.C. Joy, C. Fiori and E. Lifshin, Scanning Electron Microscopy and X-ray Microanalysis, Plenum Press, New York, 1981.

25 M. Von Ardenne, Das Elektronen-Rastermikroskop, Theoretische Grundlagen *Z. Phys.*, Vol. 109, 1938, pp. 553-572.

26 M. Von Ardenne, Das Elektronen-Rastermikroskop, Praktische Ausführung, *Z. Techn. Phys.*, Vol. 19, 1938, pp. 407-416.

27 V.K. Zworykin, J. Hillier and R.L. Snyder, A scanning electron microscope, *ASTM Bulletin*, Vol. 117, 1942, pp. 15-23.

28 C.W. Oatley, W.C. Nixon and R.F.W. Pease, Scanning electron microscopy, *Adv. Electronics Electron Physics*, Vol. 21, 1965, pp. 181-247.

29 A.D.G. Stewart and M.A. Snelling, A new scanning electron microscope, *Electron Microscopy*, Vol. 1, pp. 55-56, Prague: Czechoslovak Academy of Science, 1964.

30 I.M. Watt, The Principles and Practice of Electron Microscopy, 1985, Cambridge University Press, Cambridge, United Kingdom.

31 G.W. Stachowiak and A.W. Batchelor, Wear of roping wire and sheave under varying conditions, *Wire Industry*, Vol. 58, July 1991, pp. 380-383.

32 G.D. Danilatos, Foundations of environmental scanning electron microscopy, *Adv Electronics Electron Phys*, Vol. 17, 1988, pp. 109-250.

33 G.D. Danilatos, Introduction to the ESEM instrument, *Microsc. Res. Tech.*, Vol. 25, 1993, pp. 354-361.

34 S.L. Graindorge and G.W. Stachowiak, Changes occurring in the surface morphology of articular cartilage during wear, *Wear*, Vol. 241, 2000, pp. 143-151.

35 B.J. Griffin, A. van Riessen and L.M. Egerton-Warburton, A review of detection strategies and imaging of hydrated biological specimens in the environmental SEM, *Scanning*, Vol. 17, 1995, pp. V58-59.

36 G.W. Stachowiak and G.B. Stachowiak, Unlubricated friction and wear behaviour of toughened zirconia ceramics, *Wear*, Vol. 132, 1989, pp. 151-171.

37 A.W. Batchelor, G.W. Stachowiak, G.B. Stachowiak, P.W. Leech and O. Reinhold, Control of fretting friction and wear of roping wire by laser surface alloying and physical vapour deposition coatings, *Wear*, Vol. 152, 1992, pp. 127-150.

38 S.L. Graindorge, Wear of Articular Cartilage: Wear Mechanisms and the Anti-Wear Role of Lipids in the Joint, PhD Thesis, The University of Western Australia, 2001.

39 L. de Broglie, A tentative theory of light quanta, *Phil. Mag.*, Vol. 47, 1924, pp. 446-458.

40 E. Ruska, The development of the electron microscope and electron microscopy, *Reviews of Modern Physics*, Vol. 59 (3/1), 1987, pp. 627-628.

41 R.C. Bill and D. Wisander, Recrystallization as a controlling process in the wear of some FCC metals, *Wear*, Vol. 41, 1977, pp. 351-363.

42 D.A. Rigney and W.A. Glaeser, The significance of near surface microstructure in the wear process, *Wear*, Vol. 46, 1978, pp. 241-250.

43 J.M. Martin, J.L. Mansot, I. Berbezier and H. Dexpert, Nature and origin of wear particles from boundary lubrication with a zinc dithiophosphate, *Wear*, Vol. 93, 1984, pp. 117-126.

44 L.K. Ives, An extraction replica method for the study of surface films, *ASLE Trans.*, Vol. 28, 1985, pp. 87-90.

45 E.W. Muller and T.T. Tsong, Field Ion Microscopy - Principles and Applications, Elsevier, New York, 1969.

46 D.H. Buckley, Surface Effects in Adhesion, Friction, Wear and Lubrication, Elsevier, Amsterdam, 1981.

47 N. Ohmae, M. Umeno and K. Tsubouchi, Effect of oxygen adsorption on adhesion of W to Au studied by Field Ion Microscopy, *Trans. ASLE*, Vol. 30, 1987, pp. 409-418.

48 N. Ohmae, Field Ion Microscopy in Tribological Studies, in Surface Diagnostics in Tribology, editors: K. Miyoshi and Y.W. Chung, World Scientific, 1993, pp. 47-74.

49 G. Binnig, C.F. Quate and C. Gerber, Atomic Force Microscope, *Physical Review Letters*, Vol. 56, 1986, pp. 930-933.

50 G. Binnig and H. Rohrer, The Scanning Tunnelling Microscope, *Scientific American*, August 1985, pp. 40-46.

51 S.H. Hues, R.J. Colton, E. Meyer and H-J. Guntherodt, Scanning probe microscopy of thin films, *MRS Bulletin*, Vol. 18, January 1993, pp. 41-49.

52 H. Kumar Wickramasinghe, Scanned-probe microscopes, *Scientific American*, October 1989, pp. 74-81.

53 B. Bhushan, J.N. Israelachvili and U. Landman, Nanotribolgy: friction, wear and lubrication at the atomic scale, *Nature*, Vol. 374, 1995, pp. 607-616.

54 Handbook of Micro/Nano Tribology, editor: B. Bhushan, CRC Series Mechanics and Materials Science, CRC Press Inc., 1995.

55 A.L. Weisenhorn, P. Maiwald, H.-J. Butt and P.K. Hansma, Measuring adhesion, attraction, and repulsion between surfaces in liquids with an atomic-force microscope, *Physical Review B*, Vol. 45, 1992, pp. 11226–11232.

56 M.C. Friedenberg and C.M. Mate, Dynamic viscoelastic properties of liquid polymer films studied by atomic force microscopy, *Langmuir*, Vol. 12, 1996, pp. 6138-6142.

57 S.J. O'Shea and M.E. Welland, Atomic force microscopy at solid-liquid interfaces, *Langmuir*, Vol. 14, 1998, pp. 4186-4197.

58 Y. Kim, K.-S. Kim, M. Park and J. Jeong, Lateral force microscopy study of functionalised self-assembled monolayer surfaces, *Thin Solid Films*, Vol. 341, 1999, pp. 91-93.

59 Y. Okabe, U. Akiba and M. Fujihira, Chemical force microscopy of ---CH_3 and ---COOH terminal groups in mixed self-assembled monolayers by pulsed-force-mode atomic force microscopy, *Applied Surface Science*, Vol. 157, 2000, pp. 398-404.

60 A.J. Pidduck and G.C. Smith, Scanning probe microscopy of automotive anti-wear films, *Wear*, Vol. 212, 1997, pp. 254-264.

61 J.F. Graham, C. McCague and P.R. Norton, Topography and nanomechanical properties of tribochemical films derived from zinc dialkyl and diaryl dithiphosphates, *Tribology Letters*, Vol. 6, 1999, pp. 149-157.

62 R.S. Howland, How to buy a scanning probe microscope, Park Scientific Instruments, 1993.

63 N. Spencer, Scanning probe microscopes, entry for, Encyclopedia of Chemical Physics and Physical Chemistry, editors: J.H. Moore, N.D. Spencer and J. Moore, IOP Publishing, 2001.

64 N.A. Burnham and R.J. Colton, Force microscopy, Scanning Tunnelling Microscopy and Spectroscopy: Theory, Techniques, and Applications, editor: D.A. Bonnell, VCH Weinheim, 1993, pp. 191-249.

65 O. Wolter, T. Bayer and J. Greschner, Micromachined silicon sensors for scanning force microscopy, *Journal of Vacuum Science and Technology B* (Microelectronics, Processing and Phenomena), Vol. 9(2), 1991, pp. 1353-1357.

66 F. Ohnesorge and G. Binnig, True atomic resolution by atomic force microscopy through repulsive and attractive forces, *Science*, Vol. 260, 1993, pp. 1451-1456.

67 C.M. Mate, M.R. Lorenz and V.J. Novotny, Atomic force microscopy of polymeric liquid films, *Journal Chem. Phys.*, Vol. 90, 1989, pp. 7550-7555.

68 P.K. Hansma, J.P. Cleveland, M. Radmacher, D.A. Walters, P.E. Hillner, M. Bezanilla, M. Fritz, D. Vie, H.G. Hansma, C.B. Prater, J. Massie, L. Fukunaga, J. Gurley and V. Elings, Tapping mode atomic force microscopy in liquids, *Appl. Phys. Lett.*, Vol. 64, 1994, pp. 1738-1740.

69 S. Bec and A. Tonck, Nanometer scale mechanical properties of tribochemical films, The Third Body Concept: Interpretation of Tribological Phenomena, Proceedings of the 22nd Leeds-Lyon Symposium on Tribology, Lyon, France, 5-8 September 1995, editors: G. Dalmaz, D. Dowson, C. Taylor and T. Childs, Tribology Series 31, Amsterdam, Elsevier, 1996, pp. 173-184.

70 C.M Mate, R. Erlandsson, G.M. McClelland and S. Chiang, Atomic-scale friction of a tungsten tip on a graphite surface, *Phys. Rev. Lett.*, Vol. 59, 1987, pp. 1942-1945.

71 R. Overney and E. Meyer, Tribological investigations using friction force microscopy, *MRS Bulletin IB*, May, 1993, pp. 26-34.

72 E. Meyer, R. Overney, D. Brodbeck, L. Howald, R. Luthi, J. Frommer and H-J. Guntherodt, Friction and wear of Langmuir-Blodgett films observed by friction force microscopy, *Physical Review Letters*, Vol. 69(12), 1992, pp. 1777-1780.

73 P. Maivald, H.J. Butt, S.A.C. Gould, C.B. Prater, B. Drake, J.A. Gurley, V.B. Elings and P.K. Hansma, Using force modulation to image surface elasticity with the atomic force microscope, *Nanotechnology*, Vol. 2(2), 1991, pp. 103-106.

74 A. Rosa-Zeiser, E. Weilandt, S. Hild and O. Marti, The simultaneous measurement of elastic, electrostatic and adhesive properties by scanning force microscopy: pulsed-force mode operation, *Measurement Science Technology*, Vol. 8, 1997, pp. 1333-1338.

75 K. Topolovec Miklozic and H.A. Spikes, The use of AFM to study lubricant films, Proceedings of the 2nd World Tribology Congress, 3-7 September 2001, Vienna, Austria, (editors: F. Frank, W.J. Bartz, A. Pauschitz), 2001, pp. 179-186.

76 A. Noy, C.D. Frisbie, L.F. Rozsnyai, M.S. Wrighton and C.M. Lieber, Chemical force microscopy: Exploiting chemically-modified tips to quantify adhesion, friction, and functional group distributions in molecular assemblies, *Journal of American Chemical Society*, Vol. 117, 1995, pp. 7943-7951.

77 L. Scandella, A. Schumacher, N. Kruse, R. Prins, E. Meyer, R. Lüthi, L. Howald and H.-J. Güntherodt, Tribology of ultra-thin MoS_2 platelets on mica: Studies by scanning force microscopy, *Thin Solid Films*, Vol. 240, 1994, pp. 101-104.

78 C. Grossiord, K. Varlot, J.-M. Martin, T. Le Mogne, C. Esnouf and K. Inoue, MoS_2 single sheet lubrication by molybdenum dithiocarbamate, *Tribology International*, Vol. 31, 1998, pp. 737-743.

79 K.K. Tiong, P.C. Liao, C.H. Ho and Y.S. Huang, Growth and characterisation of rhenium-doped MoS_2 single crystals, *Journal of Crystal Growth*, Vol. 205, 1999, pp. 543-547.

80 D. Rugar, H.J. Mamin, P. Guethner, S.E. Lambert, J.E. Stern, I. McFayden and T. Yogi, Magnetic force microscopy: general principles and application to longitudinal recording media, *Journal of Applied Physics*, Vol. 68, 1990, pp. 1169-1183.

81 W. Gutmannsbauer, T. Huser, T. Lacoste, H. Heinzelmann and H-J. Guntherodt, Scanning near-field optical microscopy (SNOM) and its application in mineralogy, *Schweiz. Mineral. Petrogr. MITT.*, Vol. 75, 1995, pp. 259-264.

82 E. Synge, A suggested model for extending microscopic resolution into the ultra-microscopic region, *Philosophical Magazine*, Vol. 6, 1928, pp. 356-362.

83 D. Pohl, W. Denk and M. Lanz, Optical stethoscopy: Image recording with resolution $\lambda/20$, *Appl. Phys. Lett.*, Vol. 44, 1984, pp. 651-653.

84 T. Mathia, H. Zahouani, J. Rosseau and J.C. Le Bosse, Functional significance of different techniques for surface morphology measurement, *Int. J. Mach. Tools Manufact.*, Vol. 35, No. 2, 1995, pp. 195-202.

85 R. Castaing, Application des sondes electroniques a une methode d'analyse ponctuelle chimique et crystallographique, Ph.D. thesis, University of Paris, 1951.

86 A.W. Batchelor, A. Cameron and H. Okabe, An apparatus to investigate sulfur reactions on nascent steel surfaces, *Trans. ASLE*, Vol. 28, 1985, pp. 467-474.

87 G.W. Stachowiak and A.W. Batchelor, Engineering Tribology, Butterworth-Heinemann, 2001.

88 A.W. Batchelor and G.W. Stachowiak, Some kinetic aspects of extreme pressure lubrication, *Wear*, Vol. 108, 1986, pp. 185-199.

89 R.F. Baker and A.V. Olver, Direct observations of fretting wear of steel, *Wear*, Vol. 203-204, 1997, pp. 425-433.

90 P.M. Cann and H.A. Spikes, In lubro studies of lubricants in EHD contacts using FTIR absorption spectroscopy, *Tribology Transactions*, Vol. 34, 1991, pp. 248-256.

91 N.J. Harick, Internal Reflection Spectroscopy, Interscience Publishers, John Willey and Sons Inc., 1967.

92 S.S. Wang and S.C. Tung, Using electrochemical and spectroscopic techniques as probes for investigating metal-lubricant interactions, *Tribology Transactions*, Vol. 33, 1990, pp. 563-572.

93 Z-S. Hu, S.M. Hsu and P.S. Wang, Tribochemical and thermochemical reactions of stearic acid on copper surfaces studied by infrared microspectroscopy, *Tribology Transactions*, Vol. 35, 1992, pp. 189-193.

94 J.L. Lauer, B.G. Bunting and W.R. Jones, Investigation of frictional transfer films of PTFE by infrared emission spectroscopy and phase locked ellipsometry, *Tribology Transactions*, Vol. 31, 1988, pp. 282-288.

95 Z-S. Hu, S.M. Hsu and P.S. Wang, Tribochemical reaction of stearic acid on copper surfaces studied by surface enhanced Raman spectroscopy, *Tribology Transactions*, Vol. 35, 1992, pp. 417-422.

96 J.L. Paddy, N.C. Lee, D.N. Waters and W. Trott, Zinc dialkyl dithiophosphate oxidation by cumene hydroperoxide: Kinetic studies by Raman and ^{31}P NMR spectroscopy, *Tribology Transactions*, Vol. 33, 1990, pp. 15-20.

97 B.A. Baldwin, Relationship between surface composition and wear, an X-ray photo-electron spectroscopic study of surfaces tested with organo-sulphur compounds, *Trans. ASLE*, Vol. 19, 1976, pp. 335-344.

98 J.C. Riviere, Auger spectroscopy, Modern Physical Techniques in Materials Technology, editors: T. Mulvey and R.K. Webster, pp. 187-205, Oxford University Press, 1974, United Kingdom.

99 K. Siegbahn, C. Nordling, A. Fahlman, R. Nordberg, K. Hamrin, J. Hedman, G. Johanson, T. Bergmark, S.E. Karlsson, I. Lindgren and B. Lindberg, ESCA - Atomic, molecular and solid state structure studied by means of electron spectroscopy, Almqvist and Wiksells, Uppsala, Sweden, 1967.

100 D. Briggs (editor), Handbook of X-ray and Ultraviolet Photoelectron Spectroscopy, Heyden and Son Ltd., London, United Kingdom, 1967.

101 H. Hong, A.T. Riga, J.M. Cahoon and J.N. Vinci, Evaluation of overbased sulfonates as extreme-pressure additives in metalworking fluids, *Lubrication Engineering*, Vol. 49, 1993, pp. 19-24.

102 J.M. Martin, Th. Le Monge, C. Grossiord and Th. Palermo, Adsorption and friction in the UHV tribometer, *Tribology Letters*, Vol. 3, 1997, pp. 87-94.

103 K. Langguth and H. Ryssel, Wear of steels after implantation of aluminium and coimplantation of aluminium and oxygen ions, *Wear*, Vol. 161, 1993, pp. 127-133.

104 J.M. Martin, M. Belin, J.L. Mansot, H. Dexpert and P. Lagarde, Friction-induced amorphization with ZDDP - an EXAFS study, *ASLE Transactions*, Vol. 29, 1986, pp. 523-531.

105 E.H. Loeser, R.C. Wiquist and S.B. Twist, Cam and tappet lubrication, IV, Radio-active study of sulphur in the E.P. film, *Trans. ASLE*, Vol. 2, 1959, pp. 199-207.

9 LUBRICANT OR PROCESS FLUID ANALYSIS

9.1 INTRODUCTION

The analysis of a lubricant used during the tests is a vitally important part of experimental methodology in tribological research. A lubricant or process fluid present in the dynamic contact is subject to the high temperatures, mechanical stresses and other factors such as catalytically activated metallic surfaces. Under these conditions physical and chemical changes to the lubricant/process fluid are almost inevitable. It is therefore of practical interest to know whether the lubricant/process fluid will be degraded and to what degree. Monitoring of the changes occurring within the lubricant provides much information on its performance, e.g. the maximum contact temperature, or the presence of catalytically activated metallic surfaces. Contamination of the lubricant by e.g. airborne contaminants, water and fuel can also be monitored as this provides useful information related to the performance of the machinery, e.g. an engine or gear.

Control of the lubricant uniformity, detection of the lubricant degradation or contamination during the experiments can only be determined by its detailed analysis. During any tribological experiment there is always a number of issues related to lubricant physical and chemical characteristics to which answers can be provided by properly conducted analysis of a lubricant. For example: What is the viscosity index of a lubricant? What is the lowest operating temperature of a lubricant before viscosity becomes excessive? How big is the decline in lubricant viscosity at elevated temperatures? What is the lubricant viscosity response to pressure? Can synovial fluid provide elastohydrodynamic lubrication? Is the lubricating oil composed of napthenic, paraffinic or aromatic hydrocarbons? Does the lubricant contain any sulphur compounds? Can atmospheric pollutants introduce film formation agents to the lubricant? Can high operating temperatures during the test cause an accelerated degradation of a lubricant?

In this chapter some basic and specialized methods used in the analysis of physical and chemical properties of lubricants are described and discussed.

9.2 LUBRICANT PARAMETERS OF TRIBOLOGICAL SIGNIFICANCE

A large number of parameters are required to provide a complete description of a lubricant. Some of these parameters are of direct relevance to tribology, e.g. viscosity, oxidation stability and thermal conductivity, while many more are only of indirect relevance, e.g. smell and colour, and the remaining few have only limited significance to tribology, e.g. electrical conductivity. It should be noted, however, that even these indirectly relevant parameters, such as smell or electrical conductivity, may be important in some instances. An acrid smell may signify decomposition of a lubricant, while electrical conductivity becomes important when tribo-electrification is involved. The significance of the commonly cited lubricant parameters is summarised in Table 9.1.

Colour and odour are partly subjective parameters that influence usability considerations of lubricants and are not discussed further in this book. Each lubricant has its own characteristic hazards such as flammability and health risks. The flammability of a lubricant is defined in terms of parameters such as the 'flash point', as would be defined for any other volatile, combustible fluids.

A lubricant poses several potential health risks, which include direct toxicity (if swallowed) and allergic or inflammatory reaction of sensitive tissues in the skin and eyes. The lungs may also be at risk of inflammation if the lubricant is inhaled as a mist. A hot lubricant may release a toxic vapour of possibly carcinogenic organic compounds. Mixtures of lubricants and water may provide fertile sites for bacterial growth; this is very common in aqueous cutting fluids. If bacteria are released in an aerosol (mist), as is common in cutting, workers may sustain lung infections [1]. Lubricant additives may further complicate the toxicity issue, e.g. lead napthenate, an E.P. additive, is a classic example of heavy metal toxicity. While there is a lot of detailed descriptive data on health hazards of lubricants, a few basic parameters provide a summary view of the risks. The key parameters are the Permissible Exposure Limit (PEL), defined by the Occupational Safety and Health Administration, and the Threshold Limit Value (TLV), defined by the American Conference on Governmental Industrial Hygienists. For lubricants, the PEL and TLV are often defined in terms of g/cubic metre since oil mists are the principal hazard. There is also an oral toxicity parameter, defined as the mass of lubricant in grammes per kilogram of rat where e.g., 50% of a test rat population survive the lubricant dose. The measurement of lubricant toxicity is often defined by government regulation since the safety of workers and the public is involved [2]. Governmental agencies such as the National Institute of Occupational Safety and Health, International Agency for Research on Cancer, the Mine Safety and Health Administration, all impose legal conditions on the use of lubricants and the level of information about lubricant hazards that must be made available. The level of regulation affecting lubricants is expected to increase [3] as the health hazards of even common additives such as ZnDDP become well recognized. Lubricant toxicity tests are very specialised since they involve test animals (e.g. rats) where the data is extrapolated using various analytical models to predict the risk to humans. A useful discussion of toxicology concerns related to lubricants can be found in [4].

Table 9.1 Significance of lubricant parameters to tribological function in decreasing order.

Tribological Function	Parameter	Measurement Technique
Hydrodynamic and elasto-hydrodynamic lubrication	Viscosity	Viscometer
	Barus pressure viscosity coefficient	High pressure viscometer Elastohydrodynamic interferometry
	Temperature dependence of viscosity	Viscometer with temperature control
	Limiting shear stress, glass transition	Ultra-high shear rate viscometer
	Compressibility	Hydraulic compression
	Thermal conductivity	Heat transfer measurements
Boundary lubrication	Composition and concentration of surfactants and corrosive compounds	Infra-Red Spectroscopy
	Temperatures of friction transition or lubricant failure	Tribometer with temperature control
	Corrosivity of lubricant	Hot wire test apparatus to expose nascent surface
	Solubility of dissolved gases	Volumetric solubility test
	Diffusivity of dissolved gases	Diffusion test apparatus
Lubricant's life	Oxidation stability	Rotating bomb oxidation Differential scanning calorimetry Thin film oxidation test
	Concentration of contaminants (water) and degradation of inhibitors	Infra-Red Spectroscopy
	Acidity and Total Acid Number (TAN)	Acid neutralization test
	Smell and colour	Colour and odour comparisons with standards
	Toxicity	Tests on laboratory animals

Lubricant characteristics of more general interest, requiring measurement technology specific to tribology, are discussed in the sections below.

9.3 MEASUREMENT OF LUBRICANT CHARACTERISTICS

The fluid sample of a used lubricant can reveal much evidence of the prevailing wear occurring within the tribological system or quality of lubrication. Also, information about the oxidation or chemical decomposition of a lubricant and the depletion of additives is more easily obtained from the actual lubricant sample.

A combination of general purpose equipment and special techniques are used to measure lubricant characteristics which are specific to its tribological function. Techniques used to monitor changes occurring in the lubricant's characteristics have traditionally involved direct measurements of basic physical and chemical properties such as viscosity and acidity. To determine other lubricant parameters such as density, thermal conductivity, etc., standard techniques are applied. These standard techniques are commonly known and well described in the literature, hence they are not discussed in this book. Of more specialised interest to tribology are the measurement of lubricant viscosity at extremes of pressure, temperature and shear rate, and the oxidation of lubricants under conditions resembling those found in a wearing contact. The severity of conditions typical of a wearing contact require the application of novel and advanced measuring techniques. For example, there is little purpose in measuring viscosity at a moderate shear rate under ambient conditions if the lubricant has failed in a high speed gear system where the shear rate is extremely high. The development of methods and specialized equipment designed to obtain the required lubricants performance data, as they function in a wearing contact, remains an important research topic.

One of the more recent developments is to apply chemical analysis of used lubricants as a condition monitoring technique. Detection by chemical analysis of small traces of oxidation and degradation products can provide a much earlier warning of lubrication and wear problems than any of the traditionally used monitoring techniques.

Analysis of Chemical Changes Occurring in Lubricants and Process Fluids

Chemical changes occurring in lubricants or process fluids can be directly analyzed by Infra-Red Spectroscopy (IR spectroscopy) [5,6]. This technique is described in detail in Chapter 8. IR spectroscopy has been used, for example, to measure the accumulation of water, carbon and oxyacids in a lubricating oil during service [6]. The major practical difficulty in applying this technique is the interpretation of complex spectra produced by degraded lubricants [5,6]. Specialised numerical analysis of IR spectra is required to extract analytical information from overlapping absorption peaks. Before computers and the computer memory were available, two cells had to be used in a simultaneous analysis of the test sample and the control sample. Small differences in dimensions of either cell and misalignment nearly always prevented accurate comparison of data [5,6]. With the application of computers this problem has been eliminated so that the same IR spectroscopy cell (a vessel to hold test fluid during analysis) can be used when comparing the spectra of new and degraded lubricants. The most popular IR technique used to detect small changes in the spectra is Fourier Transform IR (FTIR) Spectroscopy.

The IR spectra are usually displayed as a standard graph of IR transmittance or absorbance within the sample versus wave number of the IR radiation. Transmittance is defined as the ratio of radiation intensity transmitted by a sample to the radiation intensity entering the sample while absorbance is a parameter that determines the attenuation of photon radiation as it passes through a material. An example of IR spectra of the same lubricating oil after varying periods of service is shown in Figure 9.1.

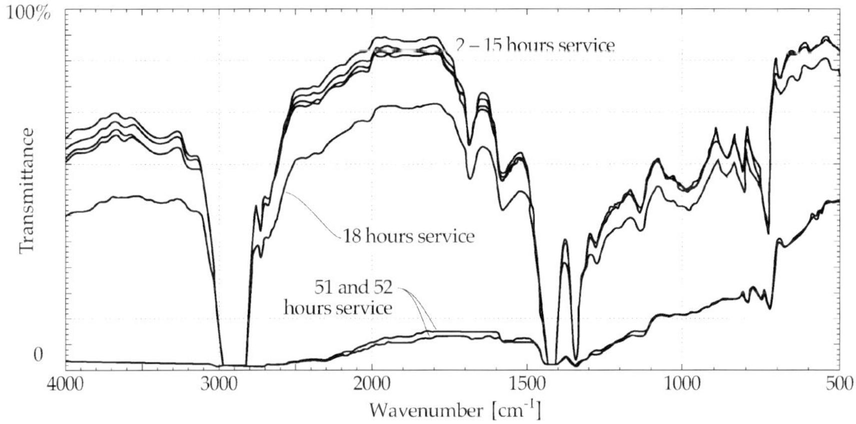

Figure 9.1 Infra-red spectra of a lubricating oil after varying periods of service [5].

As can be seen from Figure 9.1, the IR spectra of a lubricant reveal significant changes in its chemical composition occurring during service. There is a significant drop in transmittance after 51 and 52 hours probably due to build up of dispersed carbon [5]. An index or measure of lubricating oil degradation can be obtained from the IR spectra by monitoring the absorbance at specific wave numbers associated with the presence of contaminants or oxidation and decomposition products of the lubricant. For example, in one work carbon accumulation in the lubricating oil was chosen as a monitoring measure [5]. Differential absorbance at a wave number of 1900 [cm^{-1}], where the absorbance is caused by elemental carbon particles in the lubricant, was plotted as a function of service time. Differential absorbance is defined as the difference in absorbance between the control sample, which is unused lubricant, and the test sample, which is used lubricant. Graphs showing measured carbon accumulation, based on differential absorbance at 1900 [cm^{-1}], versus service time are often produced offering a clear indication of the changes occurring. An example of such graph prepared for a military truck engine is shown in Figure 9.2 [5].

It can be seen in Figure 9.2 that the differential absorbance data shows that before an oil change there is a gradual increase in carbon content with time. After the oil change a similar trend occurs for about 15 hours of operation followed by a sharp increase in carbon level indicating some sort of system failure, e.g. blocked air filter or over-choked engine [5].

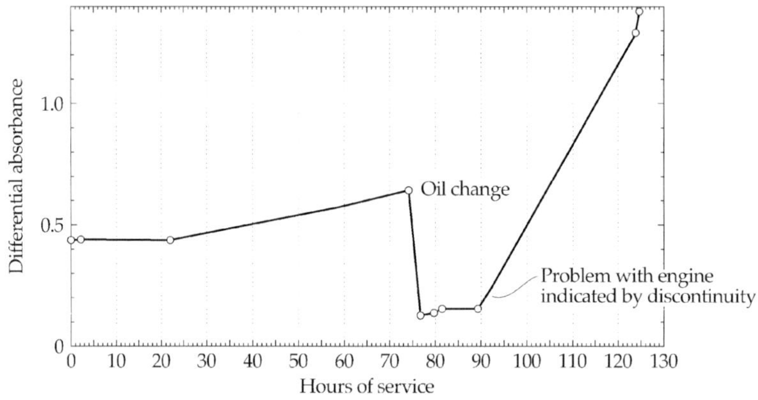

Figure 9.2 Infra-red spectroscopic measurements of carbon accumulation in a lubricating oil with varying service period [5].

Carbon measurements alone would not justify the difficulty of applying IR spectroscopy to lubricant oil monitoring. The special advantage of IR spectroscopy is the comprehensive range of information obtained about changes in the lubricating oil composition. It is possible using IR spectroscopy to simultaneously measure the depletion of antioxidants in the lubricant and the accumulation of oxidation products. Water in a lubricant can easily be detected by IR spectroscopy as well as evidence of the source of water, e.g. if glycol is also present this would indicate that the water is from a coolant leak [5]. An example of simultaneous measurement of antioxidant depletion and oxidation of a mineral oil containing a Zinc Dialkyl Dithiophosphate (ZnDDP) as an antioxidant is shown in Figure 9.3.

The concentration of ZnDDP in the oil is measured by the relative strength of IR absorption from P-S and P=O bonds present in the ZnDDP molecule. The concentration of oxyacids (oxidation products) in the oil is found from IR absorption by C=O bonds in the oxyacids. The percentage of integrated area under an absorption peak compared to either unused oil for P-S and P=O bonds or to oil at the end of the test for C=O bonds is used as a relative measure of concentration. A sharp acceleration in the rate of oil oxidation can be seen when the concentration of ZnDDP declines to less than approximately 0.1 of its level in unused oil. The results also suggest that other measurements of oil oxidation, e.g. acidity tests, may not provide good prediction of the future progress of oil oxidation as the process is effectively controlled by antioxidant concentration.

Although IR spectroscopy has the potential to offer high quality lubricant monitoring data it is still not widely used for this purpose throughout the industry because of technical difficulties.

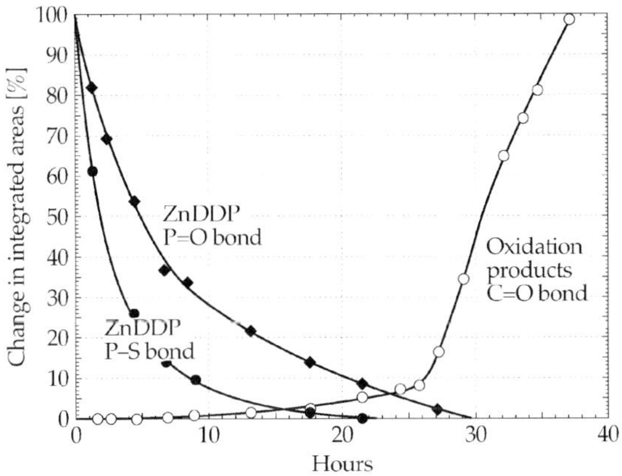

Figure 9.3 Infra-red spectroscopic measurements of ZnDDP depletion and oxidation of the carrier oil [6].

Viscometers and Characterization of Lubricant Rheology

The rheology of a lubricant is fundamental to its functioning and most of its rheological properties are evaluated using a viscometer of some kind. Viscosity tests range from general purpose tests suitable for any fluid, not just lubricants, to highly specialized tests specifically developed to measure viscosity and other rheological characteristics of tribological fluids. Lubrication of heavily loaded contacts moving at high speed, e.g. high speed gears, involves extremes of shear rate and rapid shear rate changes. In such cases, the lubricant may not have sufficient time to change its molecular structure (or packing) to accommodate a very high shear rate. Most lubricants function as a very thin film, typically between 0.1 [μm] and 10 [μm] thick, where the bulk liquid molecular structure may be distorted by the close presence of solid surfaces. In this case, the rheology of thin lubricant films can be significantly different from that of the same lubricant in bulk form. Changes imposed on lubricant rheology by the prevailing operating conditions will have a corresponding effect on lubrication mechanisms involved such as the generation of elastohydrodynamic pressure. In general, the purpose of rheological studies of lubricants is to determine how extreme conditions of shear rate affect performance of the lubricant and the mechanism of lubricating film generation.

· *Viscometers*

There are numerous viscometers available on the market. These viscometers range from the classic capillary tube which has no moving parts to advanced hydraulic systems which can measure viscosities at pressures close to 1 [GPa]. Examples of the various typical viscometers developed are illustrated schematically in Figure 9.4.

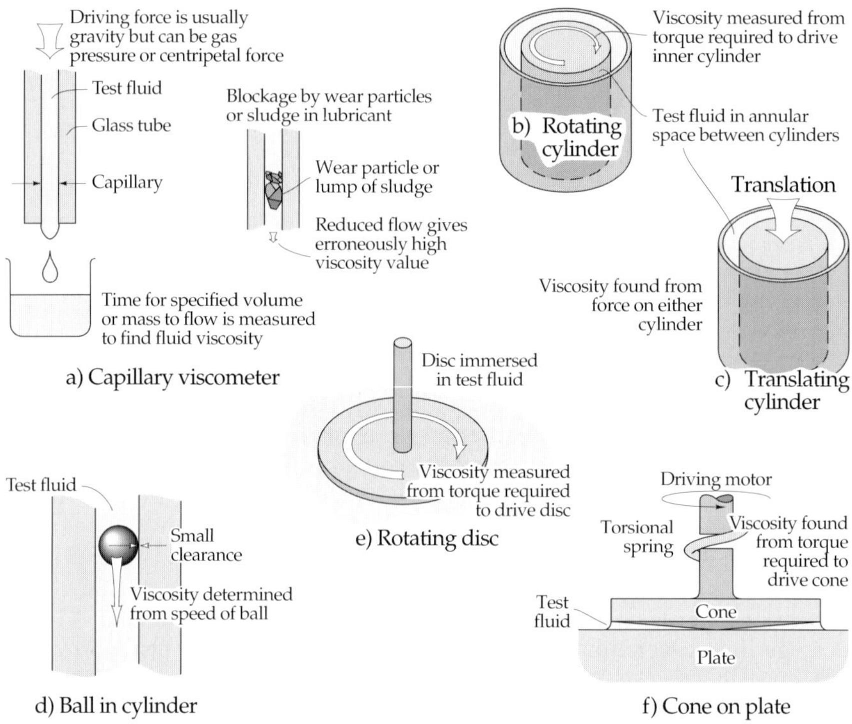

Figure 9.4 Schematic illustration of typical viscometers used in tribological studies.

The operating principles and applications of general purpose viscometers can be found in many texts on rheology or fluid mechanics and so are not discussed in this book in further detail. Unused lubricants do not usually present difficulties in viscosity measurements with most general purpose viscometers. On the other hand, used lubricants can contain wear particles which may block capillary tubes or scratch cone-on-plate viscometers. Blockage or scratching by wear particles not only damages the viscometer but also invalidates any viscosity data obtained. In a typical industrial maintenance program, viscosity measurements of used oils form the criteria upon which a lubricant condition is assessed. For example, an increase in viscosity may indicate the onset of rapid lubricant oxidation while the decrease in viscosity may indicate lubricant dilution by fuel, e.g. in diesel engines. Careful selection of the viscometer is important when measuring fluids containing particles, e.g. slurries. A capillary viscometer can be easily blocked by particles thus viscometers with relatively large clearances between rotating surfaces, such as the concentric-cylinder type, are more suitable for viscosity measurements of fluids heavily contaminated with solid particles. Even new lubricants contain particles which, in principle, can affect viscosity measurements [7].

More specialized viscometers have been developed to measure actual viscosity values of lubricants working in real industrial machinery, under conditions of high loads and shear rates. For example, a special purpose apparatus has been developed to measure the viscosity and limiting shear stress of a lubricant at high pressures comparable to those acting within an elastohydrodynamic contact [8-11]. Although the same viscosity measurement principle applies as in general purpose ambient pressure rotating cylinder viscometers, the technical demands of extreme hydraulic pressure complicate the construction of the apparatus. Concentric cylinders and translating cylinders (falling cylindrical weight in a close fitting tube) have been used in various versions of the device [8,10]. The clearance between concentric cylinders is set to approximately 1 [μm], which is sufficiently thick to suppress viscous heating during tests [8]. In addition the construction of this viscometer requires a pressure intensifier, high pressure cell and thrust bearings allowing a rotation of the cylinders while at the same time supporting a reaction force from hydraulic pressure applied. With this viscometer pressures as high as 300 [MPa] and a shear rate as high as 10^4 [s^{-1}] have been achieved [8]. The apparatus is illustrated schematically in Figure 9.5.

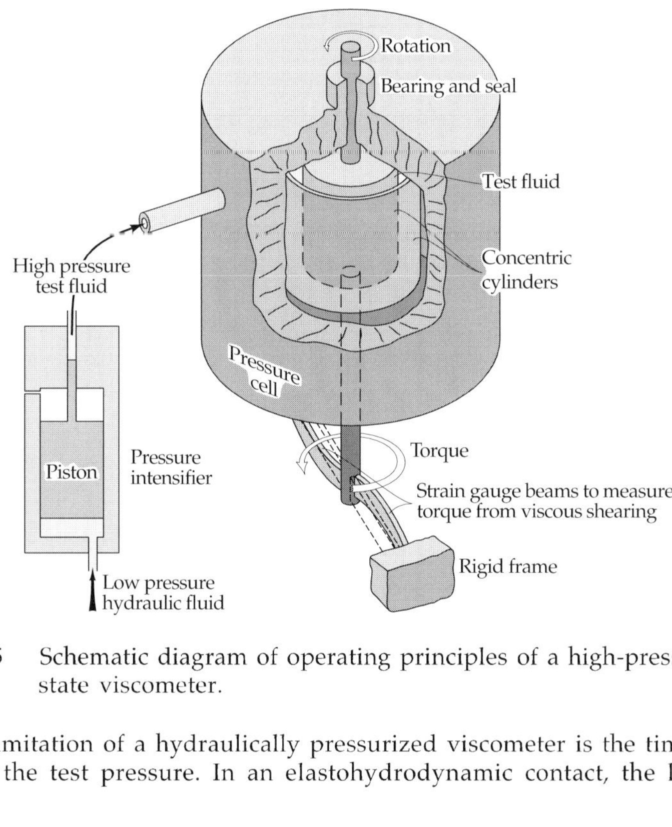

Figure 9.5 Schematic diagram of operating principles of a high-pressure steady state viscometer.

A basic limitation of a hydraulically pressurized viscometer is the time required to attain the test pressure. In an elastohydrodynamic contact, the lubricant is subjected to an extremely rapid pressure change from ambient to 1 [GPa] or more, as well as a similarly rapid rise in shear rate. In a typical contact, the shear rate can rise to 10^5 [s^{-1}] in a period as short as 100 [μs] [12]. Such a rapid increase of shear rate can, in theory, induce a change in lubricant flow from uniform shearing to discontinuous shearing in bands [12] which cannot be observed in a viscometer that is slow to attain the required test conditions. To overcome this limitation, an entirely different experimental approach was used. The test lubricant was loaded into a test cell, which was closed off by a round bar called a Kolsky bar [13]. The Kolsky bar was used to subject a lubricant sample to a pulse of shear deformation by means of a torsional oscillation. The Kolsky bar was loaded in torsion until a notched restraining bolt fractured and a torsional shear wave was released. The torsional shear wave subjected the test liquid to a sudden episode of shearing which has a duration of about 500 [μs] where the first 100 [μs] within this period elapsed before the maximum intensity of shearing was reached. With this apparatus shear rates as high as 10^4 [s^{-1}] can be achieved so that the levels of shear rate rise and shear rate are almost compatible to an elastohydrodynamic contact. The thickness of the lubricant film tested is as large as 300 [μm] which would involve viscous heating problems but for the short duration of shearing. A further advancement in this technique would probably involve pressurizing the lubricant to further refine the similarity to elastohydrodynamic conditions. The viscosity and limiting shear stress of the lubricant are measured as a function of time from the damping characteristic of the Kolsky bar. The schematic illustration of this apparatus is shown in Figure 9.6.

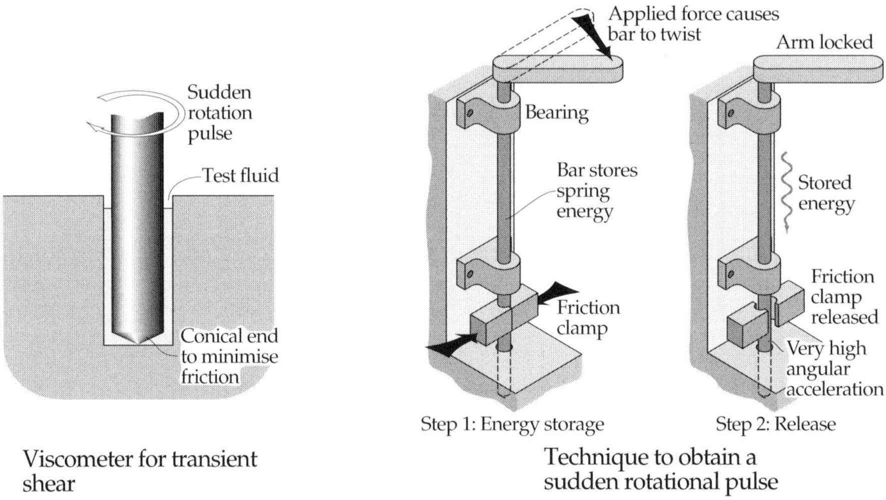

Figure 9.6 Schematic diagram of operating principles of a transient shear viscometer.

Impact viscometers allowing the measurement of lubricant viscosity at high pressures and shear rates, up to 10^6 [s^{-1}], have also been developed [14-16] and are described in Chapter 3.

· *Viscometer Selection*

A fundamental question for any researcher is what type of data is provided by each available form of viscometry. The capabilities of the various viscometry techniques developed are summarised in Table 9.2.

Table 9.2 Types of viscometry and quality of data obtained.

Type of viscometry	**Quality of data obtained**
Bulk liquid measurement techniques	
Capillary tube	Viscosity of Newtonian fluids only at low and mostly uncontrolled shear rates A limited temperature variation possible. Lubricant must not contain particles
Concentric cylinder type	Suitable for Newtonian fluids only Variable but low shear rate Temperature variation possible
Cone on plate	High and well controlled shear rate Both Newtonian and non-Newtonian fluids can be characterized Maximum temperature is usually controlled by a water heating system Lubricant must not contain particles
Techniques with particular applications to tribology studies	
Falling weight or rotating cylinder inside pressurized lubricant cell	Specifically designed to measure variation of viscosity with pressure Both Newtonian and non-Newtonian fluids can be analyzed Temperature variation possible but maximum temperature is limited by seal materials used [8] Lubricant must not contain particles
Oscillating torsional rod (Kolsky bar) inserted into a lubricant cell	Measurements at high shear rates are possible. However, the data obtained is complicated by analysis of torsional vibration Lubricant must not contain particles Usually suited for studies at ambient or low pressures unless apparatus is redesigned
Impacting or rolling ball on polished flat surface wetted with test fluid	Extremes of shear rate attained [14-16] Lubricant must not contain particles

The viscosity of lubricants can be determined with a high degree of precision but such measurements can be very costly. Hence the most advanced techniques should not be applied to crude measurements or simple tests of lubricant type since the discrimination between different lubricants according to viscosity only requires the simplest type of viscometer. For example, viscosity measurements needed for the analysis of hydrodynamic lubrication require only low-pressure viscosity data. On the other hand, viscosity data for the analysis of elastohydrodynamic lubrication should ideally be obtained from apparatus designed for high values of shear rate, pressure and rate of application of both. In practice, however, such apparatus is still not feasible and prediction of the elastohydrodynamic characteristics of a lubricant is often derived from formulae that use ambient pressure viscosity data. At present it is possible to obtain viscosity and limiting shear stress measurements at high pressures and shear rates but at a rate of rise in pressure and shear that is much slower than actually occurs in an elastohydrodynamic contact [10,12]. Viscosity dependence on pressure and shear rate can be estimated from elastohydrodynamic film thickness measurements conducted, for example, on the falling ball apparatus. However, these estimations always involve assumptions about the characteristics of the lubricant in an elastohydrodynamic contact. For instance, it is difficult, if not impossible, to determine the exact rheological conditions of shear rate and temperature inside an elastohydrodynamic contact without assuming that, for example, the lubricant shears in a uniform manner. Measurements subject to unproved assumptions are always unsatisfactory which is the underlying reason for the expense and effort directed to construct more sophisticated viscometers, e.g. the high pressure and shear rate viscometers described above.

9.4 LUBRICANT OXIDATION TESTS

A basic problem with hydrocarbon lubricants is that they oxidize almost as readily as other hydrocarbons such as kerosene, gasoline and fuel gas. The form of oxidation involved in lubricants is a low temperature acidification, which is much slower and milder than the oxidation in direct combustion but it still causes problems of excessive lubricant viscosity and corrosive wear.

From the industrial view point it is important to know what is the remaining useful life of lubricant (RUL) before the 'uncontrolled' oxidation of the base oil occurs. This can be determined by conducting laboratory lubricant oxidation stability tests described in this section. In these tests the lubricant is oxidized under various conditions of temperature, pressure and the presence of catalysts. At the completion of the test, lubricant characteristics such as Total Acid Number, viscosity, antioxidant depletion, sludge content or induction time are used to measure oil degradation and determine the oxidation stability and the remaining useful life of a lubricant tested.

Laboratory oxidation tests are widely used to evaluate lubricating oils for their oxidation stability. There are many variations of these tests ranging from proprietary tests of individual commercial organizations, where the tests are usually empirical in nature, to carefully controlled experiments designed to reproduce in detail the oxidation conditions occurring in a wearing contact. A typical example of the empirical type of test is a standard oxidation test used by United States railroad companies. In this test the catalytic effect of metals on oxidation is modelled by immersing a bronze bushing, a steel bolt and ball into a 350 [ml] oil sample [17]. There is a large variety of empirical tests and they are not discussed here in detail as these tests are not easily applied to investigations that differ from the context in which the test was originally formulated. Oxidation tests specified by industrial standards organizations such as the American Society for Testing and Materials are of general significance as they are applicable to most studies and facilitate the comparison of data between different laboratories. More advanced oxidation tests designed to include every known feature of lubricant oxidation usually remain as specialized research experiments.

It should be mentioned that lubricating oil oxidation is related to important industrial problems such as lubricant degradation in service, sludge and lacquer formation. Sludge is a loose deposit of insoluble material while lacquer is a hard adhering coating, which forms on hot surfaces in contact with lubricant. In the majority of cases a standard oxidation test apparatus can be used for tests of lacquer formation while in some special cases as, for example, in tests of aircraft lubricants specialized apparatus are used [18].

In general, all the lubricant oxidation stability tests fall into three major categories:

- bulk oxidation tests,
- micro-scale oxidation tests and
- non-standard tests.

The bulk oil oxidation test is used to determine whether excessive oil oxidation occurs when a lubricating oil is stored in the sump or storage tank of a machine. Sump temperatures range from 75°C to 150°C in most machinery [19], which usually is sufficient to cause oxidation problems. In many cases the lubricating oil spends most, but not all, of its lifetime in the bulk form of several litres or more.

However, when a lubricating oil is actually functioning as a friction reducer, it is generally present as a thin film. Oxidation of a thin oil film is subject to much less severe limitations of oxygen diffusion in oil, than in the case of bulk oxidation, and the metallic surface which the oil film rests on may catalyze oil oxidation [19,20]. The temperature of a thin oil film close to the source of frictional or combustion heat is often higher than that inside a remote oil sump and may reach 200°C or possibly even 350°C [19]. This temperature difference may cause a change in the controlling oxidation mechanism of the oil and the oxidation inhibitors, which effectively control the lower temperature bulk oxidation, can become ineffective at the higher temperatures. Thus to effectively predict the remaining useful life (RUL) of the lubricant in machinery, both a thin layer oxidation test and a bulk oxidation test need be conducted [19].

It needs to be mentioned that within these basic categories there are various specific test methods and some of them are briefly described in the following sections.

Bulk Oil Oxidation Tests

Bulk oil oxidation tests are conducted according to carefully developed methodologies which are specified by industrial standard organizations. Commonly used standardized oxidation tests are listed in Table 9.3.

As can be seen from Table 9.3 all the standard oxidation tests use oxygen and a catalyst to increase the speed of reaction and represent as closely as possible the operating conditions of the particular oil. The exception is IP 48 which uses air as an oxidising agent and no catalyst. The data in Table 9.3 also shows that the test duration is exceedingly long for the majority of standard oxidation tests. The exceptions are the Rotating Bomb Oxidation Test - RBOT (ASTM D 2272 or IP 229) and the Thin Film Oxygen Uptake Test - TFOUT (ASTM D 4742) which is a modified RBOT for automotive oil evaluation, both of which are generally completed within 10 hours.

Table 9.3 Commonly used standard oil oxidation tests (adapted from [21]).

Test Method		Description	Origin of sample	Type of Catalyst	Oxidising Agent	Temp °C	Test Parameters	Test Duration	Sample Size
IP	ASTM								
48	-	Determination of oxidation characteristics of lubricating oil	Base oils	No catalyst	Air at 15 [L/hr]	200	Kinematic viscosity at 37.8°C Ramsbottom carbon residue	Two 6 [hrs] periods	40 [ml]
157	D 943	Isothermal bulk oil oxidation test	Steam turbine oils	Iron & copper coils	Oxygen at 3 [L/hr]	59	Acidity increase	1000 [hrs] or until TAN varying from 0.25 to 2.0 [mg KOH/g]	300 [ml]
229	D 2272	Determination of relative oxidation stability of mineral turbine oils by rotating bomb	Steam turbine oils	Copper coil	Initial oxygen pressure at 0.62 [MPa]	150	Induction time	Until 0.175 [MPa] drop 2 - 30 [hrs] [22]	50 [g]
280	-	Determination of oxidation stability of inhibited mineral turbine oils	Turbine oils	Iron & copper naphthenate solution	Oxygen at 1 [L/hr]	120	Volatile acids Soluble acids Sludge	164 [hrs]	30 [g]
306	-	Determination of oxidation stability of inhibited mineral turbine oils	Straight mineral oil	Tube 1 - Copper coil Tube 2 - No catalyst	Oxygen at 1 [L/hr]	120	Volatile acidity Soluble acidity Sludge Total oxidation products (TOP)	48 [hrs]	25 [g]
307	-		Insulating oil	Copper coil	Oxygen at 1 [L/hr]	100	Soluble acidity Sludge	164 [hrs]	25 [g]
335	-		Inhibited mineral oil	Copper coil	Oxygen at 1 [L/hr]	120	Induction time	236 [hrs]	25 [g]
-	D 4742	Thin film oxygen uptake test	Engine oil	Oxidised fuel & soluble metal catalysts	Initial oxygen pressure at 0.62 [MPa]	160	Induction time	Until pressure drop 0.7 - 10 [hrs] [23]	25 [g]

Acronyms: IP = Institute of Petroleum, ASTM = American Society for Testing and Materials.

· *Open Vessel Tests*

This is the simplest type of bulk oil oxidation test during which the bulk oxidation of oil at room temperature and elevated temperatures is measured. A sample of oil is held in a glass vessel, which is placed in a thermostatic bath. Air or another gas such as oxygen is then bubbled through the oil sample via a tube to initiate oil oxidation as schematically illustrated in Figure 9.7.

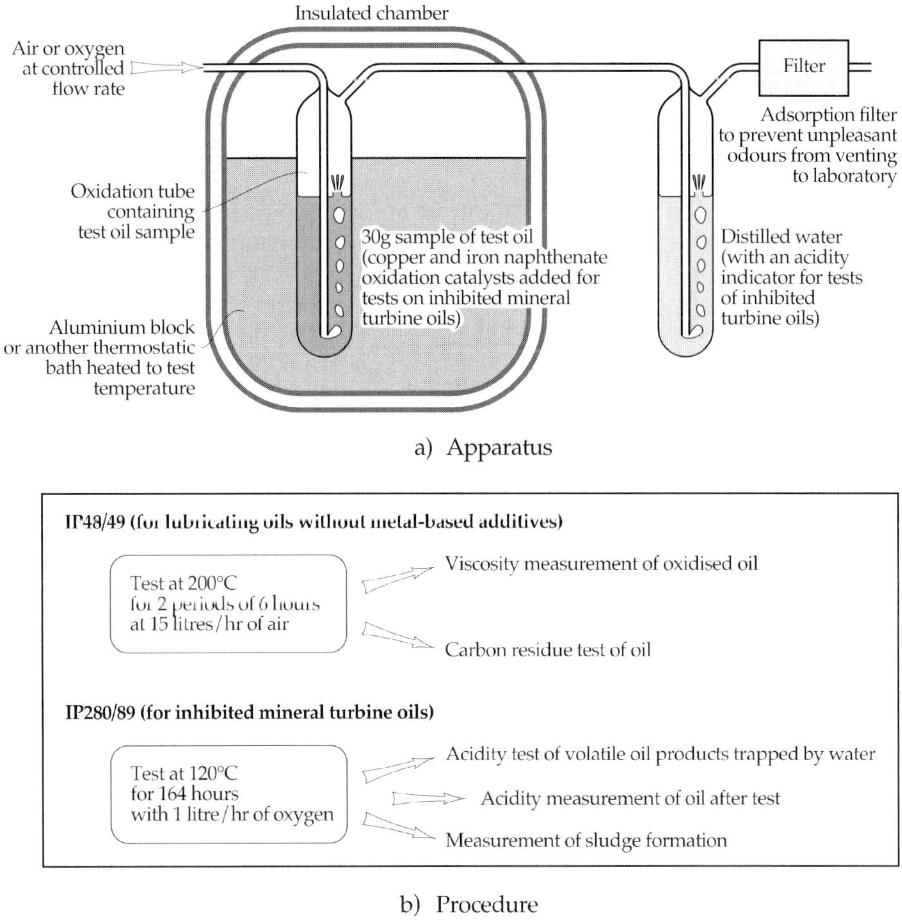

Figure 9.7 Schematic illustration of oxidation test at atmospheric pressure.

The vessel containing the oil is usually covered and the waste gas is passed through a second vessel containing water. The water absorbs any volatile oil oxidation products, which would otherwise escape into the atmosphere. The acidity of the oil sample is measured by neutralisation with potassium hydroxide solution after a standard period of oxidation which can be as long as 1500 hours

[24]. The length of time required to reach a standard level of acidity is taken as a measure of the oxidation resistance of the oil. A widely accepted acidity level is about 2 mg KOH/gram of oil, where acidity is denoted by the amount of KOH (potassium hydroxide) required for neutralisation of the oxidized oil. It is found by experience that beyond this level of acidity the remaining useful life of the oil is short.

Open vessel test procedures are specified in the Institute of Petroleum Standards IP 48/49 and IP 280/89. IP 48/49 is a faster but less realistic test since the oxidation temperature is 200°C compared to 120°C used in the latter version of this test, i.e. IP 280/89. IP 48/49 is used to determine the extent of oxidation by measuring the oil viscosity changes. This is an obsolete method since it does not indicate oxidation until the chemical reactions are so well advanced that a significant portion of the oil is already oxidized. IP 280/89 also includes an acidity measurement of the water from the second vessel. This standard enables the determination of the acidity of volatile oxidation products. Measurements of acidity change correspond more closely to observed changes in real lubricants than the extreme viscosity changes measured in IP 48/49. A copper and iron napthenate catalyst is also added to the oil in IP 280/89 to simulate the catalytic effect of organo-metallic compounds on oil oxidation, i.e. copper napthenate simulates metal contamination of a lubricating oil in actual machinery. The main disadvantage of this test is the amount of time required to its completion since several working days might be needed to complete the testing of just one oil sample. This rate of testing is too slow to be useful in a condition monitoring program where rapid answers are required and therefore it can only be used to determine the quality of the oil during its formulation [24].

· *Bomb Oxidation Tests*

Oxidation of oil can be accelerated by substituting oxygen for air and performing the test under elevated pressure, which increases the solubility of oxygen in oil. A 'bomb' is the term used to describe a mechanically strong enclosed vessel specifically designed for containing the mixture of oxygen and oil during a test. Since the reaction products are largely vapour or gas and the oxygen is pressurised, a robust vessel is required to withstand the stresses that may be generated by an unexpectedly reactive oil.

An apparatus used to determine the oxidation stability of greases is illustrated schematically in Figure 9.8 while the test procedures are specified in IP 142/85. The vessel is pressurized with air or oxygen in order to raise the severity of the test and shorten its duration. The relative oxidation data is obtained from the decline in gas pressure as the grease sample consumes oxygen to form non-gaseous oxidation products. With most lubricants, there is an 'induction period' where very little visible oxidation occurs. Once the induction period is passed, then oxidation appears to proceed rapidly and the pressure in the bomb declines sharply. The time required for the specified pressure loss to occur is used as a measure of the oxidation stability of greases.

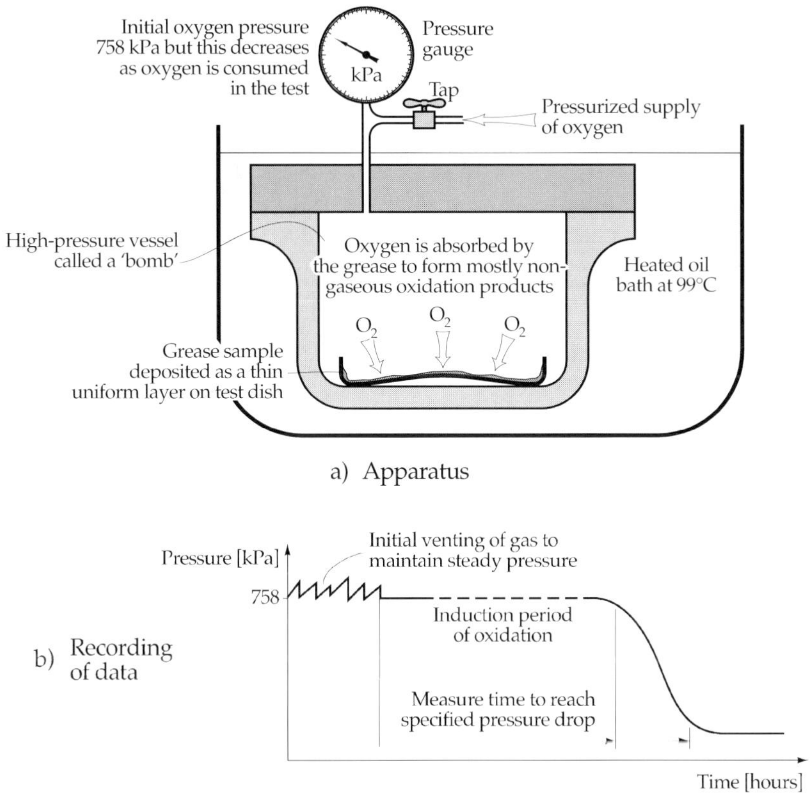

Figure 9.8 Schematic illustration of the 'bomb oxidation test' for greases.

To determine an oxidation stability of lubricating oils, a more sophisticated test, called the 'rotating bomb oxidation test', is used. The details of the test are described in IP 229/93 and ASTM D 2272 standards. The rotating bomb apparatus consists of a glass vessel inclined at an angle of 30° to the horizontal while being rotated by an electric motor. The purpose of this arrangement is to maintain a continuous circulation of the lubricant around a metal catalyst while being surrounded by a pressurized oxygen or air atmosphere. The metal catalyst is in the form of a wire coiled around the inside of the glass vessel. The circulation of lubricant ensures uniform access to oxygen and catalyst by every part of the lubricant sample. During the test, the oil and oxygen are added to the glass vessel ('bomb') and the oxygen pressure is raised to 620 [kPa]. The bomb is then heated to 150°C by immersion in a thermostatic bath and rotated at 100 [rpm]. Water and a solid copper catalyst are also added into the bomb to provide a closer simulation of operating conditions for a real lubricant. The induction time, manifested by a rapid drop in oxygen pressure, is used as a relative measure of oxidation stability of lubricating oils. The test is considered complete when the pressure drops by 175

[kPa]. The rotating bomb oxidation apparatus is illustrated schematically in Figure 9.9.

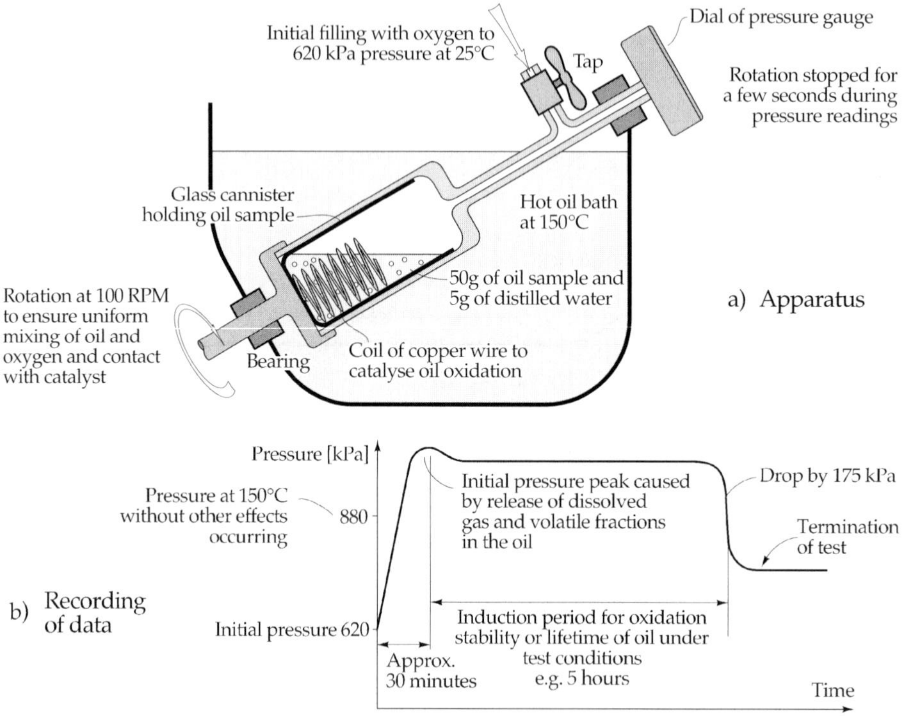

Figure 9.9 Schematic illustration of the 'rotating bomb oxidation test' for lubricating oils.

The rotating bomb oxidation test can typically require 2 to 30 hours for completion [22] and while much quicker than the tests at atmospheric pressure, this length of time still poses some difficulties for the wider applications of this test in machine condition monitoring. Furthermore, complex formulated oils such as engine oils containing Zinc Dialkyl Dithiophosphate (ZnDDP) do not show an easily determined, distinct decline in pressure and so are difficult to assess by this technique. The test is found to be most effective for oils containing aromatic amines and phenolic amines as antioxidants, which are typically used for turbine oils and this is where the test is most widely used [24].

Micro-Scale Oil Oxidation Tests

As is suggested by the name, micro-scale oxidation tests involve small amounts of oil. There are several reasons for using smaller rather than larger amounts of oil during the tests:

- ease of obtaining an oil sample since not all mechanical systems contain a large quantity of lubricating oil,
- need to accurately simulate oxidation of oil when it is present as a thin layer upon a hot metal surface, and
- need to reduce the testing time, i.e. making the technique more cost effective and more reliable, e.g. for applications in machine condition monitoring.

- *Thin Layer Oxidation Test*

The basic principle behind a thin layer oil oxidation test is to apply a small volume of test oil over a comparatively large dish so that all parts of the oil are in close contact with atmospheric oxygen and catalytic material of the dish [25-27]. During this test a greatly reduced volume of oil is used compared to bulk oxidation tests, i.e. 40 [μl] compared to 40 [ml] as specified in IP 48/49. The oil is contained in an open dish as a layer of approximately 0.8 [mm] deep and 8 [mm] in diameter. The dish is placed in an enclosed vessel with special ducts directing a stream of oxygen, air or an inert gas over the sample. Inert gas, such as nitrogen, can be used to determine the evaporation rates of the lubricant. The enclosed vessel is then placed in a thermostatic bath and heated to the temperature required. The apparatus and the test procedure are illustrated schematically in Figure 9.10.

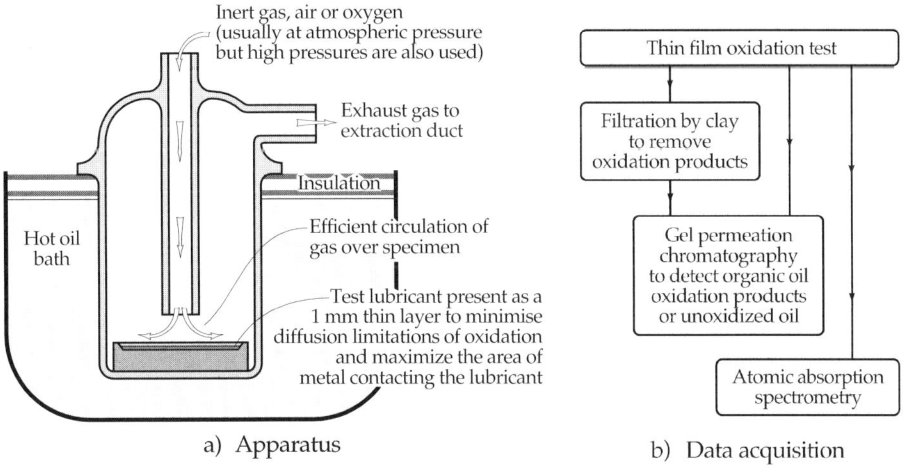

Figure 9.10 Schematic illustration of the 'thin layer' lubricating oil oxidation test.

The dish holding the sample can be made of various materials, e.g. copper or iron, in order to compare their catalytic effect. The catalytic activity of the dish can affect both the oxidation and evaporation rates. A high pressure version of the 'thin layer' apparatus has also been developed for tests where a high gas pressure is considered critical in experimental simulation [28].

Oxidation data is obtained by subsequent analysis of the test sample by standard techniques. For example, in one work the oil sample was analyzed before and after a test by Gel Permeation Chromatography (GPC) in order to find the molecular weight distribution of the unoxidized and oxidized oils [25]. The extent of evaporation, which is a problem during this type of test, is determined by filtering half of the oxidized sample through clay to remove all remaining oxidized components. The clay filtered sample is then analysed in GPC. The area between the untested lubricant curve and the clay-filtrated curve, as shown in Figure 9.11, indicates the total amount of lubricant converted by oxidation and/or lost by evaporation. The areas between the oxidized lubricant curve and the clay-filtered curve show the high-molecular-weight products (A) and 'same' unchanged molecular products (B) [25].

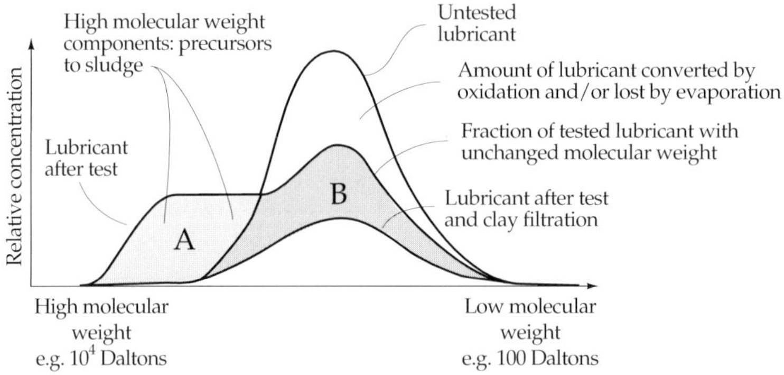

Figure 9.11 Schematic illustration of the procedure to determine the degree of oil oxidation after 'thin layer' oxidation test and Gel Permeation Chromatography.

An alternative test to that described above has also been developed and involves a 1.5 [g] sample of oil which is used in a modified rotating bomb oxidation test [23,25]. Often 0.5 [ml] of water and a catalyst are added. The oil is held in a metal dish, which may be of copper, steel, aluminium or other metal in order to simulate any catalytic effects. The test oil is subjected to the same level of oxygen pressure, rotation and inclination to the horizontal as in the standard rotating bomb oxidation test (ASTM D 2272). Temperature is, however, increased from 150°C to 160°C [24]. This test is found to show good correlation with engine lubricant specification tests [23] and has been used in the development of lubricants [29]. The time required for each thin layer test is significantly less than that in the rotating bomb oxidation test.

Non-Standard Tests

There is also a large number of other, non-standard oxidation stability test methods. All these methods have been developed with two principal aims:

- to reduce the test duration in-comparison to standard techniques, and
- to ensure that the technique developed can be used to evaluate the oxidation stability of a range of lubricants.

A useful test should be applicable to, for example, mineral, synthetic, automotive and turbine oil formulations, as well as to assist in product development work. Some of these methods are briefly described below.

- *Differential Scanning Calorimetry Tests*

The basic problem in lubricant monitoring is to determine the remaining time before excessive lubricant oxidation occurs. An accelerated oxidation test of lubricants using Differential Scanning Calorimetry (DSC) has been proposed [30]. DSC is commonly used in the characterisation of materials' thermal properties by measuring the specific heat of a material versus a change in the material's temperature. DSC has also been widely used to analyze the composition of oils and other hydrocarbons [31,32] but its application to measurements of lubricant degradation is relatively recent. Two stainless steel pans mounted on a heater and a thermocouple sensor are used. One pan contains a quantity of test sample while the other pan, which is empty, acts as a reference [24]. Both pans are simultaneously heated to the same temperature and the amount of power required to maintain the sample pan at the same temperature as the reference pan when heated is recorded. Sophisticated temperature controlling devices allow measurements of the minute heat fluxes from exothermic and endothermic reactions occurring inside the test sample to be made. The reference sample provides a datum temperature for the differential measurement of heat fluxes from the test material. A schematic illustration of the DSC operating principle is shown in Figure 9.12.

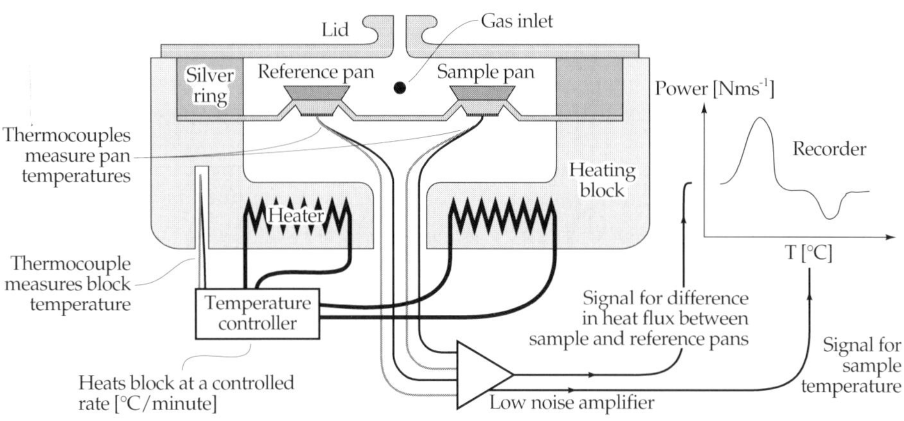

Figure 9.12 Schematic illustration of the Differential Scanning Calorimetry operating principle [33].

The principle behind the application of DSC and its related technique, Differential Thermal Analysis (DTA), to the assessment of a lubricant's oxidation stability is the acceleration of chemical reactions by elevated temperature where these reactions are detectable from the heat flux changes occurring when the sample is heated [34]. These heat flux changes represent exothermic, e.g. oxidation, or endothermic, e.g. evaporation, reactions occurring in the oil sample. Antioxidant depletion and oxidation reactions that may only progress slowly over many hours of lubricant service can be modelled in a few minutes at high temperature in a DSC system. With the DSC a remaining service life of a lubricant can be accurately predicted by a quick, about 45 [min], test which requires only a very small sample of less than 1 [ml] volume.

In the early work it has been found that using conventional DSC apparatus equipped with open pans to oxidize mineral oils leads to inaccurate results. At temperatures greater than 170°C the lubricant evaporates, affecting the magnitude and shape of the oxidation curve, which in turn gives imprecise values of the induction time. The use of a pressurised environment has therefore been suggested in order to suppress the evaporation of the lubricant [30]. Elevated pressures reduce sample volatility and evaporation signal interference and increase the peak magnitude and sharpness [34]. This can be achieved through the modification of the existing DSC chamber to incorporate high pressures [34], the use of a separate pressurised cell [35-39] or the use of an oxygen purged sealed pan or capsule [40]. The sealed capsule DSC technique (SCDSC) appears to be the least developed method, yet it has the advantages of a lower instrument cost and an easier operating technique in comparison with its high-pressure counterparts [24,30].

As with other analytical techniques such as IR spectroscopy, the application of SCDSC to lubricant degradation requires a carefully managed experimental procedure. Usually a test oil sample of about 2-3 [μl] volume is placed in a threaded stainless steel test capsule. The insertion of oil into the capsule is performed in an oxygenated glove box to ensure that the oil is exposed to oxygen during the test. It has been reported that the capsule can be pressurized with oxygen to approximately 0.7 [MPa] [30]. The capsule is then sealed to constant torque with a gold plated copper seal and placed in the DSC apparatus. First the system is equilibrated at 50°C and then the temperature is ramped at 10°C/minute to 140°C. A slower heating rate of approximately 3°C/minute is then imposed until an exothermic heat pulse is detected [41]. As the temperature increases, the levels of natural and added antioxidants in the lubricant deplete. During this period, the DSC trace is essentially horizontal. When the antioxidants are consumed, uncontrolled autocatalytic oxidation of the base oil occurs. This reaction is detected by the release of heat from exothermic oxidation reactions, manifested as a sharp deviation from the baseline. This exothermic heat peak is characteristic of oxidation reactions and signifies that the oil has begun rapid oxidation and has therefore reached the end of its test life. The period of time between the commencement of heating from 50°C and the exothermic heat pulse is called the induction time, usually about 30 to 45 minutes, and is used as a measure of oil oxidation stability. An example of the heat flux versus time record and illustration of the method used to determine the induction time from the DSC data is shown schematically in Figure 9.13.

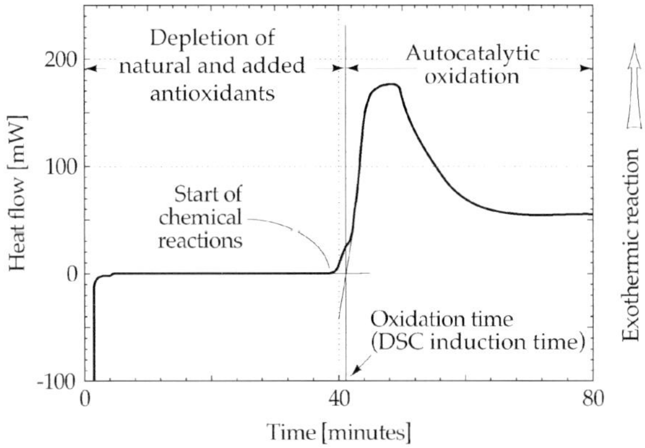

Figure 9.13 Schematic illustration of the detection of heat release from oxidized oil and the method used to determine the induction time, i.e. level of oil oxidation and its oxidation resistance, from the DSC data (adapted from [24]).

It was found that if the time for the heat pulse in DSC to occur is plotted against the lifetime of the oil as determined by conventional acidity tests, an approximate relationship results as shown in Figure 9.14.

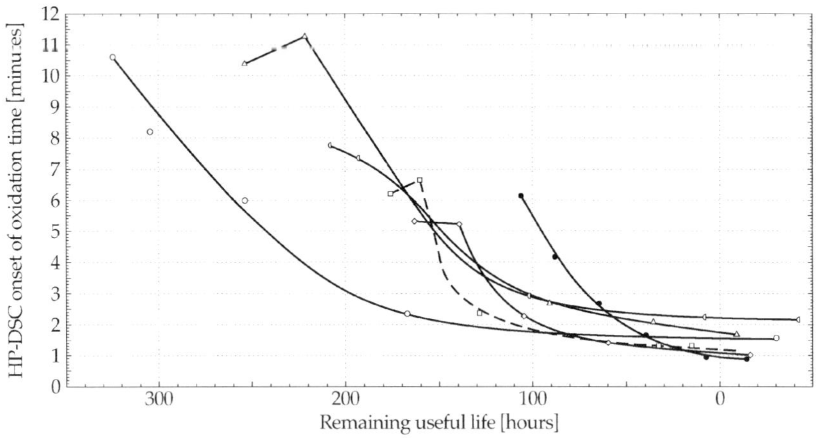

Figure 9.14 Relationship between period before heat pulse as measured in a DSC test and residual service life of a lubricating oil [30].

More detailed studies have shown a very good correlation between oxidation data from DSC and data from much slower, large sample, conventional oxidation tests, e.g. rotating bomb oxidation tests [41,42]. It was found that the quality of DSC data from the tests with stainless steel pans is not significantly diminished by complications such as suspended water and metal wear debris [41]. However, wear debris contamination may corrupt the results obtained when aluminium pans are used in high pressure DSC [41] as the aluminium exhibits a lower catalytic activity than iron [25]. In some isolated cases the accuracy of the DSC data in terms of oil life prediction could be affected by varying content of soluble iron catalysts such as iron napthenate.

The problem with the DSC test method is that the delay period before release of heat is subject to many influences, not just oil degradation. A significant problem is the reduction in predicted service life (i.e. reduced delay period) caused by fuel contamination of the lubricant [43]. This erroneously low predicted life is due to the more rapid oxidation of fuel compared to lubricant [43]. In cases like this instead of hazarding a guess that fuel contamination is occurring and more lubricant service life remains than indicated by the test, it is more practical to directly test the lubricant for fuel contamination using, for example, IR spectroscopy.

However, with carefully executed DSC tests, most of these problems can be reduced and the DSC method provides reliable oil oxidation data. For example, it was shown that the DSC data can be used to produce oil oxidation 'maps' that enable predictions of the remaining useful life of a lubricant to be made as illustrated in Figure 9.15 [41].

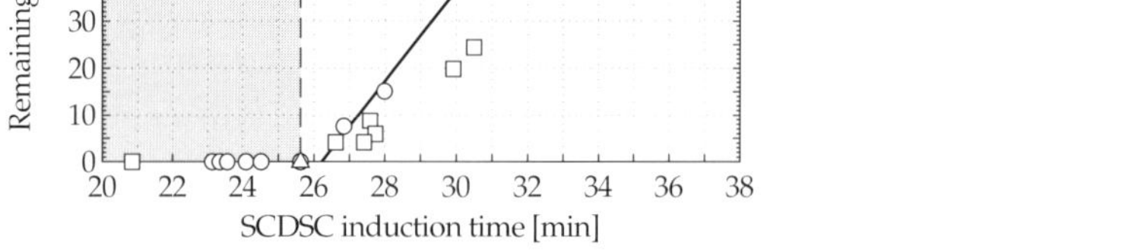

Figure 9.15 Example of the map allowing determination of the percentage of the remaining useful life of laboratory and industry oxidized turbine oil samples [41].

The DSC technique has been successfully applied in evaluation of mineral [35-37,42] and synthetic oils [35,37,38,42,44], diesel [39], turbine oils [41,42], aviation oils [30,45] and greases [36]. DSC has also been applied to the prediction of antioxidant degradation times for lubricants under simulated engine conditions [34], the evaluation of perfluoropolyalkylether fluids [37], postulating a kinetics model for tricresyl phosphate oxidation [46], design of optimum additive systems and concentrations for inhibiting oxidation in lubricating oils, aluminium and lithium greases [36], ester lubricants [21,44], antioxidants performance evaluation [47] and many other applications [24].

· *Differential Thermal Analysis (DTA)*

Differential Thermal Analysis (DTA) has been used to determine both oxidation and thermal stability characteristics of lubricating oils and organic products [28]. The technique is similar to DSC. A sample pan containing up to 50 [mg] of oil together with a reference pan are placed in a furnace, which is continuously purged with oxygen while both pans are heated under controlled conditions up to 800°C. As the temperature is increased, the reactions occurring in the oil sample are detected from the temperature difference 'ΔT' recorded between the sample pan and reference pan. The positive 'ΔT' represents exothermic, e.g. oxidation, reaction while negative 'ΔT' represents endothermic, e.g. evaporation, reaction occurring in the sample. When the oil oxidises usually two exothermic peaks are observed [24]. The first peak corresponds to oil oxidation while the second represents the combustion of oxygenated compounds into different by-products as illustrated in Figure 9.16.

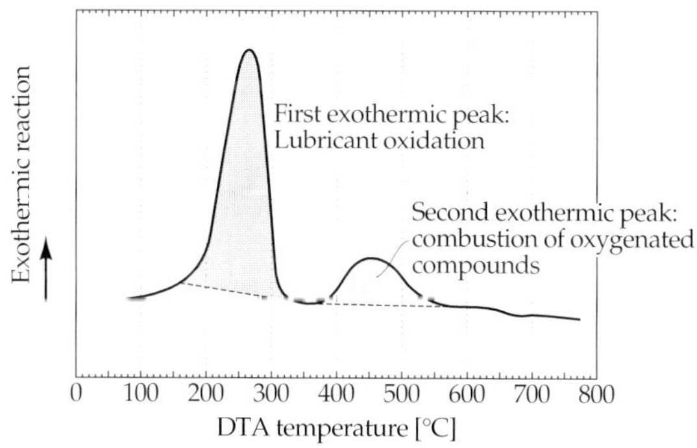

Figure 9.16 Typical DTA oil oxidation plot (adapted from [48]).

The extrapolated temperature of the first peak could be used to assist in defining the remaining useful life of the lubricant, i.e. the lower the temperature, the lower the oxidation life [24].

· *Thermogravimetric Balance*

The weight changes occurring in the sample pan due to thermo-oxidation can be continuously monitored by a thermogravimetric balance (TG) technique and the results used to determine the oxidation stability of an oil sample. The weight loss curve defines both the temperature at which the formation of higher molecular weight oxygenated compounds begins (oxidation point) and the weight of these compounds. It has been found that the amount of oxygenated compounds remaining is inversely proportional to the oxidation stability of the oil [49]. An example of TG weight loss curve is shown in Figure 9.17.

In general, the use of DTA/TG techniques in the determination of oxidation stability of oils has been limited. For example, the effect of base oil volatility on the oxidation stability of lubricating oils has been evaluated [50], however, remaining life studies of a lubricant have received little attention. Instead, these techniques have found application in thermal stability studies of petroleum products [51].

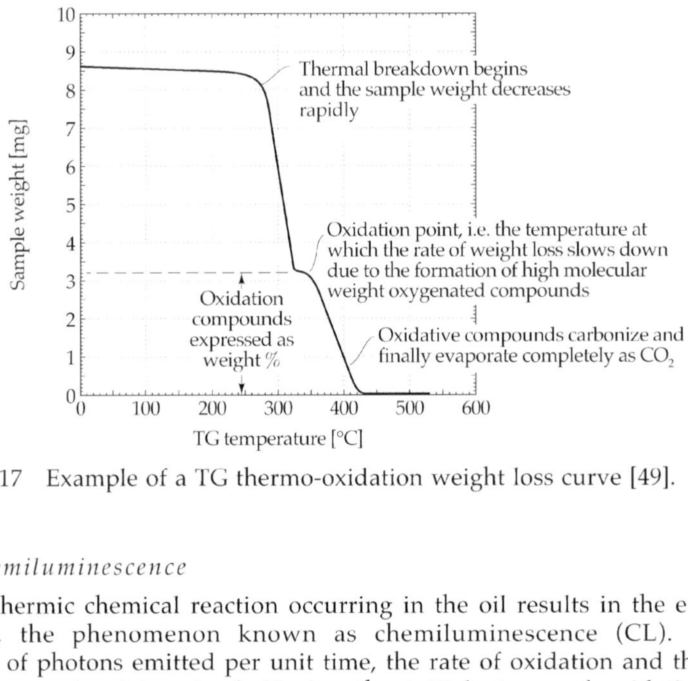

Figure 9.17 Example of a TG thermo-oxidation weight loss curve [49].

· *Chemiluminescence*

The exothermic chemical reaction occurring in the oil results in the emission of photons, the phenomenon known as chemiluminescence (CL). From the intensity of photons emitted per unit time, the rate of oxidation and the onset of oxidation can be determined. During the initial stages of oxidation photon emission is low due to the consumption of antioxidants. When the antioxidants are consumed, autocatalytic oxidation of the base oil occurs resulting in a dramatic increase in the intensity of emitted CL photons. The time taken to observe this increase in CL photons relates to the remaining oxidation life of the lubricant, i.e., the longer the time, the longer the remaining oxidation life. The CL technique has the advantages of being rapid, versatile and non-intrusive and can be used to monitor both low and high rates of oxidation [52]. More information on CL can be found in [53-55].

Other Non-Standard Methods

There would be enormous practical benefits if a simple device or technique could be developed that consistently measures the extent of lubricant degradation. Many inventors have proposed various devices which subject an oil sample to heat or electric fields in order to determine its remaining useful life (RUL). For example, one device contains two parallel plates immersed in the test lubricant while subjected to an electric potential. Measured variations in capacitance and resistance are claimed to allow detection of suspended metals, water and micelles formed by detergent additives [56]. Another device uses a semiconductor detector to analyze volatile oxidation products released from degraded lubricants [57]. The semiconductor detector was, in fact, originally developed for alcohol breathalysers [58]. Oxidation of a test lubricant was successfully monitored, but the detector requires a precise adjustment for a particular lubricant [57]. This effectively renders this device unsuitable for routine testing of unspecified lubricants. The most commonly used non-standard methods are summarized in Table 9.4.

Table 9.4 Commonly used non-standard oxidation stability tests (adapted from [21]).

Test	Oils Evaluated	Test Conditions	Test Duration
Universal Oxidation Test	Base oil doped with phenolic antioxidants [59]	Temp: 50 to 375°C. Water, soluble metal catalysts and metal coupons can be added. Glassware allows the return of volatiles. Different gases and flow rates can be used. Measures time to a TAN of 2.0 [mg KOH/g].	20 to 250 [hrs]
Chemiluminescence (CL)	Mineral oil [53] Oil fractions [52] Automotive oils [54] Vegetable oils [60]	Temp: room to 350°C. Heating rate 1 to 15 [°C/min]. Thin film test - small oil volume. Different gases and flow rates can be used. Measures light from exothermic reactions.	1 to 3 [hrs] [53] < 10 [min] [52]
Penn State microoxidation test - atmospheric version	Vegetable oils [61] Diesel oils [62] Copper additives [63]	Temp: ambient to 275°C. Sample vol. 40 [μl]. Catalyst: mild steel coupon. Air at 20 [ml/min]. Measures oxidation products after set period.	Up to 40 [min]
Penn State microoxidation test - pressurised version	Mineral and synthetic base oils [64,65]	Same conditions as atmospheric version. 1 atm of air + 4 atm of nitrogen. Measures oxidation products after set period.	Up to 1 [hr]
Thermogravimetric balance (TG)	Mineral oil [49]	Temp: ambient to 600°C at 5 [°C/min]. Sample size: 10mg. Oxygen at 20 [ml/min]. Measures sample weight loss with temp.	Up to 2 [hrs]
Differential Thermal Analysis (DTA)	Mineral oils [48,50]	Temp: ambient to 800°C. Sample size: 40 [mg]. Oxygen atmosphere. Measures temperature at which the exothermic reaction occurs.	Up to 1.5 [hrs]
Differential Scanning Calorimetry (DSC)	Mineral [35,37,44], synthetic oils [35, 37,38,46,47], diesel [39], aviation oils [30,45] and greases [36]	Temp: ambient to 300°C. Sample size: 0.7-0.8 [mg] [39] to 20 [mg] [37]. Oxygen atmosphere. Measures time to exothermic reaction.	Up to 1 [hr]

The techniques listed in Table 9.4 have both advantages and disadvantages in relation to determining the oxidation stability of lubricating oils. The main disadvantage common to the majority of these techniques is related to the availability of the test equipment, which in turn appears to be related to the volume of published literature. For example, the Universal Oxidation Test was developed by the ASTM Committee D02.09-D-3 but since the initial development work the test has been rarely cited [21]. The research, as indicated by available literature, on Penn State microoxidation test, developed in the late 70s, appear to be confined to the Penn State University [21]. Studies on Differential Thermal Analysis and Thermogravimetric balance of oil oxidation are also limited. Instead, these combined techniques generally find application in thermal stability rather than oxidation stability studies of petroleum products [21,51].

9.5 SUMMARY

The characterisation of a lubricant is a fundamental part of any study of lubricated wear and friction. Various techniques of rheometry and chemical analysis are utilized as lubricant characterisation involves both chemical and physical parameters. There are many specialised techniques developed to provide data where general purpose equipment is inadequate. With each advance in understanding of tribology, the range of information required for lubricant characterisation increases. Where previously values of viscosity were sufficient to describe the lubricant rheology, it is now necessary to consider the pressure-viscosity coefficient and limiting shear stress at extreme shear rates. Early lubricant degradation tests merely considered temperature while now the catalytic effect of a metal surface is included in the test.

Oil oxidation tests have progressively developed from simple combinations of prolonged oxidative stress followed by a physical measurement to a detailed analysis using the latest techniques in analytical chemistry. A basic problem to be resolved in any oxidation test is how to efficiently condense the complex indications on oxidation into one or two direct parameters. Parameters such as viscosity or acidity have some serious limitations or disadvantages. The limitations of trying to correlate oxidation with acidity, for example, have led to a more exact analysis based on the direct detection of the oxidative reactions occurring in the oil. This is the reasoning behind oil testing using Differential Scanning Calorimetry, which yields the unique combination of accurate data and rapid testing. For the development of new lubricants and for some detailed analysis of oil failures, it is necessary to know how the oil degrades. Advanced chemical techniques such as Fourier Transform Infra-Red Spectroscopy are required to detect, distinguish and quantify the multiple interacting chemical reactions that control oil oxidation. In summary, careful planning and execution of lubricant characterisation is a necessary part of any tribological investigation.

REFERENCES

1 C.J. Peppiatt and G. Shama, Ultraviolet treatment of microbially contaminated metal working fluids using a thin film contactor, *Lubrication Engineering*, Vol. 56, No. 12, 2000, pp. 11-17.

2 S. Kalhan, S. Twinning, R. Denis, R. Marano, R. Messick and B. Johnston, Shear stable mist suppressants for aqueous metal working fluids: Development and field evaluations, *Lubrication Engineering*, Vol. 56, No. 9, 2000, pp. 27-33.

3 W.J. Bartz, Ecological and environmental aspects of cutting fluids, *Lubrication Engineering*, Vol. 57, No. 3, 2001, pp. 13-16.

4 J.M. Perez and D.I. Hoke, Toxicology, environment, safety and health, Tribology Data Handbook, editor: E.R. Booser, CRC Press, 1997.

5 J.P. Coates and L.C. Setti, Condition monitoring of crankcase oils using computer aided infrared spectroscopy, *SAE Paper No. 831681*, SP-558, 1983, pp. 37-50.

6 J.P. Coates and L.C. Setti, Infrared spectroscopic methods for the study of lubricant oxidation products, *ASLE Transactions*, Vol. 29, 1986, pp. 394-401.

7 G.W. Stachowiak, T.B. Kirk and G.B. Stachowiak, Ferrography and fractal analysis of contamination particles in unused lubricating oils, *Tribology International*, Vol. 24, No. 6, 1991, pp. 329-334.

8 S. Bair and W.O. Winer, A new high-pressure, high shear stress viscometer and results for lubricants, *Tribology Transactions*, Vol. 36, 1993, pp. 721-725.

9 E. Höglund and B. Jacobson, Experimental investigation of the shear strength of lubricants subjected to high pressure, *Journal of Lubrication Technology, Trans. ASME*, Vol. 108, 1986, pp. 571-578.

10 S. Bair and W.O. Winer, The high pressure high shear stress rheology of liquid lubricants, *Journal of Tribology, Trans. ASME*, Vol. 114, 1992, pp. 1-13.

11 B. Jacobson, Rheology and elastohydrodynamic lubrication, Elsevier, Amsterdam, 1991.

12 R. Feng and K.T. Ramesh, The rheology of lubricants at high shear rates, *Journal of Tribology, Trans. ASME*, Vol. 115, 1993, pp. 640-649.

13 K.A. Hartley, J. Duffy and R.H. Hawley, The torsional Kolsky (Split-Hopkinson) Bar, ASM Metals Handbook, 9th edition, ASM International, Vol. 8, pp. 218-228.

14 G.R. Paul and A. Cameron, The ultimate shear stress of fluids at high pressures measured by a modified impact microviscometer, *Proc. of the Royal Society of London, Series A*, Vol. 365, 1979, pp. 31-41.

15 P.L. Wong, S. Lingard and A. Cameron, The high pressure impact viscometer, *Tribology Transactions*, Vol. 35, 1992, pp. 500-508.

16 P.L. O'Neill and G.W. Stachowiak, High shear rate impact microviscometer, *Tribology International*, Vol. 29, No. 7, 1996, pp. 547-557.

17 R.D. Stauffer and J.L. Thompson, Improved bench oxidation tests for railroad diesel engine lubricants, *Lubrication Engineering*, Vol. 44, 1988, pp. 416-423.

18 E. Jantzen, The simulation of aircraft engine oil deposits under static conditions and their influencing factors, *Lubrication Engineering*, Vol. 44, 1988, pp. 319-323.

19 E.E. Klaus, Status of new direction of liquid lubricants, New Directions in Lubrication, Materials, Wear and Surface Interactions, editor: W.R. Loomis, April 1983, NASA Lewis Research Center, USA, Noyes Publ., New Jersey, USA, 1985, pp. 355-375.

20 W.R. Jones, Thermal and oxidative stabilities of liquid lubricants, New Directions in Lubrication, Materials, Wear and Surface Interactions, editor: W.R. Loomis, April 1983, NASA Lewis Research Center, USA, Noyes Publ., New Jersey, USA, 1985, pp. 403-427.

21 B.F. Bowman, The Effect of Oil Oxidation on Scuffing, PhD thesis, The University of Western Australia, 1997.

22 T.M. Warne and P.C. Vienna, High-temperature oxidation testing of lubricating oils, *Lubrication Engineering*, Vol. 40, 1984, pp. 211-217.

23 C-S. Ku and S.M. Hsu, A thin-film oxygen uptake test for the evaluation of automotive crankcase lubricants, *Lubrication Engineering*, Vol. 40, 1984, pp. 75-83.

24 W.F. Bowman and G.W. Stachowiak, Review of techniques used for evaluating the oxidative condition of lubricating oils, *Trans. Mechanical Engineering, Institution of Engineers, Australia*, Vol. ME 23, No. 1, 1998, pp. 19-26.

25 D.B. Clark, E.E. Klaus and S.M. Hsu, The role of iron and copper in the oxidation degradation of lubricating oils, *Lubrication Engineering*, Vol. 41, 1985, pp. 280-287.

26 E. Cvitkovic, E.E. Klaus and F. Lockwood, A thin-film test for measurement of the oxidation and evaporation of ester-type lubricants, *ASLE Transactions*, Vol. 22, 1979, pp. 395-401.

27 E.E. Klaus, L. Cho and H. Dang, Adaptation of the Penn State micro-oxidation test for the evaluation of automotive lubricants, Soc. Automotive Eng. Technical Paper No. 801362, SAE Sp. Pub. 80-473, 1980, pp. 83-92.

28 S. Gunsel, E.E. Klaus and J.L. Duda, High temperature deposition characteristics of mineral oil and synthetic lubricant basestocks, *Lubrication Engineering*, Vol. 44, 1988, pp. 703-708.

29 J.E. Davis, Oxidation characteristics of some engine oil formulations containing petroleum and synthetic base-stocks, *Lubrication Engineering*, Vol. 43, 1987, pp. 199-202.

30 R.E. Kauffman and W.E. Rhine, Development of a remaining useful life of a lubricant evaluation technique, Part I: Differential scanning calorimetric techniques, *Lubrication Engineering*, Vol. 44, 1988, pp. 154-160.

31 F. Noel, The characterization of lube oils and fuel oils by DSC analysis, *Journal of the Institute of Petroleum*, Vol. 57, 1971, pp. 354-358.

32 C. Giavarini and F. Pochetti, Characterization of petroleum products by DSC analysis, *Journal of Thermal Analysis*, Vol. 5, 1973, pp. 83-94.

33 H.H. Willard, L.L. Merritt, Jr., J.A. Dean and F.A. Settle, Jr., Instrumental methods of analysis, 7th edition, Wadsworth Publishing Company, Belmont, California, 1988.

34 S.M. Hsu, A.L. Cummings and D.B. Clark, Differential scanning calorimetry test method for oxidation stability of engine oils, Proc. Conf. Measurements and Standards for Recycled Oil-IV, National Bureau of Standards, Gaithersburg, Washington, DC, 1982, pp. 195-207.

35 K.I. Papathomas and A.D. Katani, Oxidative stability of lubricants for dry circuit sliding contacts, *Journal of Thermal Analysis*, Vol. 35, 1989, pp. 153-161.

36 R.J. Smith and G.J. Gowlett, Rapid assessment of lubricant oxidation by differential scanning calorimetry, Proc. International Tribology Conference, Melbourne, The Institutions of Engineers, Australia, National Conference Publication No. 87/18, 1987, pp. 325-330.

37 J.R. Barnes and J.C. Bell, Laboratory screening of engine lubricants for high temperature performance, *Lubrication Engineering*, Vol. 45, 1989, pp. 549-555.

38 C.E. Snyder, L.J. Gschwender and O.L. Scott, Characterisation of model perfluoropolyalkylethers by miniaturised thermal oxidative techniques - Part II: Pressure differential scanning calorimetry, *Tribology Transactions*, Vol. 38, 1995, pp. 733-737.

39 Y. Zhang, P. Pei, J.M. Perez and S.M. Hsu, A new method to evaluate the deposit forming tendency of liquid lubricants by differential scanning calorimetry, *Lubrication Engineering*, Vol. 48, 1992, pp. 189-195.

40 W.F. Bowman and G.W. Stachowiak, Determining the oxidation stability of lubricating oils using sealed pan differential scanning calorimetry (SPDSC), *Tribology International*, Vol. 29, No. 1, 1996, pp. 27-34.

41 W.F. Bowman and G.W. Stachowiak, Application of sealed capsule differential scanning calorimetry – Part I: Predicting the remaining useful life of industry-used turbine oils, *Lubrication Engineering*, Vol. 54, No. 10, 1998, pp. 19-24.

42 W.F. Bowman and G.W. Stachowiak, New criteria to assess the remaining useful life of industrial turbine oils, *Lubrication Engineering*, Vol. 52, No. 10, 1996, pp. 745-750.

43 S. Schwartz, Discussion to [30], *Lubrication Engineering*, Vol. 44, 1988, pp. 160-161.

44 J. Yao and J. Dong, Antioxidation synergism between alkali metal salts and arylamine compounds in synthetic lubricants, *Tribology Transactions*, Vol. 39, 1996, pp. 498-500.

45 D.G. Pachuta, J.A. Thich, R.W. Knipple and D.A. Stephanic, Experiences with an analytical monitoring program designed for commercial flight tests of advanced Mil-L-23699 turbine fluids, Turbine Oil Monitoring ASTM STP 1021, editors: W.C. Young and R.S. Robertson, ASTM, Philadelphia, 1989, pp. 54-76.

46 S. Shankwalker and D. Placek, Oxidation kinetics of tricresyl phosphate (TCP) using differential scanning calorimetry (DSC), *Lubrication Engineering*, Vol. 50, 1994, pp. 261-264.

47 W.F. Bowman and G.W. Stachowiak, Application of sealed capsule differential scanning calorimetry, Part II: Assessing the performance of antioxidants and base oils, *Lubrication Engineering*, Vol. 55, No. 5, 1999, pp. 22-29.

48 H.H. Abou El Naga and A.E.M. Salem, Testing the thermooxidation of lubricating oils via differential thermal analysis, *J. Therm. Anal.*, Vol. 32, 1987, pp. 1401-1413.

49 H.H. Abou El Naga and A.E.M. Salem, Base oil thermooxidation, *Lubrication Engineering*, Vol. 42, 1986, pp. 210-217.

50 H.H. Abou El Naga and A.E.M. Salem, The effect of base oil volatility on lubricating oil oxidation stability, *Lubrication Engineering*, Vol. 44, 1988, pp. 931-938.

51 H.H. Abou El Naga and A.E.M. Salem, Testing thermal stability for base oils via thermogravimetric balance and differential thermal analyser, *Lubrication Engineering*, Vol. 41, 1985, pp. 470-476.

52 P. Pei, S.M. Hsu, S. Weeks and R. Lawson, Chemiluminescence instrumentation for fuel and lubricant oxidation studies, *Lubrication Engineering*, Vol. 45, 1989, pp. 9-15.

53 L. Zlatkevich, New chemiluminescence apparatus and method for evaluation of thermal oxidative stability of lubricants, *Lubrication Engineering*, Vol. 44, 1988, pp. 544-552.

54 S.K. Ivanov, Z.D. Kalitchin and V.S. Aleksandrov, Use of chemiluminescence in determining the thermal and oxidative stability of motor oils, *Lubrication Science*, Vol. 4, 1992, pp. 135-153.

55 D.B. Clark and S.M. Hsu, Chemiluminescence of fuels and lubricants - A critical review, *Lubrication Engineering*, Vol. 39, 1983, pp. 690-695.

56 T. Kato and M. Kawamura, Oil maintenance tester: A new device to detect the degradation levels of oils, *Lubrication Engineering*, Vol. 42, 1986, pp. 694-699.

57 R.N. Bolster and H. Wohltejen, Electronic detection of synthetic lubricant oxidative breakdown, *ASLE Transactions*, Vol. 29, 1986, pp. 377-382.

58 T. Seiyama, A. Kato, K. Fujiishi and M. Nagatani, A new detector for gaseous components using semiconductive thin films, *Analytical Chemistry*, Vol. 34, 1962, pp. 1502-1503.

59 T.M. Warne and P.C. Vienna, High-temperature oxidation testing of lubricating oils, *Lubrication Engineering*, Vol. 40, 1984, pp. 211-217.

60 B.W. Matthaeus, Determination of the oxidative stability of vegetable oils by Rancimat and conductivity and chemiluminescence measurements, *J. Am. Oil Chem. Soc.*, Vol. 73, 1996, pp. 1039-1043.

61 S. Asadauskas, J.M. Perez and J.L. Duda, Oxidative stability and antiwear properties of high oleic vegetable oils, *Lubrication Engineering*, Vol. 52, 1996, pp. 877-882.

62 V. Palekar, J.L. Duda, E.E. Klaus and J. Wang, Evaluation of high-temperature liquid lubricants using the Penn State micro-oxidation test, *Lubrication Engineering*, Vol. 52, 1996, pp. 327-334.

63 S. Gunsel and F.E. Lockwood, The influence of copper-containing additives on oxidation and corrosion, *Tribology Transactions*, Vol. 38, 1995, pp. 485-496.

64 S. Gunsel, E.E. Klaus and J.L. Bailey, Evaluation of some poly-alpha-olefins in a pressurised Penn State microoxidation test, *Lubrication Engineering*, Vol. 43, 1987, pp. 629-635.

65 S. Gunsel, E.E. Klaus and J.L. Duda, High temperature deposition characteristics of mineral oil and synthetic lubricant basestocks, *Lubrication Engineering*, Vol. 44, 1988, pp. 703-708.

10 WEAR PARTICLE ANALYSIS

10.1 INTRODUCTION

During the operation of machinery and during almost every tribological experiment wear debris is produced. An integral part of tribological experimental methodology is to collect samples of used lubricant/process fluid containing the wear debris. Several specialized methods of extracting and analyzing wear debris from used lubricants/process fluids have been developed to become a fertile source of experimental data.

Wear particles are a valuable source of information about wear processes since their composition, size and morphology, i.e. their shape and surface topography, are indicative of the wear mechanism involved in their formation. Accurate data on the wear particle characteristics is also relevant to assessments of 'health' status of the machinery and reliability in mechanical systems that are sensitive to particulate contamination.

The size and number of wear particles determine whether closely conforming components can continue to slide freely, e.g. piston against cylinder in combustion engine, articulating surfaces in artificial implants or various devices in biomedical applications which require stringent control of wear particles. For example, it was found that in the artificial heart valves, the most wear resistant material may produce excessively large wear particles compared to more rapidly wearing materials [1]. For artificial heart valves, a maximum allowable size of wear debris is necessary to prevent obstructions from forming in capillaries, so the size of wear debris dictates the choice of heart valve material. In artificial joints, e.g. in total knee replacements, it is thought that submicron size ultrahigh molecular weight polyethylene (UHMWPE) particles may introduce more severe cellular reaction, osteolysis and aseptic loosening than large wear particles [2].

There appears to be a trade-off between the reduction in wear severity and the production of small submicron size wear particles which increase the cellular response leading to late aseptic loosening of total knee replacements [2]. For almost all machines in use, an increasing number and size of wear particles usually indicate impending machine failure. This latter aspect of wear particle analysis is fundamental to machine condition monitoring which has become an essential technique for the maintenance of machinery.

Wear particle analysis has been transformed from early tests of sludge at the bottom of oil sumps to a precise experimental technique which enables the resolution of wear mechanisms sometimes even without the need to interrupt an experiment or machinery under test. The non-invasive nature of the investigation is a further advantage of wear particle analysis in tribological experimentation which can often provide data which is not obtainable by any other means.

The investigator or an engineer needs to know how wear particles present in the tribological system can be detected or captured and how, based on their composition, quantity, size and morphology, information about the status of the machinery or the wear mechanisms occurring during the experiment can be obtained. In this chapter some basic and specialized methods used in wear particle analysis are described and discussed.

10.2 TECHNIQUES USED IN WEAR PARTICLE ANALYSIS

Wear particle analysis has gained commercial interest and there is continuous improvement in the various techniques developed. There are two basic types of techniques of wear particle analysis widely used in laboratories and throughout the industry. One technique, which is the oldest, is based on the analysis of the chemical composition of the lubricant. Most lubricants are made of hydrocarbons, while wear particles usually contain iron, copper, tin and lead (e.g. bronzes and babbitts) so that chemical data from the lubricant analysis easily reveals the presence and quantity of metallic wear particles in the lubricating oil. Another technique, more recently developed, involves an extraction of wear particles from a lubricating oil sample and then measurement of their number, their size, shape and assessment of their morphology. These characteristics of the wear particle population in a used lubricating oil reveal much about the nature and severity of wear processes occurring and also 'health' status of the machinery.

Chemical Analysis of Particles Present in Lubricant or Fluid Sample

The most common method used to analyse the chemical composition of particles present in lubricant or process fluid is 'flame spectroscopy'. The technique is based on the emission or absorption of a characteristic colour by each chemical element when heated in a gas flame or electric arc. Common examples of light emission at high temperatures are the orange colour of sodium when the kitchen salt is sprayed into a flame or the green colour when copper sulphate particles are burnt in a flame. It is apparent that if a sample of lubricating oil were vaporised by a similar flame then any copper present in the wear particles suspended in the oil from worn bearings would be detectable by a green colour in the flame. Systematic analysis of all the colours produced by exposure of a lubricant sample to severe heat gives a comprehensive data of the chemical elements present in this sample. The concentration of various elements such as iron, copper, lead, zinc, calcium, silicon, sulphur and phosphorus present in the oil sample can also be assessed. Two forms of spectroscopy utilizing this principle have evolved and are in general use, i.e. Atomic Absorption Spectroscopy (AAS) and Atomic Emission Spectroscopy (AES) [e.g. 3-5].

The operating principles of AAS and AES techniques are illustrated schematically in Figure 10.1 In the AAS a small amount of a diluted oil sample is fed into an acetylene-oxygen or nitrous oxide mixture flame [3]. A hollow cathode lamp emits a beam of light, via the chopper, corresponding exactly to the energy needed to excite the element to the elevated energy state. The atoms from the vaporised oil sample absorb light at a specific wavelength characteristic for each element present. Since each cathode lamp emits light of a specific wavelength a separate lamp is required for each element [3]. An alternating absorbed signal from the flame travels to the monochromator, tuned to the desired wavelength, together with the continuous reference signal from the lamp. The absorbed light is then related to the concentration of the element in the oil.

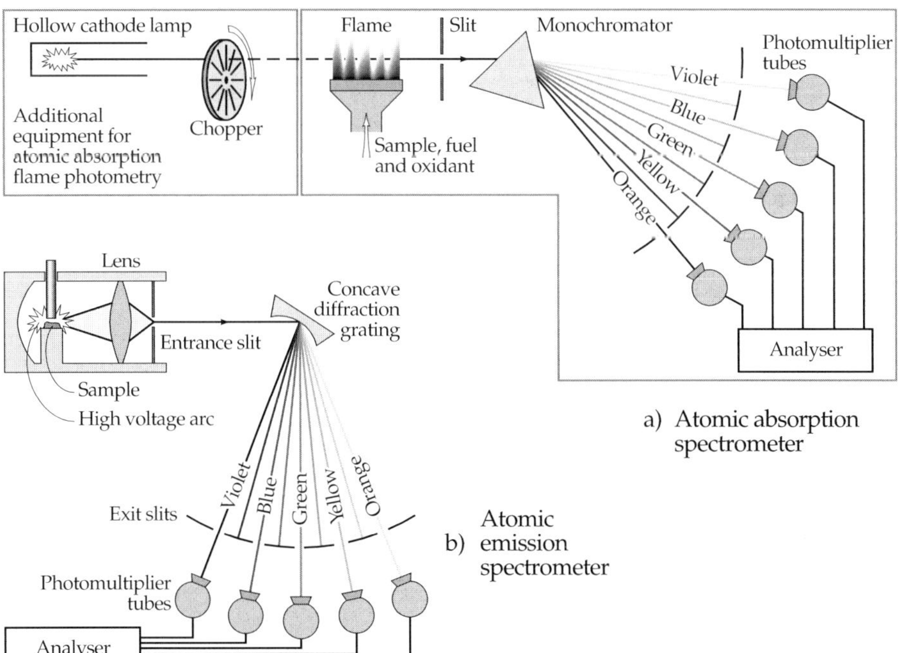

Figure 10.1 Schematic illustration of the operating principles of flame spectroscopy; a) atomic absorption spectrometer and b) atomic emission spectrometer.

Most AES instruments are either of the rotating disc electrode type (RDE) or the inductively coupled plasma (ICP) type. In RDE/AES spectrometer the rotating disc electrode brings the oil sample into a gap between the disc and a stationary electrode. An applied high voltage current results in an electric arc causing the individual atoms in the oil sample to give off light or radiant energy [3]. The ICP/AES technique employs an electrode free plasma created by passing argon gas through a plasma torch located inside a radio frequency induction coil connected to a radio frequency alternating current generator. An ionized argon gas (plasma) becomes very hot and reaches temperatures between 8,000-10,000°C. The oil sample is aspirated through the centre of the torch into the plasma where the atoms are excited to give off radiant energy [3]. Photomultiplier tubes detect and convert the radiant energy to a signal which is related to a concentration of elements present in the oil sample.

The technique of flame spectroscopy was for the first time commercially employed in condition monitoring by United States railroad companies in the mid-20th century as a means of detecting diesel engine component failures in locomotives [6]. This program has been known as Spectrographic Oil Analysis Program or more commonly SOAP. Nowadays the abbreviation SOAP is used in machine condition monitoring to indicate the oil analysis by either AAS or AES technique.

These two techniques, AAS and AES, have their relative merits and disadvantages. The AES provides rapid sample analysis since all elements present in a sample are detected simultaneously while AAS, although relatively slower, offers greater sensitivity to low concentrations of elements and requires a smaller sample size [3,7]. Also the AAS technique is relatively simple to use and the instrumentation is least costly, however, it requires a different light source for each element [3]. The RDE/AES spectrometer is usually preferred for its simplicity of operation and reliability. It can also analyse larger particles present in the used oil sample than ICP/AES or AAS techniques, i.e. it is inefficient for particles > 10 [μm] while ICP/AES or AAS techniques are inefficient for particles > 5 [μm] [3,8].

Information on the concentrations of chemical elements in the oil is useful in the assessment of either the severity of wear occurring during the experiments or the 'health status' of the machinery. Unfortunately useful information on concentrations of lighter elements such as carbon and oxygen cannot be obtained by these techniques. AAS and AES therefore cannot provide information on either the accumulation of soot in engine oils or increases in oxygen content due to the formation of oxyacids from oil oxidation.

For all ferrous sliding systems, the most important data is, in most cases, the concentration of iron since this usually relates to the wear particle population in the lubricating oil. There are, however, exceptions where iron content is not controlled by wear debris concentration. For example, in mining and mineral processing machinery exposed to iron ore dust, an elevated iron concentration may be due to iron oxide (dust) contamination of the lubricating oil. The copper concentration reveals wear of bearing babbitts and bronzes or else corrosion of copper components by excessively reactive Extreme Pressure additives present in the lubricating oil. Concentrations of lead and zinc may depend on either the wear of babbitts and bronze components or may relate to the concentrations of lubricant additives.

Most crankcase lubricating oils (e.g. motor vehicle engine oils) contain zinc dialkyldithiophosphate (ZnDDP) as a lubricating additive. The zinc concentration will therefore be related to the concentration of the additive and unlike the other metal concentrations, a minimum concentration of ZnDDP and hence zinc is essential for proper functioning of the lubricant. However, in leaded fuels a substitution reaction between the fuel and zinc dialkyldithiophosphate takes place resulting in lead dialkyldithiophosphate [9]. Thus in some systems an elevated concentration of lead may be due to this substitution reaction. Usually the concentration of lead should ideally be small like the concentrations of iron and copper.

The measured concentrations of zinc and calcium can be reduced by the presence of viscosity improvers in the lubricating oil [10,11]. It is believed that viscosity improvers disrupt the flow of oil from the capillary tube outlet of an Atomic Absorption Spectrometer. Abnormally large droplets of oil are believed to form from linkage between smaller droplets by strands of polymer viscosity index improver. These polymer bonded droplets are too large for effective vaporisation in the AAS [10]. Thus a regression formula has been developed and used to calculate the true concentration of zinc and calcium if the concentration of viscosity index improver is known [11]. However, the effect of polymer viscosity index improvers on AAS data of other metals does not appear to have been evaluated in detail.

A detectable silicon concentration indicates in nearly all cases that the lubricating oil is contaminated by silica, which is the main constituent of sand or other geological minerals. If the lubricating oil is a silicone or silahydrocarbon, then silicon data cannot be used to provide an indication of contamination by sand and other abrasives. Apart from the silicone lubricating oils, silicon is an undesirable element in a lubricating oil. Calcium indicates either the presence of clay contaminants or else detergent lubricating additives such as calcium sulphonate. The concentration of water has to be carefully monitored as water has detrimental effects on the performance of machinery. Usually the water percentage is included in SOAP reports even though it is determined by other methods.

In a practical situation, SOAP is used to specify whether to continue operating a machine or else the type of maintenance required. Decisions on operating status and maintenance are based on whether the concentrations of contaminant metals and water in the oil exceed acceptable limits. These limits for contaminant metals and water are based on experience and can vary depending on the type of industry. Typical example of such limits is shown in Table 10.1.

SOAP can also be used in tribological experiments on lubricated wear to determine whether an oil is functioning at conditions resembling those found in actual machinery.

Table 10.1 Limits in ppm for trace metal concentrations in used crankase oils [12].

Fe	100	Cr	40	K	50
Pb	100	Al	40	Si	20
Cu	40	B	20	Ag	6

SOAP is usually applied repetitively over certain periods of time to allow trend analysis of the various metal concentrations in lubricating oils. These concentrations usually show a steady increase with time so that it is possible to predict when maintenance is required. When there is rapid increase in metal concentration this usually indicates an accelerated wear which may lead to some more serious problems or even failure of the entire system. In such cases the machinery is scheduled for maintenance. SOAP allows the machine operator to replace expensive regular maintenance periods with cheaper failure prevention maintenance. For example, data from a crankcase lubricating oil plotted against time is shown in Figure 10.2.

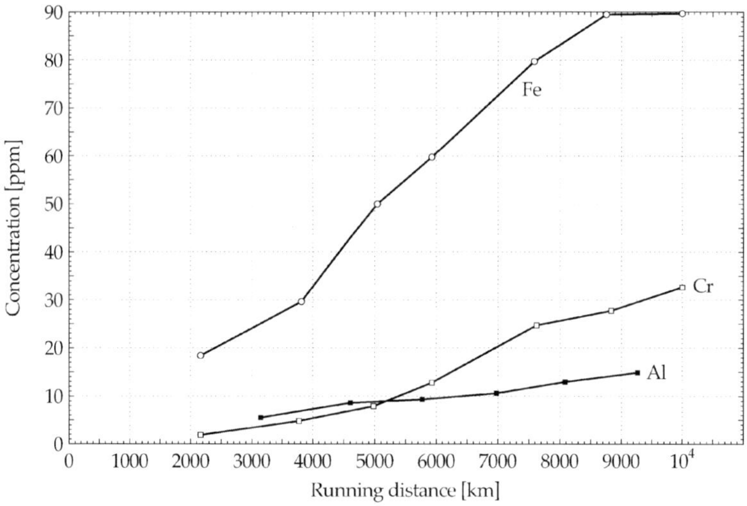

Figure 10.2 Example of trend analysis of metal concentrations in lubricating oil [13].

It can be seen from Figure 10.2 that there is a gradual increase in iron content resulting from cumulative wear. The increase in chromium concentration indicates the deterioration of piston rings which have chromium coated surfaces. Aluminium concentration indicates wear level of engine bearings which, as can be seen from Figure 10.2, is lower than that in the piston rings in the overall engine operation [13].

The limitation of SOAP analysis is that many metals, including iron, fail to vaporise completely in the oxy-acetylene flame or electric arc during spectroscopy [14]. Another limitation is that SOAP analysis provides accurate data only when very small, suspended in the oil, wear particles are present in the system. As the size of the particles increases the SOAP accuracy becomes progressively worse. In general SOAP tends to be insensitive to particles larger than 10 [μm] [3,8,13-16].

The larger particles are removed by centrifuging prior to spectroscopy to ensure consistent data. Sometimes large metal wear particles are converted into fluid reaction products by mixing the lubricating oil sample with hydrofluoric acid and aqua-regia which dissolve those particles [15]. The shape of wear particles appears to influence the SOAP results as lamellar wear particles have been found to be reliably detected at larger particle sizes than spherical particles [16]. Higher surface area to volume ratio of lamellar particles compared to spherical particles allows them to be vaporised more efficiently. Refinements in the technology of SOAP such as the replacement of the oxy-acetylene flame with a high temperature argon plasma in ICP/AES allows for the increased quality of data [17]. The high temperature plasma also allows much higher sensitivity, with iron concentrations as low as 0.004 [ppm] becoming detectable. This is sensitive enough to detect rolling contact failures where only a very small amount of metal is released into the lubricating oil before a much larger flake of metal is detached from a pit or spall on a rolling surface [17].

Analysis Based on Separation of Wear Particles From Lubricating Oil or Process Fluid

Once spectrographic oil analysis or SOAP had become established as a routine technique for lubricating oils, its limitations as a source of information about wear became readily apparent. Practical experience of machine failures quickly revealed that data from SOAP contained only some of the required evidence to predict machine operating status. SOAP provides a good general indication of lubricant additive depletion, contamination of oil by dirt and the accumulation of wear metals in the oil, but very little information about the type of wear occurring. As the SOAP analysis allows the detection of only very small wear particles that are suspended in the oil, it can completely miss out certain types of machine failures during which large particles are generated, e.g. scuffing or contact fatigue.

As a response to the inadequacies of SOAP, in the 1970s Seifert and Westcott developed a technique called 'ferrography' [18]. Ferrography is an analytical technique, which allows wear particles to be separated from the lubricating oil according to their size, thus allowing their individual examination. The advantage of ferrography is that it offers a solution to the practical difficulties normally impeding close viewing of wear particles [18,19]. Ferrography is based on the application of a strong magnetic field to the separation of wear particles. Although other separation methods have been developed more recently as substitutes for ferrography the basic objective of viewing wear particles remains the same.

Collection and Separation of Wear Particles by Ferrography

Ferrography is based on allowing a sample of lubricating oil or oil diluted by solvent to flow over a glass slide, which is suspended above a magnetic field. The glass slide is inclined at an angle to the axis of the magnetic field so that the oil sample, as it flows down the glass slide, is subjected to a large magnetic field gradient. The glass slide has a non-wetting barrier painted on one surface to constrain the test fluid, delivered at a certain rate by a pump, to flow in one direction. The fluid sample flows by gravity, along the slide but within the non-wetting barrier, and finally is picked up by a drain tube. Strong magnets are used to produce a magnetic field of large magnitude which must be strong enough to capture the majority of wear particles present in the oil sample. The glass slide or the glass tube is placed immediately above the poles of the magnet. The combination of viscous and magnetic forces acting on particles allows for the separation and settlement of wear particles on the glass slide in the following manner:

- large particles of iron and steel are strongly attracted by the magnetic field and are deposited close to the entry region of the glass slide, i.e. the position where the oil sample is poured onto the slide, and
- fine particles are deposited further down the slide, i.e. close to the exit, where the magnetic field is particularly intense.

Once all the fluid in an oil sample has run across the slide a washing and fixing process is applied to remove the residual oil and hold the particles in a fixed position. Common solvents, e.g. petroleum ether, can be used for this purpose. At this stage the slide becomes a permanent record of the lubricant's condition and is commonly known as a 'ferrogram'. The wear particles present on the ferrogram can be further studied using various microscopy techniques, e.g. optical, SEM, etc. A schematic illustration of the ferrography principle is shown in Figure 10.3.

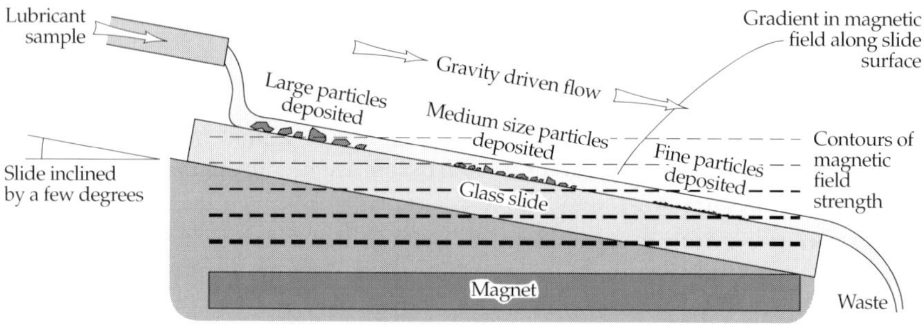

Figure 10.3 Schematic illustration of wear particle collection by ferrography.

It is also possible to collect wear particles of non-magnetic materials in some instances. Either the particles are sufficiently heavy and large to sink through the flowing layer of sample fluid or the particles can be artificially magnetized. For example, in studies of wear particles from synovial joints, chelation by erbium ions of protein material is exploited [20,21]. Erbium ions are magnetic and it has been found that sufficient chelation between cartilage and erbium ions occurs to allow for effective ferrography of synovial joint wear particles [21,22]. The early models of ferrographs were manually operated [18,19] and demanded care in their operation and interpretation of data. Semi-automated models, which do not require skilled technicians to obtain valid data, have also been developed [23].

After ferrography has been successfully performed, a slide containing a thin streak of wear particles, illustrated schematically in Figure 10.4, results.

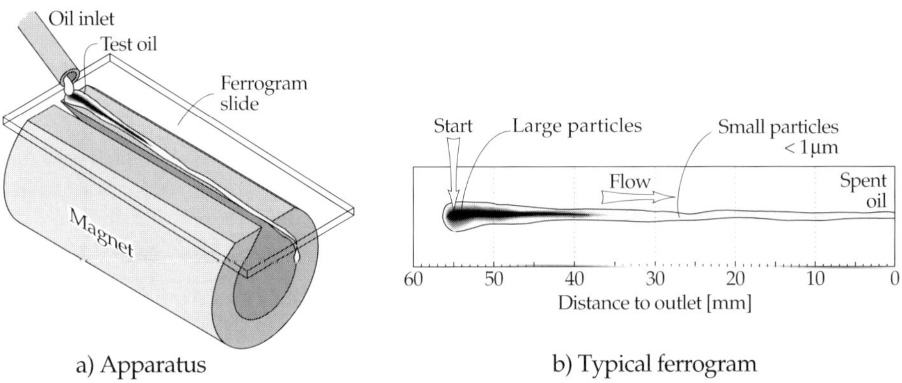

Figure 10.4 Schematic illustration of wear particle collection by ferrography.

The two ways in which data can be obtained from this slide are called:

- analytical ferrography, and
- direct reading ferrography.

The first method is to examine the slide under a microscope in order to study the morphology of specific wear particles while the other is to measure the transmission of light through the wear particles deposited in a glass tube.

- *Analytical Ferrography*

This technique is based on the direct observation of collected wear particles to diagnose wear mechanisms occurring and to predict the imminent type of machine failure. Analytical ferrography allows the investigator to sort and display the wear particles according to their size in a manner for convenient examination [24]. Using this technique it is possible to distinguish between the wear particles produced by various wear mechanisms and to diagnose accordingly. This method can reveal a wide range of information about wear mechanisms operating in a lubricated contact but the quality of information obtained totally depends on the expert's knowledge of wear. Despite this limitation this technique can provide valuable wear monitoring data [e.g. 24-28].

In many practical applications wear particle morphology is assessed using optical or electron microscopy techniques requiring high level of expertise. In the majority of cases an optical microscope is used as it is more accessible and much cheaper that a Scanning Electron Microscope (SEM). Although the resolution of optical microscope is much lower than that of SEM it often provides an adequate detail of wear particles [24]. The advantage of microscopy examination is that the human skill is exploited resulting in quick and accurate assessment of particle morphology. The disadvantages are personal observer's bias and the need for trained personnel. As there are literally millions of lubricated tribological systems that require wear monitoring the labour costs of manual wear particle observation tend to be excessively high. Therefore analytical ferrography, despite its many benefits, is not applied to all but only to a few critical lubricated contacts. Examples of SEM images of typical ferrograms and wear particles are shown in Figures 10.5 and 10.6.

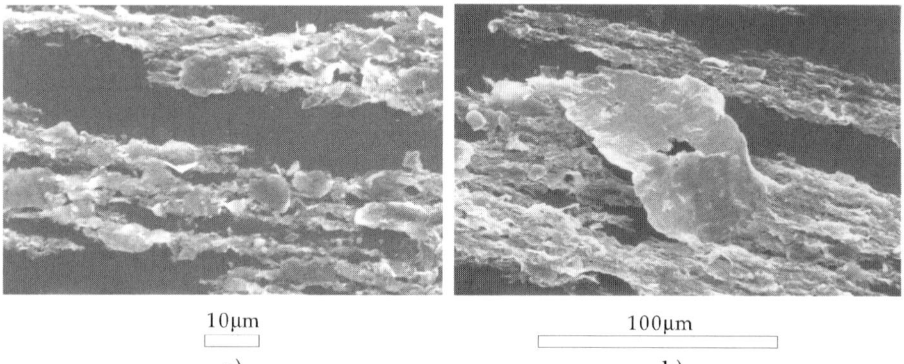

Figure 10.5 Examples of SEM images of typical ferrograms; a) normal sliding wear with characteristic small plate-like particles, b) abnormal sliding wear with large delamination wear particles.

It can be seen from Figures 10.5 and 10.6 that wear particles exhibit specific shape and morphology which are characteristic to a wear process involved in their formation. For example, abrasive wear often produces wear particles resembling miniature cutting tool swarf while flat lamellar particles between 20 and 100 [μm] in size and about 1 μm thick are often the result of delamination wear. Irregularly shaped chunky particles are caused by adhesive wear or severe wear. Small spherical particles, 1 to 5 [μm] in diameter, originate during fatigue wear from surface fatigue pits but larger spherical particles are probably welding debris. SEM, sometimes Field Emission SEM (FESEM), combined with Energy Dispersive Spectroscopy (EDS), described in detail in Chapter 8, can be used to investigate the fine detail of particles and determine their chemical content. In particular, FESEM facilitates very detailed examination of minute wear particles, e.g. below 1 [μm] in major dimension. The examples of small wear particles imaged by FESEM are shown in Figure 10.7. Images of typical wear particles have been compiled to create the wear particle atlases and to aid assessment of wear mechanisms occurring [27,29].

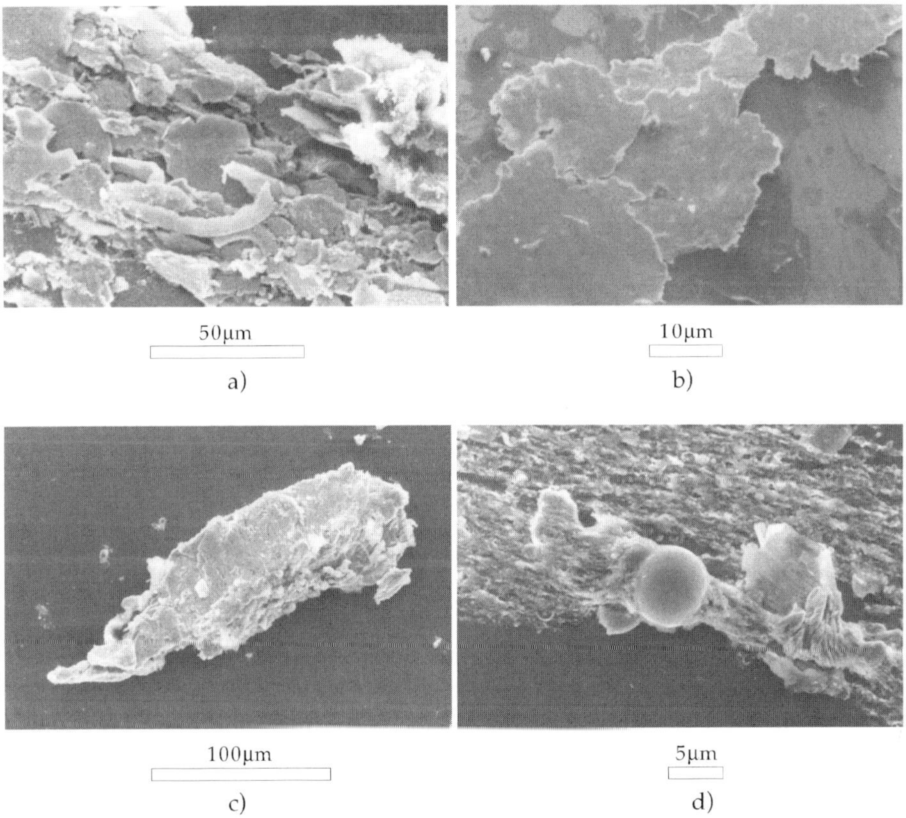

Figure 10.6 Examples of SEM images of typical wear particles; a) abrasive wear, b) delamination wear, c) adhesive or severe wear and d) fatigue wear.

Analytical ferrography is particularly useful in the monitoring of contact fatigue of rolling bearings, gears and other industrial equipment. In one study, a number of spherical particles released from developing pits on rolling surfaces were detected. A period of 23 hours of bearing operation elapsed between detection of these particles and failure of the bearing [26]. Ferrography has also been used to monitor wear of other industrial equipment such as combustion engines, pumps and turbines. When correctly applied this technique can provide an adequate warning of approaching machine failure, hence the technique is very useful in the early detection of wear problems.

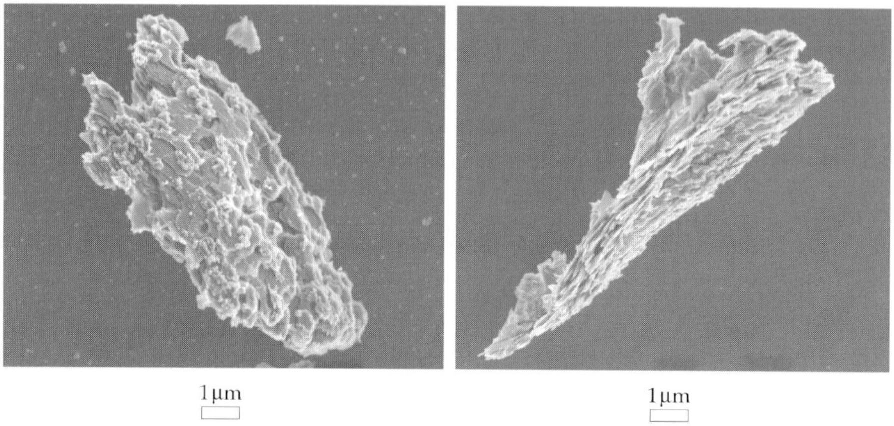

Figure 10.7 Examples of small wear particles obtained from sheep synovial joints and imaged by FESEM.

· *Direct Reading Ferrography*

Another method of obtaining data suitable for routine condition monitoring is to measure the changes in light transmission through the streak of wear particles deposited in the glass tube. In this method the glass slide is replaced by a glass tube and the method is often called a 'direct reading ferrography'. When there are a lot of wear particles present in the glass tube, light transmission is poor, and conversely, when there are very few wear particles, most of the light is transmitted. A set of light detectors is used to measure the changes in the optical density occurring during the wear particle deposition. The light detectors are positioned along the glass tube. One detector is positioned at the beginning of the tube where the large particles are deposited while the other is positioned further down the tube where the small particles are deposited. Semi-empirical formulae were developed to define the severity of wear from light transmission measurements [29]:

$$\mathbf{WPC} = (\mathbf{A}_L + \mathbf{A}_S)/\mathbf{C}$$

$$\mathbf{PLP} = (\mathbf{A}_L - \mathbf{A}_S)/(\mathbf{A}_L + \mathbf{A}_S) \times 100\%$$

$$\mathbf{WI} = (\mathbf{A}_L + \mathbf{A}_S)(\mathbf{A}_L - \mathbf{A}_S) = \mathbf{A}_L^2 - \mathbf{A}_S^2$$

where:

WPC is the wear particle concentration [dimensionless];

PLP is the percentage of large particles [dimensionless];

WI is the wear severity index, also sometimes called intensity of wear [dimensionless];

$\mathbf{A_L}$ is the digital reading of opaqueness at the end of the wear particle deposit closer to the start of flow by the oil sample [dimensionless],

$\mathbf{A_S}$ is the digital reading of opaqueness at the end of the wear particle deposit further from the start of flow by the oil sample [dimensionless], and

C is the dilution coefficient [dimensionless].

An example of wear severity index measurements from ferrography of wear particles from a gas turbine oil is shown in Table 10.2 [30].

Table 10.2 Example of wear data from direct reading ferrography [30].

Sample N°	Flying time [h]	DR readings		Total wear	Severity of wear	Severity of wear index
		D_L	D_S	$D_L + D_S$	$D_L - D_S$	$WI = D_L^2 - D_S^2$
Test bed	–	3.5	1.3	4.8	2.2	11.0
1	183	10.1	1.6	11.7	8.5	99.0
2	230	6.8	4.8	11.6	2.0	23.0
3	267	9.7	2.5	12.2	7.2	88.0
4	309	9.3	4.8	14.1	4.5	63.0
5	360	7.2	3.5	10.7	3.7	39.0
6	430	73.2	9.8	83.0	63.4	5262.0
7	492	35.0	4.0	39.0	31.0	1209.0
8	546	56.9	13.0	69.9	43.9	3068.0
9	579	17.1	2.7	19.8	14.4	285.0
10	648	12.1	4.8	16.9	7.3	123.0
11	679	47.2	5.2	52.4	42.0	2200.0

It can be seen from Table 10.2 that the value of parameter '**WI**' tends to rise sharply at periodic intervals during operation of the turbine. The excessive values of '**WI**' were used as a warning of required maintenance. When the gas turbine was stopped for maintenance and the lubricating oil replaced, the value of '**WI**' was observed to decline. It was also observed that a rise in the '**WI**' value usually preceded a severe damage to the gas turbine, so providing a warning period for preventive action [30]. Direct reading ferrography is now routinely used for machine condition monitoring throughout industry.

Collection and Separation of Wear Particles by Rotary Particle Separators

Rotary particle separators have been proposed as a means of wear particle collection without limitations of particle magnetization [31]. The basic principle of a rotary particle separator is the application of centrifugal force to extract the heavier wear particles from the lighter lubricant. The speed of separation critically depends on the wear particle size, according to the balance between mass-dependent centrifugal forces and size-dependent viscous forces. Under suitable test conditions, effective wear particle separation can be obtained. Wear particles are segregated according to their size and density on a slide mounted on a rotating disc. The slide is then examined under the microscope and the wear particle morphology is assessed.

Collection of Wear Particles by Membrane Filters

A simple and cheap way of extracting wear particles from lubricant or process fluids is based upon passing a liquid through the filter in a form of a membrane. As the pores of the membrane are quite small, about 0.8 [μm], most of the wear debris are retained on the membrane [28]. SEM or an optical microscope can then be used to examine the particle morphology.

Particle Discrimination by Analytical Techniques and Prediction of Wear

The size range of particles analyzed by spectrographic oil analysis (SOAP) is entirely different from the size range of particles retrieved by ferrography or rotary particle separators. This difference in size has a significant effect on the quality of information obtained when using these analytical techniques. Detection efficiency versus particle size for SOAP, ferrography and a chip detector is illustrated in Figure 10.8 [32]. Detection efficiency is a comparative measure of the proportion of wear particles (or wear particle mass) that are revealed by the analytical technique in use.

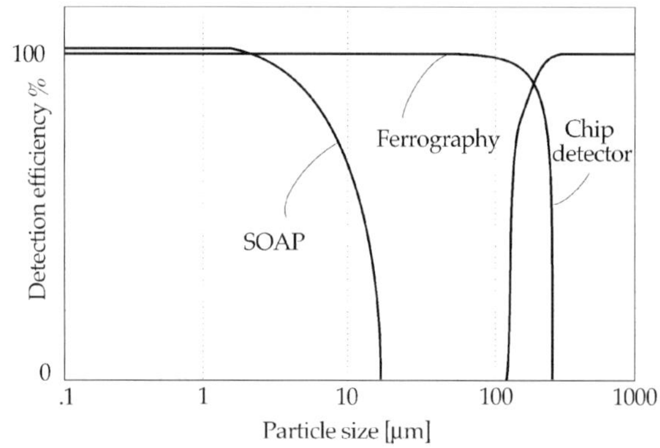

Figure 10.8 Detection efficiency versus particle size for spectrometer, ferrography and chip detectors (adapted from [32]).

Reliable wear particle detection by SOAP is limited to particle sizes less than 10 [μm] and this restriction prevents monitoring of most wear particles by this technique [14-17]. Research on raising the wear particle size limit of SOAP is proceeding but a practical modification to SOAP that enables reliable detection of large particles has not yet been developed [33,34]. Since most wear particles of interest in machine condition monitoring are between 10 and 100 [μm] in size [35] ferrography, as it exhibits a suitable size range for these particle types detection, appears to be well suited for wear particle analysis. Chip detectors or magnetic plugs are intended for the collection of extremely large particles, which are usually produced in the final stages of extreme wear.

A further advantage of the bias by magnetic particle separation towards larger particles is the quality of wear monitoring data obtained. Wear monitoring is based on finding unequivocal signals of changes in machine operating condition. It has been shown that ferrography provides a sharper indication of abnormal wear than SOAP [36]. Data from SOAP tends to show a steady increase in wear metal concentration versus time, giving a very little warning of undesirable changes in wear rates occurring. On the other hand, ferrography provides data which either tends to a value characteristic of safe machine operating condition or else the data shows a rapid rise in wear particle concentration which is a warning of imminent wear problems. The reason for this difference between SOAP and ferrography lies in the size of particles analyzed by either technique. The extremely small particles detected by SOAP tend to accumulate as a suspension in the lubricant and the concentration of these particles gradually increases as wear progresses. The larger particles sampled by ferrography remain suspended in the lubricant for a relatively short period of time. In this larger size range of particles, there is usually a fixed concentration of particles in the lubricant during normal wear. The concentration is determined by the balance between the rate at which wear particles are released into the lubricant by wear and the rate of wear particle precipitation. The difference in wear particle retention characteristics of the lubricant is illustrated schematically in Figure 10.9.

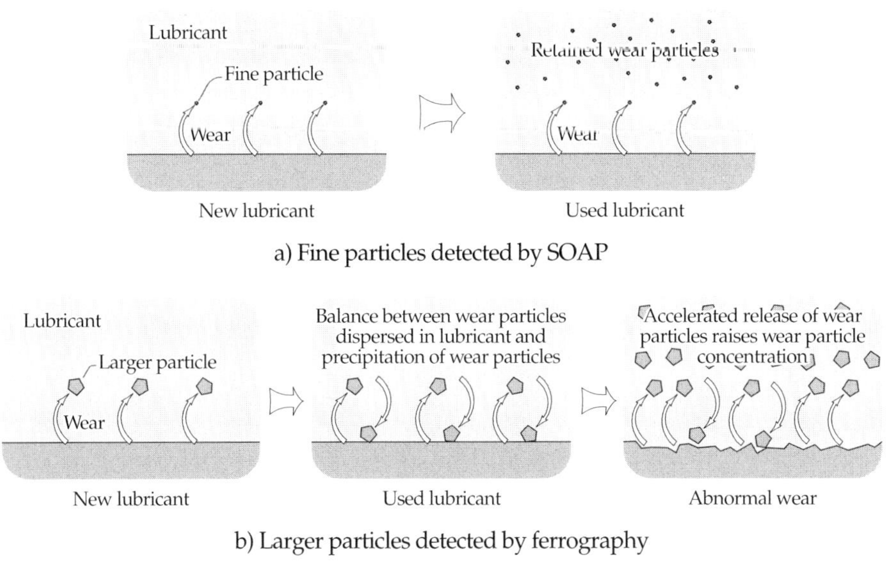

Figure 10.9 Lubricant wear particle retention characteristics for large and small particles.

A transition from normal to severe wear causes an increase in wear particle formation so that the concentration of wear particles in the 10-100 [μm] size range increases. This increase in concentration causes a sharp change in the wear severity index measured by direct ferrography.

Concentration of wear particles will be greatly affected by an oil change. This effect is different for SOAP and ferrography. A schematic diagram of fluctuations in wear particle concentration with an oil change for both SOAP and ferrography is shown in Figure 10.10.

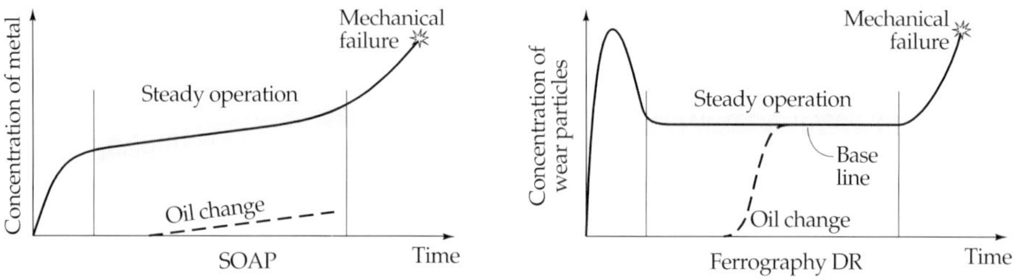

Figure 10.10 Schematic illustration of changes in wear particle concentration induced by oil change monitored by SOAP and ferrography [25,32,36].

It can be seen from Figure 10.10 that for SOAP data there is a continuous rise in wear metal particle concentration, which gradually accelerates as mechanical failure approaches. It can also be seen from this figure that wear monitoring based on SOAP and limits of wear particles concentration can be easily compromised by replenishment or change of lubricant. Replacement of the lubricant causes the wear metal concentration to decline without any full recovery of the wear metal concentration to the original level. However, the wear trend (slope of the curve) remains the same.

Ferrography data of wear particle concentration shows an initial peak during 'running-in' or conditioning of the moving components followed by a 'plateau' of steady wear. A long period of steady wear particle concentration is established after component conditioning is completed and the wear particle concentration only rises when mechanical failure occurs. A change of lubricating oil will cause the wear particle concentration to fall to zero before a rapid rise to the original wear particle concentration level symptomatic of normal wear.

Particle Counting

In systems where wear rates are low and clearances between interacting surfaces are very small, e.g. hydraulic systems, particle counting is used as a condition monitoring measure [3]. Particle counters are based on light blocking or scattering principles. In light blocking sensors a momentary decrease in light intensity produced by a particle passing through a 'view volume' is detected, resulting in a pulse proportional to a particle size [3]. In light scattering sensors light reflected or scattered by a particle is measured and related to a particle quantity and particle size. The measurements give the number of particles in specific size categories which can be used as a condition monitoring measure.

These two techniques have their relative merits and disadvantages. Light blocking sensors are more common as they are more effective in counting dark particles, handle larger concentrations of wear or contaminate particles better and have longer service life. On the other hand, light scattering sensors can detect smaller particles [3].

10.3 ANALYSIS OF THE SHAPE OF WEAR PARTICLES

One of the basic problems in wear debris analysis is the accurate assessment of wear particle morphology. In order to avoid the use of human experts and subsequent bias errors there is a need for a development of practical methods allowing comprehensive data from the wear particles collected to be obtained and the particles objectively classified. In the quantitative methods presented so far, all the complexity of wear particle morphology is condensed into a few simple parameters such as wear metal concentration or concentration of large and small wear particles. This approach to wear analysis inevitably leads to omissions of many critical features of the prevailing wear processes.

A useful fact in wear studies is that the shape and surface morphology of wear particles contain evidence of how the wear particles are formed, i.e. the particle morphology is strongly related to wear processes involved in their formation. A wear analysis technique, which explicitly assesses the shape of the wear particles could provide much information about the causes of wear. From a practical viewpoint, the significance of this approach is that it allows, for example, corrosive and adhesive wear to be distinguished. The routine assessment of the shape of wear particles, not just their size and number, is beginning to be realized.

Image Analysis of Wear Particles

In machine condition monitoring the visual microscopy inspection of wear particles provides useful information about the 'health' status of the machinery. The particles, captured by ferrography, are inspected on a regular basis and if something unusual is observed then the machine is shut down and examined before any more serious damage occurs. The same particle examination can be conducted for lubricants used in tribological testing with the aim to determine the wear processes occurring. Currently, the morphological assessment of wear particles depends on an expert examining them under the microscope. However, visual examination of wear particles is not particularly popular as it is slow and costly, i.e. it requires a certain level of expertise to differentiate between various types of wear particles. The level of information required to distinguish between the various types of wear particles is far more sophisticated than is provided for by standards on particle size measurement as detailed in e.g. British Standard 2955 [37].

Sometimes the interpretation of wear particle images can be enhanced or facilitated by the application of artificial colour. Colour helps the human observer to resolve the fine detail of the image, such as evidence of gouging, cracking or pitting on the wear particle.

At present, computerized wear particle assessment systems assist the user to distinguish between different shapes of wear particles by visual comparison between a library of standard shapes and the wear particle under examination [38]. The basic principle employed in these systems is subjective evaluation where an example would be 'this particle resembles cutting swarf so abrasive wear is probably occurring'. In most cases, however, wear particle shapes do not give such a distinctive indication of the prevailing wear mechanism.

Another approach involves the application of Fourier analysis to give some measure of the sharpness and curvature of the wear particles' outlines [39]. It has been suggested that particles originating from different wear mechanisms could be distinguished by this technique [39]. Recently image analysis techniques and fractal methods have been applied to the numerical characterization of wear particle morphology. Ideally, these techniques may eventually overcome the subjectivity problem involved in microscopic examination of wear particles and allow the whole analysis process to become automated. An automated image analysis of wear particles could provide reliable interpretation of the wear particle shape and surface morphology. In other words, the image analysis techniques could provide most of the information obtainable from analytical ferrography, without the need for experts.

Wear particle morphology can be fully described by two sets of parameters, i.e.:

- particle boundary descriptors, and
- particle surface topography descriptors.

Wear Particle Boundary Descriptors

The boundary of the projected particle image is usually described by a set of numerical shape descriptors, such as aspect ratio, shape factor, convexity, elongation, curl, roundness, etc. [e.g. 40-43]. The calculation methods and the influence of measurement conditions on these parameters such as the effects of accelerating voltage, filtering, noise, etc., are described in detail in [43-45]. It has been observed that wear particle boundaries often resemble chaotic shapes. These shapes can be analyzed and described by fractal geometry to reveal their underlying similarities and differences [46]. Fractals are commonly associated with chaos. Wear particles are generated by processes which are almost certainly chaotic so that fractals can be used to describe their shapes and also their surface topographies. The shapes of these particles are in turn closely associated with the wear process that formed them. If big chunky particles with irregular outlines are found then we know that surface fatigue, which is a very destructive form of wear, is probably occurring. On the other hand, if only fine debris with smooth outlines is produced then normal mild wear takes place [37].

Recently, fractal geometry has been used in the description of the complex shapes of particle boundaries and their surface topographies. A parameter describing the complexity of a particle boundary, called fractal dimension, has been developed together with various strategies for its calculation [43,46-57]. It has been shown that fractals are useful in describing powders [51,58-60], pigments [58] and wear particles [e.g. 47,48,54]. Fractal techniques have also been employed to characterize surface irregularities [61-67], surface anisotropy and directionality [69,70] and surface topography [64-67,70,71].

· *Characterization of Wear Particle Boundary by Fractal Methods*

Originally the potential of fractal geometry has been demonstrated on the example of a coastline of Britain. If one imagines the measurement of the coastline of e.g. Australia, Britain or Madagascar, a basic consideration is the step-length of measurements [46]. When the perimeter length of the coastline is quoted a certain step-length of measurement is assumed, e.g. 1 [km], 1 [m], etc. If the step-length of measurement is as long as 100 [km] then the measured coastline will be very roughly approximated and, as a result, will be much smaller than if a much finer step-length of e.g. 1 [m] was chosen. With a very short step-length, much finer details, i.e. every visible indentation in the coastline, are included in the measured perimeter [46] while with the long step-length these fine details are omitted. Fractal geometry encapsulates information about the complexity of a specific shape using a relationship between the variation in measured perimeter and the step-length. A parameter called 'fractal dimension' is used to express this relationship in the following manner [46]:

$$\mathbf{D = 1 - m}$$

where:

D is the fractal dimension [dimensionless] and

m is the slope of the plot of log(perimeter length) versus log(step-length) as illustrated in Figure 10.13.

This principle equally applies to a coastline or to the wear particle boundary as observed using a microscope.

A technique commonly used to determine the complexity of particle boundary is known in the literature as a 'structured-walk' technique or as a 'Richardson's method' [e.g. 50,53,72-74]. The boundary of a particle projected image is 'walked' around at different step sizes. After completion of each 'walk' the perimeter length is estimated by summing the Euclidean length of the steps around the particle boundary. The variation of length of measured perimeter versus step-length is obtained by a series of measurements analogous to tracing a pair of dividers around a particle boundary [75]. As the edges of wear particle images are often diffuse, edge enhancement techniques have to be applied in order to obtain a perimeter suitable for analysis [47]. The average step-length is found by dividing the perimeter estimate by the number of steps. The difference between the step size and step-length is illustrated schematically in Figure 10.11 [62] while the process of evaluating the variation of perimeter length measured as a function of step-length is illustrated in Figure 10.12.

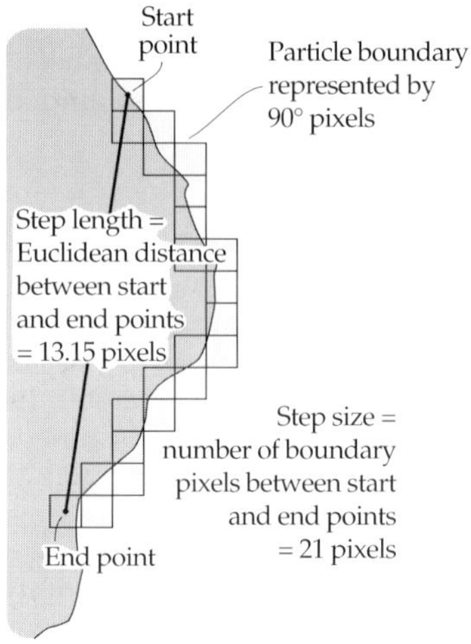

Figure 10.11 Schematic illustration of a difference between step size and step-length shown on a section of digitized particle boundary (adapted from [62]).

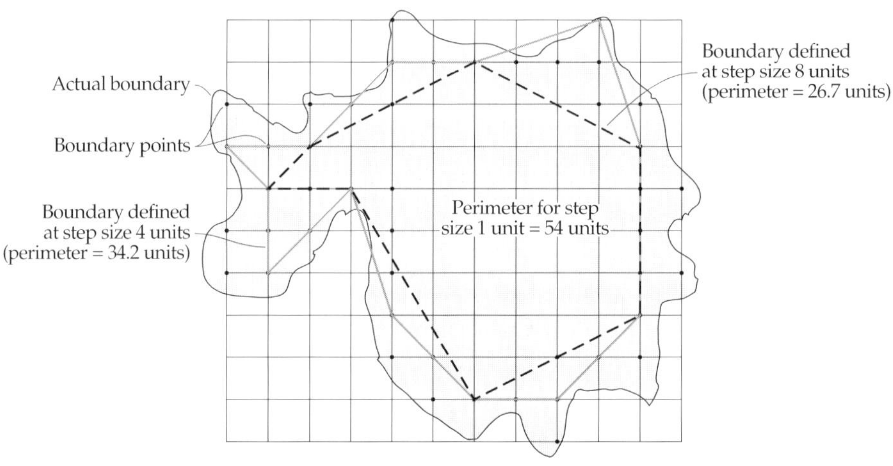

Figure 10.12 Schematic illustration of the variation of measured perimeter with step-length for a wear particle (adapted from [75]).

The perimeter estimates obtained at different step-lengths are then plotted on a log-log scale. The resulting plot is often called a 'Richardson's plot'. From the slope of the line of best fit to the plot a fractal dimension '**D**' is calculated as **D = 1 - m**, where '**m**' is the slope of the line. A plot of the perimeter length versus step-length data for the 'particle boundary' shown in Figure 10.12 is shown in Figure 10.13.

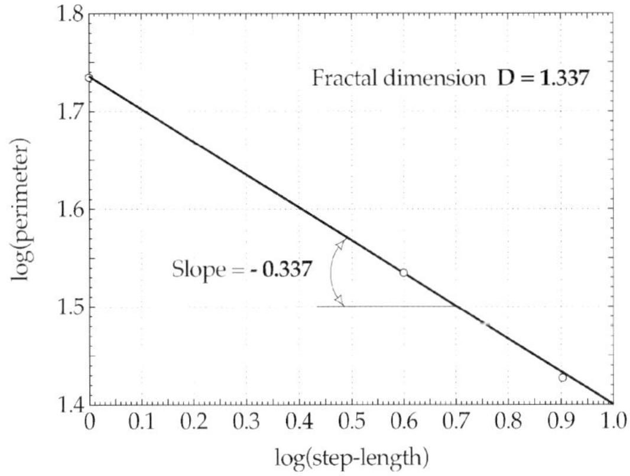

Figure 10.13 Example of perimeter versus step-length plot from which the fractal dimension is calculated (adapted from [75]).

The fractal dimension of 1.337 obtained from this plot can be used in wear monitoring data. For wear particle boundaries '**D**' varies between 1 and 2, i.e. a more complex boundary exhibits a higher fractal dimension. This is illustrated in Figure 10.14 where two wear particles and their corresponding Richardson's plots are shown.

It can be seen from Figure 10.14 that the boundary of the particle 'a' is more complex than the boundary of the particle 'b'. This visual feature is confirmed in the slopes of the corresponding Richardson's plots obtained, i.e. the slope of curve 'a' is steeper than the slope of curve 'b', and this is in turn reflected in fractal dimensions calculated for those particles [43]

Richardson's plots often exhibit bi-modal characteristics which have been attributed to different features of particle boundary. For example, the upper part of the plot at small step-lengths has been associated with 'textural' characteristics of the boundary while the lower part of the plot at large step-lengths has been associated with boundary 'structural' characteristics [e.g. 76] as shown in Figure 10.15.

More detailed work revealed that although in some cases, e.g. soot agglomerates, particle boundaries can exhibit two distinct physical scales [76], in the majority of cases this bi-modal behaviour can only be attributed to an artefact produced by the limited data size. At large step-lengths any object tends to behave as a Euclidean object exhibiting finite data errors [50,73]. At large step-lengths the perimeter is increasingly underestimated due to longer line segments spanning the object. Therefore, the interpretation of the bi-modal behaviour of a Richardson's plot must be treated with caution [43].

Various fractal methods have been developed for the analysis of particle boundaries [e.g. 57,72]. There are some problems associated with the selection of any of these methods for analysis due to the variation in step-length, the difficulties associated with line fitting into the Richardson's plot and the sensitivity of the individual method to measurement conditions such as noise and focusing [78]. Commonly used fractal methods were investigated in order to find which of them gave the most accurate and repeatable results when applied to wear particle boundaries [78]. It has been shown that fractal analysis by the estimation normalized approach method, developed by Kennedy and Lin [79], gives the most accurate results when applied to a wear particle boundary [78]. The complexity of a wear particle boundary can now be reliably described [67].

Fractal boundary descriptors have been successfully applied to wear particles found in tribological systems [73,80], joint replacements [81] and synovial joints [82,83].

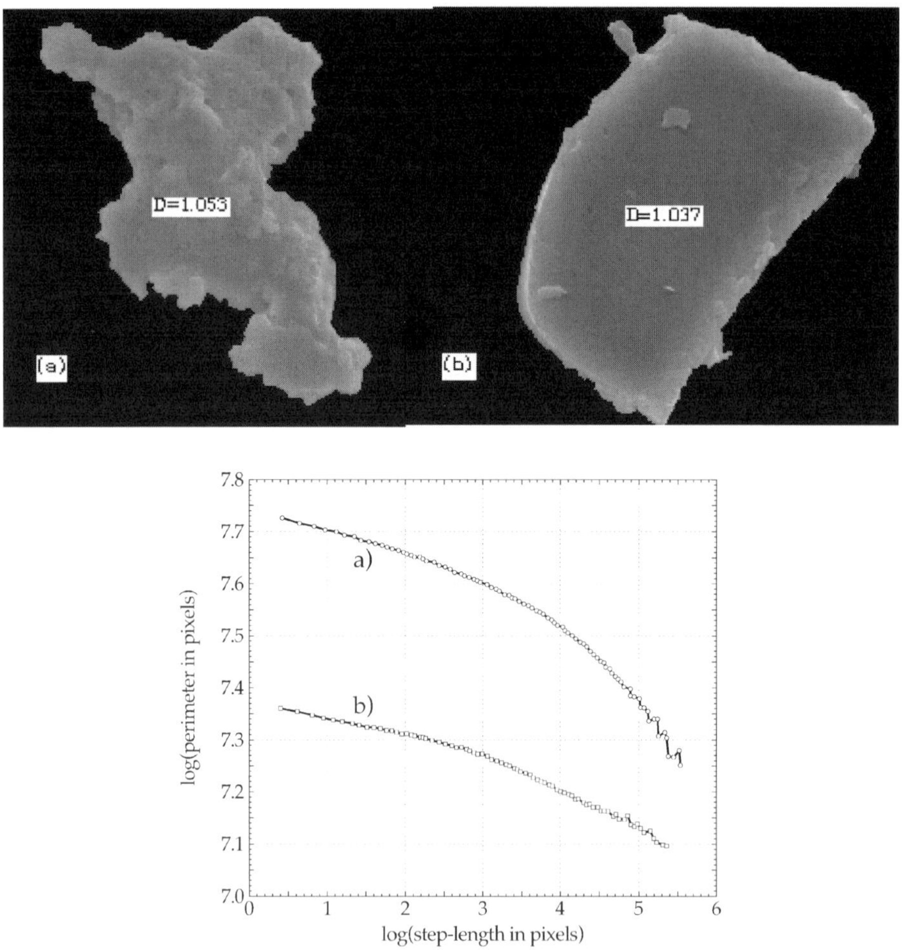

Figure 10.14 SEM micrographs and corresponding Richardson's plots of wear particles obtained from total hip replacement prosthesis operating in human body (adapted from [43]).

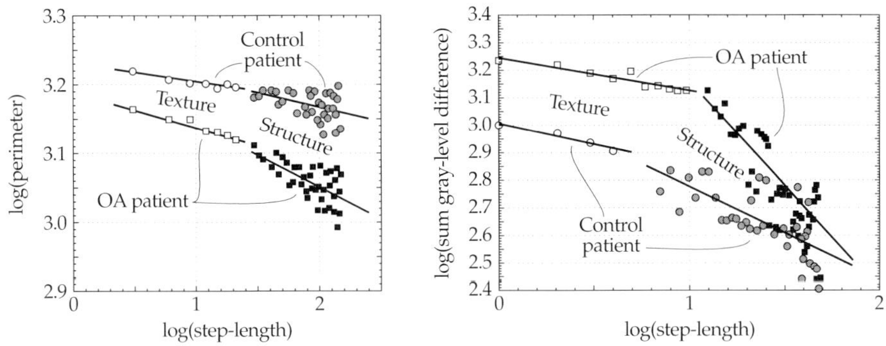

Figure 10.15 Fractal analysis of wear particles from healthy and osteoarthritic (OA) synovial joints (adapted from [77]).

Characterization of Wear Particle Surface Topography

Characterization of the spatial properties of particle surfaces is a much more difficult problem than characterization of particle boundaries. The evidence from microscopy observations seem to indicate that wear particle surfaces, in general, appear to exhibit a fractal nature. Although the reasons for this are not well understood, it seems that the particle surfaces are generated by wear processes which probably dissipate energy chaotically over many different dimensional scales [64,69]. In the majority of cases wear particle surfaces exhibit details at every scale, i.e., they are multiscale objects, and a part of a particle surface at a large scale is similar to a part of the surface at a small scale, i.e., it is 'self-similar'. The latter feature is a fractal property called 'piece-wise self-transformability'. The self-transformability means that a surface contains transformed copies of parts of itself [64].

One of the problems in the characterization of wear particle topography is the acquisition of accurate three dimensional (3-D) data of a particle surface. There are a number of techniques that could be used to acquire 3-D data from a particle surface, e.g. atomic force microscopy (AFM), laser scanning confocal microscopy (LSCM), interferometric microscopy (IM) and scanning electron microscopy (SEM) stereoscopy [87]. Although AFM has excellent vertical (~0.1 [nm]) and horizontal (~0.2 [nm]) resolutions its maximum vertical range is limited to about 4 [μm] [43,84]. As the majority of the wear particles exhibit surface features with height relief much greater than 4 [μm] and are of average size greater than 10 [μm] the AFM is not suitable for acquiring their 3-D surface topography data. The average size of wear particles is too small to be resolved by commercially available 3-D imaging techniques, including LSCM [85] and IM [43,86] as the lateral resolution provided by these techniques is poor [67]. For example, the lateral resolution of LSCM and IM, at their best, is about 0.4 [μm] and 0.2 [μm], respectively. This means that for a particle size of 10 [μm] the best LSCM provides 25 surface scans while IM provides 50 scans [87]. As both of these techniques are based on the light reflected from the surface under examination, a moderate reflectivity of the specimen surface is required. Hence these techniques tend not to work well with rough surfaces and cannot be used in the analysis of particles with complicated surface topography [87].

It has been shown that, amongst various techniques that could be used for the 3-D data acquisition of wear particle surface topography, SEM stereoscopy gives the best results [87,88]. SEM is routinely used in the examination of wear particle morphology and a stereoscopic pair of SEM images can be used to obtain information about 3-D features of the surface [87,89]. The principles of SEM stereoscopy are described in Chapter 6.

The sizes of wear particles range from very large, i.e., above 500 [μm], to very small, i.e., below 1 [μm]. Images of the larger particles, i.e., above 10 [μm], can be acquired in the SEM, while the smaller particles images can be acquired in the field emission SEM (FESEM). FESEM gives excellent results in imaging small particles due to its high signal to noise ratio which can be obtained for a small probe size at low accelerating voltages [87,88].

In SEM stereoscopy two images of the particle taken under different angles are processed by a stereoscopic matching computer program [87]. The tilt angle, which can vary between 5° to 15° depending on the particle morphology, is chosen to ensure that the corresponding points can be clearly recognized on the two images. The accuracy of the particle surface data obtained depends on the magnification used and the pixel size. The horizontal resolution for FESEM particle images can be as high as 0.005 [μm] while the vertical resolution is usually lower by a factor of six or eight [87,88]. The example of two images of the same particle, taken under different angles, for SEM stereoscopy analysis is shown in Figure 10.16.

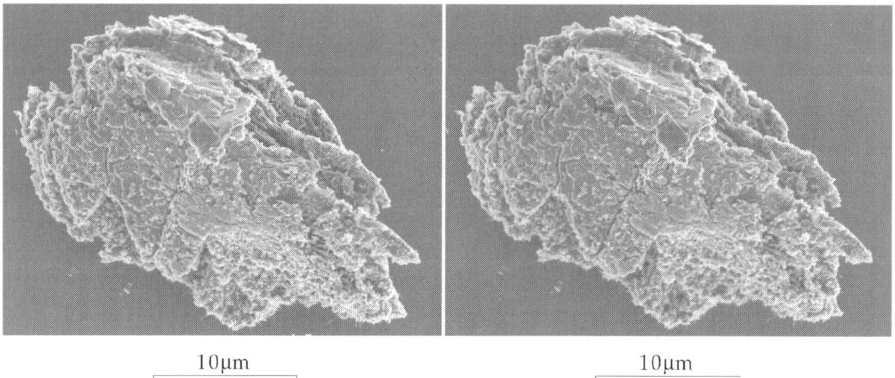

Figure 10.16 Two FESEM images of cartilaginous wear particle obtained from a sheep synovial joint taken under two tilt angles of 0° and 10°. The wear particle was identified as a cartilaginous by a strong sulphur signal detected by energy dispersive X-ray analysis (EDS).

The 3-D particle surface topography data obtained from the SEM or FESEM stereoscopy can be processed by different techniques and also encoded in different formats, i.e. colour stereo, perspective view, contour map and range image for further analysis [88]. It was found that the particle surface data represented by range images is in the most useful form for further analysis [70,88]. The usefulness of other forms of 3-D surface topography representation, i.e. a perspective view, colour stereo, or contour map image, for further numerical analysis is rather limited [70,88]. The examples of the different ways of surface data representation of the wear particle, i.e. in the form of perspective view, contour maps and range image, are shown in Figure 10.17.

a) b) c)

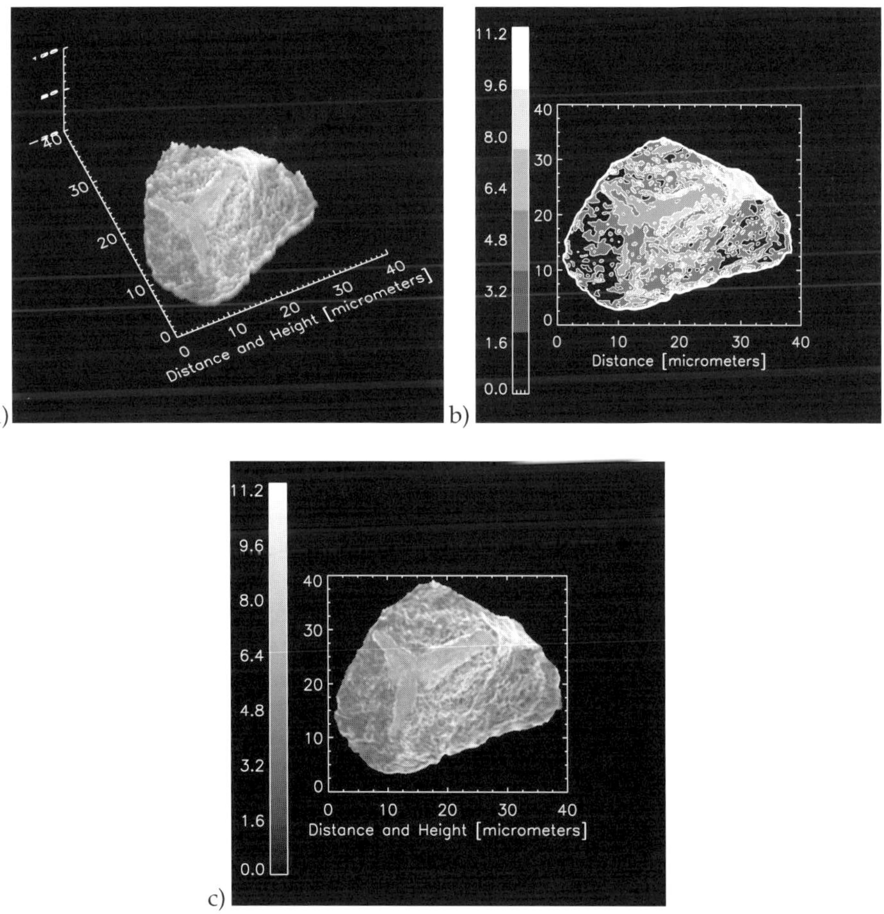

Figure 10.17 Examples of the wear particle surface data representation in the form of; a) perspective view, b) contour maps and c) range image.

In range image the surface elevation data is encoded into an image containing 256 different grey-scale levels. On these images the pixel brightness value represents a value of single elevation data point while the horizontal and vertical axes represent a location of the data point in an 'x-y' plane [88]. This means that the brightest pixel, depicted by the grey level of '255', represents the highest elevation point on the particle surface, while the darkest pixel, depicted by the grey level of '0', represents the lowest elevation point on the particle surface [70,88].

In the early attempts of characterizing the wear particle surface texture grey-scale profiles taken across the particle surface were used. A line, usually along the longest dimension of the wear particle, was drawn and the grey-scale variation measured. The sums of the differences between grey levels were plotted against the distance between neighbouring data points on logarithmic graph scales, resulting in a linear relationship from which a fractal dimension was then calculated [e.g. 75]. The process is schematically illustrated in Figure 10.18.

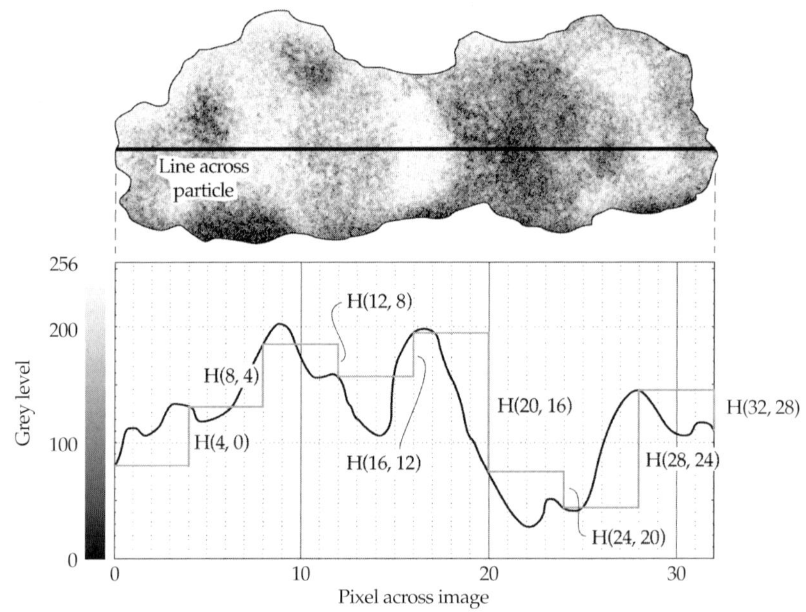

Figure 10.18 Schematic illustration of the technique used to characterize the surface profiles of wear particles.

However, it was soon discovered that this way of calculating fractal dimension from a surface profile was not particularly accurate as the method used involved the similar principles applied in calculating a particle boundary fractal dimension. As there are some basic differences between the particle boundaries and surface profiles they cannot be treated in the same way.

Mandelbrot [90] indicated that 'the compass method' used in calculation of fractal dimensions from particle boundaries should not be used in calculation of fractal dimension from profiles. The reason for this is that the particle boundary is self-similar while the profile is self-affine [62]. Self-affine profiles scale by different amounts in the vertical (z) and horizontal (x) directions and, also, the units of the measurements may not be the same, i.e. the profile can be treated as the time series, where the elevation varies with time [43].

A number of techniques have been developed to evaluate the fractal dimension from a profile, e.g. horizontal structuring element method (HSEM) [91], correlation integral [92,93], fast Fourier transform (FFT) [e.g. 68] and Richardson technique [62]. The performance of these techniques has been tested with artificially generated profiles [61]. The results obtained showed that the problem of choosing any particular algorithm for the calculation of fractal dimension from a profile is not a simple one [61]. An algorithm, which works well with artificially generated profiles, may not, necessarily, give the best results when applied to the 'real' data. Since there is no way of knowing the 'true' or even 'nominal' dimension of the surface profile under consideration, the use of any of these techniques must be treated with caution [43].

Recently a new method, called a Partition Iterated Function System (PIFS), has been developed and applied to the characterization of 3-D surface topography [64-67,70]. The PIFS method is based on the postulate that, since most of the complex structures observed in nature can be described and modelled by a combination of simple mathematical rules [e.g. 94,95], it should, in principle, be possible to describe a surface by a set of such rules. Once this set of rules is found surface topography could be uniquely described in detail with a precision which would supersede all other methods developed and used so far [67].

It has been shown that a wear particle image exhibits a certain degree of 'self-transformability', i.e. it is composed of image parts which can be converted to fit approximately other parts located elsewhere in the image [64,96]. This is illustrated in Figure 10.19, which shows an image of wear particle with the 'self-transformable' parts marked by the squares.

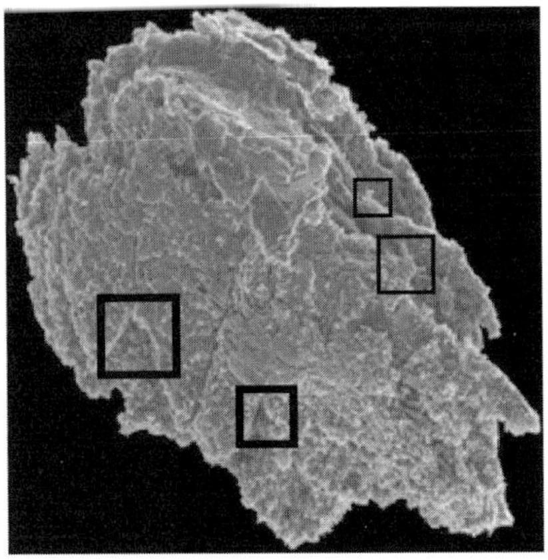

Figure 10.19 Range image of a wear particle with marked 'self-transformable' parts.

For an image of wear particle the affine transformations, needed to convert parts of an image onto other parts of the same image, are found. This process is illustrated in Figure 10.20 where the larger part of the image transforms to a smaller part of the image using mathematical transformation containing information on scale, translation, rotation, etc. Sets of these transformations constitute PIFS file encapsulating the whole information about the surface and particle boundary [64-67,71].

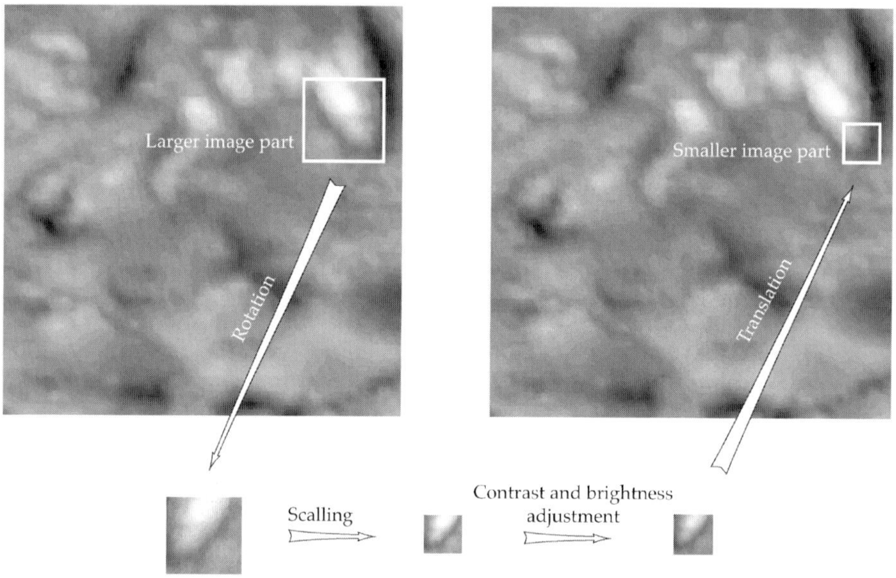

Figure 10.20 Schematic illustration of PIFS construction.

When these transformations are iteratively applied into any initial image the result is a sequence of images which converges to the original particle image. This is illustrated on the example shown in Figure 10.21 where sets of rules were found for the range image of wear particle containing the 3-D surface topography information encoded into the pixel brightness values and applied to some starting image, i.e. black square.

It can be seen from Figure 10.21 that almost an exact replica of the original image has been obtained after only 12 iterations of PIFS transformations. The advantage of this method is that the affine mappings that constitute the PIFS data are scale invariant [96], i.e. the reconstruction of an image from PIFS transformations can be done at any scale [70].

The limitation of this technique is, however, that it describes the surface at all scales at once. Particle surfaces are multi-scale objects (i.e. they exhibit different scales of surface features) and the surface data exhibit's non-stationary characteristics (i.e. features which are superimposed on each other and located at different positions on a surface). This problem was solved by applying a combination of fractals and wavelets to describe the surface topography [97-100]. The fractal methods allow the characterization of surface topography over the widest achievable range of scales while the wavelet methods provide the characterization of surface topography at individual scales. First, wavelets decompose the surface data into different scale components, e.g. roughness, waviness and form, and also into smoothed and detailed images at different scales. Fractal methods, i.e. PIFS transformations, are then applied to describe the resulting images. The method, based on the hybrid fractal-wavelet transforms, allows the multi-scale and non-stationary nature of wear particles and tribological surfaces to be described [97-100].

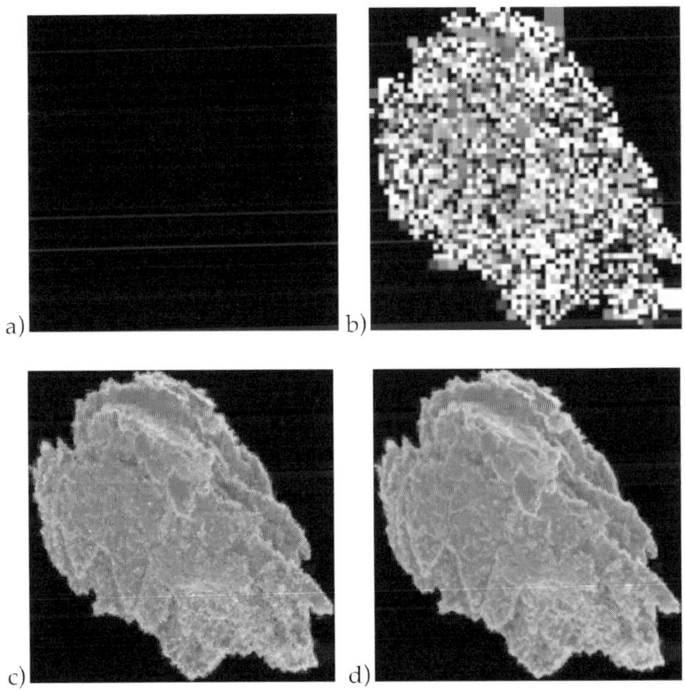

Figure 10.21 The application of the PIFS data obtained for the wear particle image shown in Figure 10.19; a) initial image, b) one iteration, c) four iterations and d) 12 iterations.

Classification of Wear Particles

Classification of wear particles is usually based on visual inspection, requiring experts which is time consuming, costly and not always reliable. Research efforts are directed towards making the classification process computer-based and automated. There have been various attempts to automate the wear particle classification process mostly using neural networks [e.g. 101,102]. These methods are based on representing the particle morphology by a discrete number of values, called pattern features, e.g. area, length, shape factor, Fourier descriptors and co-occurrence texture descriptors, forming the base of a feature vector [e.g. 101-105]. The problem is that numerical descriptors used in these feature vectors, such as area, length, shape factor, boundary fractal dimension, etc., are not scale-, rotationally- and translation-invariant, i.e. they do not uniquely characterize wear particle or surface morphology.

To overcome this problem a pattern recognition method has been developed based on the PIFS technique [66,70,71]. The objective is to assign an unclassified wear particle or surface into a specific group/class according to a predefined criterion such as, for example, the wear mechanism, surface texture, etc. It has been shown that this method allows one to assign wear particles to specific groups without the need for experts [70,106]. The method has been applied to three classes (groups) of wear particles obtained from sheep synovial joints and to an unclassified wear particle [106]. Examples used in this study are shown in Figure 10.22 [106].

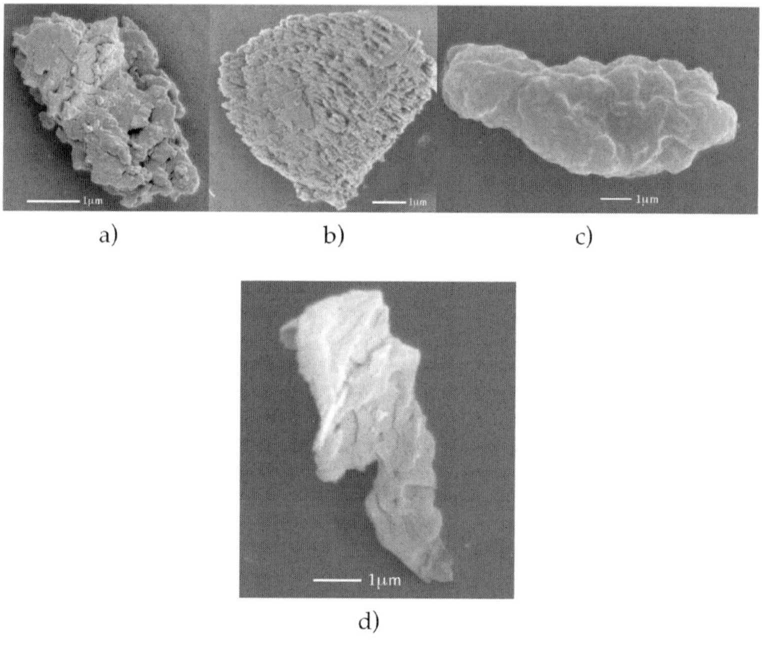

Figure 10.22 Examples of FESEM images of wear particles belonging to three particle groups; a) cleavage type particle, b) spongy type particle, c) cartilaginous particle, d) unknown particle [106].

PIFSs were constructed for images of all particles, i.e. 20, 18 and 15 particles in the first, second and third group respectively. The PIFSs constructed for the particles in those three groups were individually decoded using the unknown particle image shown in Figure 10.22d as the initial image. For the decoding, only one iteration was used. The images obtained after the decoding of the PIFSs constructed for particles depicted in Figure 10.22a-c are shown in Figure 10.23b-d. The peak-to-peak signal-to-noise ratio (PSNR) values were calculated between the decoded images and the unclassified particle image shown in Figure 10.23a for each group of particles. The highest PSNR value was obtained for the third group of particles indicating that the unclassified particle belongs to the group of cartilaginous particles [106].

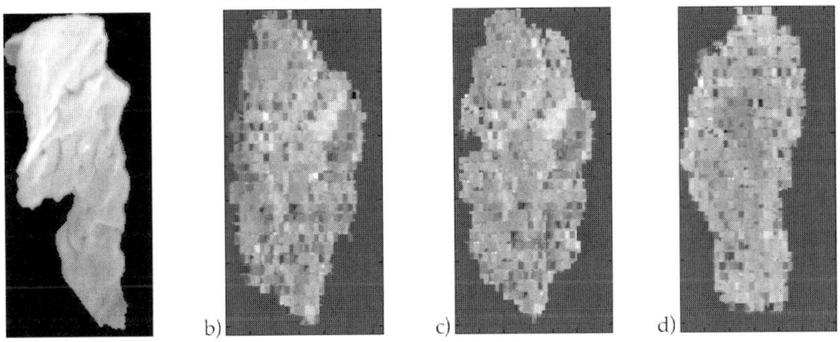

a) b) c) d)

(54-55dB) (20 particles) (53-55dB) (18 particles) (55-56dB) (15 particles)

Figure 10.23 Range image of the unclassified particle (a), images obtained after the decoding of the PIFS transformations constructed for; b) the cleavage type wear particle shown in Figure 10.22a, c) the spongy type particle shown in Figure 10.22b and d) cartilaginous particle shown in Figure 10.22c using the unclassified particle as the initial image [106].

Recently, based on a hybrid fractal-wavelet method, a pattern recognition system has been developed and applied to the classification of tribological surfaces [98-100]. The basic principles of this system are described in Chapter 11.

The research work on the development of classification systems continues. It is thought that eventually a classification system will emerge allowing, for example, the maintenance engineer to submit the images of wear particles or even of failed engineering surfaces on line, e.g. via Internet. The analysis and diagnosis will then be conducted in an automated manner, i.e. no experts will be required [70].

10.4 IMAGE ANALYSIS OF ABRASIVE GRITS

The size and shape of abrasive wear grits is known to be a significant factor in controlling abrasive and erosive wear [107-110]. The size of a grit is usually defined as the minimum size of a sphere which encloses the entire particle. This quantity is measured by sieving particles through a series of sieves of known hole size.

Finding an adequate particle shape parameter that would describe particle angularity has been a more difficult matter as abrasive grits exhibit both chaotic features visible as irregularity in particle boundary and regular geometrical features such as a cube or pyramid. In the early work invariant Fourier descriptors [111] have been used to distinguish between abrasive particles of differing angularities [112]. The shape descriptors were derived from Fourier coefficients describing the 'unwound' particle boundary [112]. The empirical equation relating the angularity of particles, in terms of Fourier descriptors, to abrasion rates has also been proposed [112].

Attempts have also been made to describe abrasive particle shapes using various conventional numerical descriptors and to correlate particle shape with its ability to abrade or erode [e.g. 43,107,110,113]. For example, common abrasives such as SiC, Al_2O_3 and SiO_2 have been characterized using aspect ratio (width/length) and perimeter^2/area shape parameters [113]. It was found that the erosion rate increased with increasing P^2/A and decreasing W/L for these three types of abrasive particles [113]. In another work roundness factor (Perimeter^2/4πArea) and statistical parameter R_{ku} (Kurtosis) describing edge detail have been employed to differentiate among five groups of abrasives [40]. Recently progress in the area has been greatly accelerated with the application of numerical methods and image processing techniques resulting in the development of new numerical parameters describing the angularity of particles [107-109,113-116,118].

In the 90s two new particle angularity parameters, i.e. 'spike parameter - linear fit' (SP) and 'spike parameter - quadratic fit' (SPQ) have been introduced and related to abrasive and erosive wear rates [107-110].

SP parameter is calculated in a manner similar to the boundary fractal dimension, i.e. particle boundary is 'walked' around at a fixed step size [72,107-110,119]. The particle boundary between the start and the end point at each step is then represented by a set of triangles constructed at different scales [107] as illustrated in Figure 10.24a [107,108,119]. The method is based on the assumption that the sharpness and size of these triangles are directly related to particle abrasivity, i.e. the sharper (smaller apex angle) and larger (perpendicular height) the triangles are the more abrasive the particle is [107]. A numerical parameter describing the sharpness and size of these triangles, called the 'spike value' (SV), is defined as:

$$\mathbf{SV} = \cos\left(\frac{\theta}{2}\right)\mathbf{h}$$

where:

h is the perpendicular height of the triangle, and

θ is the apex angle as shown in Figure 10.24a [119].

The method involves calculating the SVs for the largest and sharpest triangles at each step around the particle boundary and then combining them into a 'spike parameter - linear fit' expressed by the following formula [107,108]:

$$\mathbf{SP} = \frac{\Sigma\left[\Sigma(\mathbf{sv}_{\mathbf{max}}/\mathbf{h}_{\mathbf{max}})\right]/\mathbf{m}}{\mathbf{n}} \qquad (10.1)$$

where:

$\mathbf{SV}_{max}$ is $\mathbf{max}\left[\mathbf{cos}\left(\frac{\theta}{2}\right)\mathbf{h}\right]$ for a given step size;

$\mathbf{h}_{max}$ is the height at '$\mathbf{SV}_{max}$';

m is the number of valid '**SV**' for a given step size;

n is the number of different step sizes used.

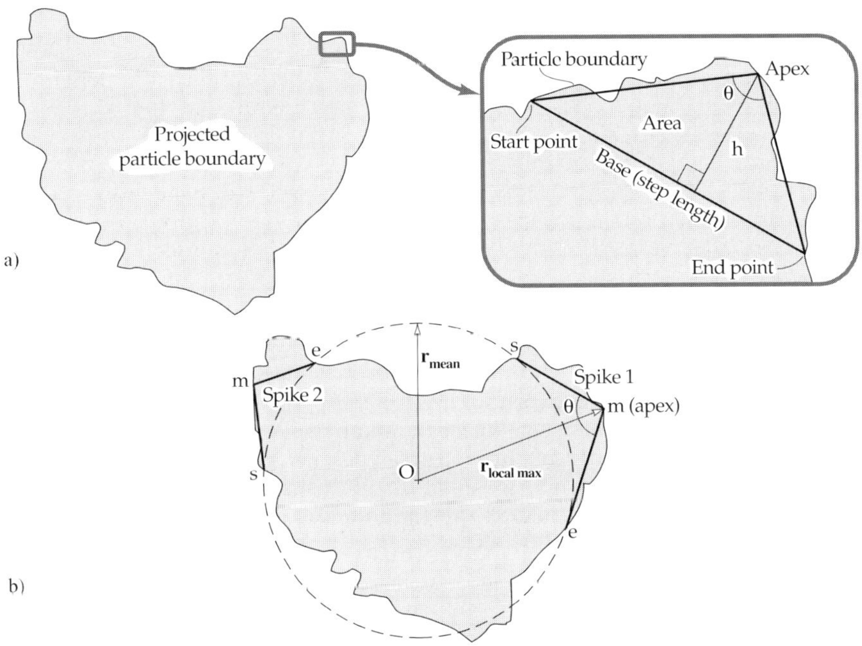

Figure 10.24 Schematic illustration of particle angularity calculation methods of; a) 'spike parameter - linear fit' (SP) and b) 'spike parameter - quadratic fit' (SPQ) (adapted from [119].

When calculating an SPQ parameter a particle boundary centroid 'O' is first located and the average radius circle found [109] as schematically illustrated in Figure 10.24b [119]. Protrusions outside the circle are called 'spikes'. For each protrusion the local maximum radius is found and this point is treated as the spike's apex [109]. The sides of the 'spike' between the points 's-m' and 'm-e' (Figure 10.24b), are then represented by fitting quadratic polynomial functions which are then differentiated at point 'm' yielding the apex angle 'θ' and the spike value 'SV', defined as **SV=cosθ/2**. From the spike values obtained SPQ is then calculated according to the following formula [109]:

$$\mathbf{SPQ} = \mathbf{SV}_{average} \qquad (10.2)$$

The advantage of SPQ over SP is that it considers only the boundary features, i.e. protrusions, which are likely to come in contact with the opposing surface [119].

It has been shown that both SP and SPQ correlate well with abrasive wear rates, i.e. two body, three-body abrasive and erosive wear [108-110]. This is illustrated in Figure 10.25 where the abrasive wear rates obtained with chalk counter-samples are plotted against various abrasive grits characterized by different angularity parameters [119].

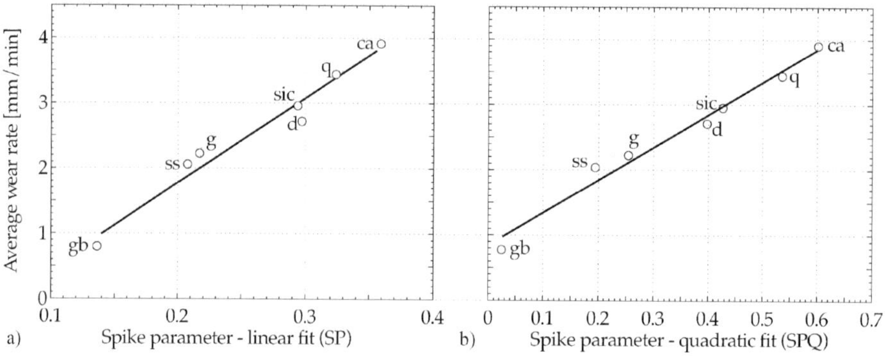

Figure 10.25 Relationship between wear rates and particle angularity described by; a) 'spike parameter - linear fit' and b) 'spike parameter - quadratic fit' (SPQ) for different abrasive grit types, i.e. 'gb' - glass beads, 'ss' - silica sand, 'g' - garnet, 'd' - natural industrial diamonds, 'sic' - silicon carbide, 'q' - crushed quartz and 'ca' - crushed sintered alumina (adapted from [119]).

A recently developed new particle abrasivity parameter is based on the ratio of the groove area '$\mathbf{A}_g$' produced by the particle to projected penetration area '$\mathbf{A}_p$', as illustrated in Figure 10.26a. It is assumed that these values depend on the particle orientation and penetration depth '**h**' and are directly related to load and wear rates. The values of groove area and projected penetration area can be determined using a conical approximation as schematically illustrated in Figure 10.26b. For a specific abrasive grit sample, i.e. silica sand, quartz, alumina, etc., of relatively uniform size, these values are calculated and averaged yielding a parametric relationship between groove area and projected penetration area as a function of the penetration depth [115,116]. Because of the cone approximation used, the technique is called Cone-Fit Analysis (CFA) [115,116].

It has been shown that the projected penetration area, when plotted against groove area at various loads, correlates well with two-body abrasive wear rates, as illustrated in Figure 10.27 [115,116]. The gradient of the curves obtained is directly related to particle abrasivity or alternatively, the ability of particles to remove material.

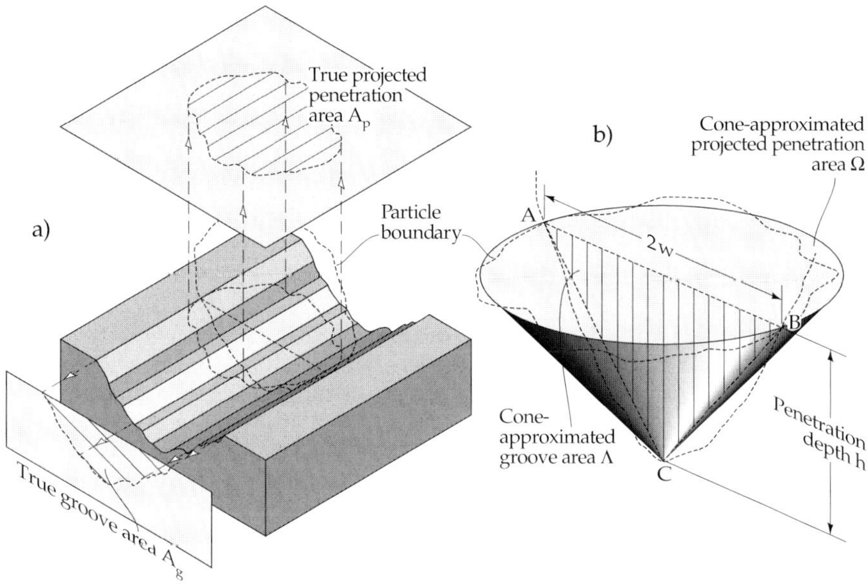

Figure 10.26 Schematic illustration of the cone approximation technique used to determine the particle abrasivity.

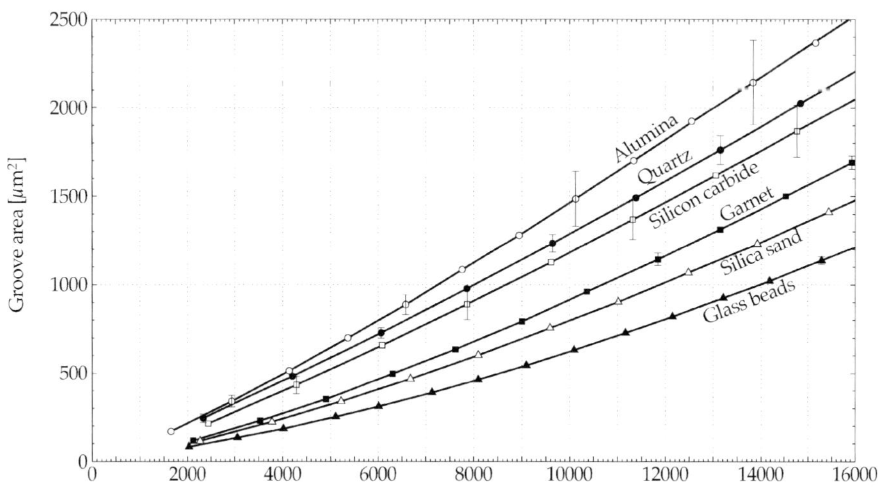

Figure 10.27 Relationship between projected penetration area and groove area at various loads (adapted from [116]).

The curves depicted in Figure 10.27 not only show the order of abrasivity of different grits but also show the relationship between load and wear rate. Note that the projected penetration area is related to load while the groove area is related to wear rate. It has also been shown, for SiC (brittle fracture type particle), that, statistically, abrasive particle shape does not change with particle size. Consequently, the well known particle size effect in abrasion appears to be more attributable to a combination of abrasive particle clogging and material strengthening (according to the strain gradient theory [117]) at small particle sizes [115,116]. This problem is not trivial since finding a solution would involve studies of particle fracture under load and plastic deformations of the abraded material.

In more recent work, the cone-fit algorithm [115,116] has been improved to account for the fact that particle asperities are generally not conical. The modified technique, called 'sharpness analysis', calculates the 'groove area' by numerical integration of the particle boundary [118,120,121]. Additional experiments conducted with quartz and alumina particles revealed no clear relationship between particle shape and size, thus supporting the argument that particle shape is not directly responsible for the particle size effect. Research is now directed towards investigating the relationship between grit shape and size and the geometrical properties of corresponding abrasive surfaces (e.g. abrasive paper, grinding wheels, etc.).

10.5 SUMMARY

Analysis of wear particles from used lubricants (or process fluids) can reveal a wide range of data about wear and lubrication mechanisms occurring within a functioning machine or during tribological test. Analytical techniques range from simplified methods designed for routine industrial application to research tools which are used to analyze the fine detail of wear debris. The two basic types of wear particle analysis available, Spectrographic Oil Analysis (SOAP) and ferrography, each have their own advantages and limitations. SOAP provides a rapid assessment of both quantity and composition of wear particles suspended in lubrication oil while analytical ferrography provides more detailed information about the wear mechanisms taking place. Visual analysis of retrieved wear particles, although required for reliable wear monitoring is, however, more expensive and time consuming. There is a strong research interest in the monitoring of wear particle formation, their recognition and classification. New techniques such as image analysis of wear particles combined with the latest pattern recognition methods are being introduced to extend the range and quality of machine condition monitoring data. There is also a strong interest in the characterization of particle ability to abrade.

REFERENCES

1 S.H. Teoh, S.C. Lim, E.T. Yoon and K.S. Goh, New method for in-vitro wear assessment of materials used in mechanical heart valves, Proceedings of the Symposium on Biomaterials' Mechanical Properties, Sep 14-18 1992, Pittsburgh, PA, USA, ASTM Special Technical Publication, No. 1173, 1994, pp. 43-52.

2 M. Kuster and G.W. Stachowiak, Factors affecting polyethylene wear in Total Knee Arthroplasty, *Orthopedics*, Vol. 25 (2), 2002, pp. 235-242.

3 M. Lukas and D.P. Anderson, Laboratory used oil analysis methods, Tribological Data Handbook, editor: E.R. Booser, CRC Press LLC., 1997 pp. 897-914.

4 A. Walsh, The application of atomic absorption spectra to chemical analysis, *Spectrochimica Acta*, Vol. 7, 1955-1956, pp. 108-117.

5 B.J. Russell, J.P. Shelton and A. Walsh, An atomic-absorption spectrophotometer and its application to the analysis of solutions, *Spectrochimica Acta*, Vol. 8, Issue 6, 1957, pp. 317-318.

6 A.R. Lansdown, Lubrication, A Practical Guide to Lubricant Selection, Pergamon Press, Oxford, 1982.

7 M. Lukas, Lubricating oil analysis, Sawyer's Turbomachinery Maintenance Handbook, editor: J.W. Sawyer, Vol. III, Chapter 8, Turbomachinery International Publications, Norwalk, CT, 1980, pp. 1-20.

8 W.E. Rhine, C.S. Saba and R.E. Kaufmann, Metal particle detection capabilities of rotating disc emission spectrometers, *Lubrication Engineering*, Vol. 42(12), 1986, pp. 755-761.

9 M. Kawamura, K. Fujita and K. Ninomiya, The lubricating properties of used engine oil, *Wear*, 1982, Vol. 77, pp. 195-202.

10 H.H. Abou El Naga, M.M. Mohamed and M.F. El Meneir, Determination of zinc and calcium in multigrade crankcase oils, *Wear*, Vol. 140, 1990, pp. 49-61.

11 T.L. Oliphant, J.R. Terry and E.E. Klaus, The effect of viscosity index improvers on the determination of zinc using atomic absorption spectroscopy, *SAE Reprint 860548*, 1986 (Society of Automotive Engineers).

12 D.F. Hays, Diesel, dual-fuel, and gas engines, Handbook of Lubrication, Theory and Practice of Tribology, editor: E.R. Booser, Volume I, Application and Maintenance, ASLE, CRC Press 1983, pp. 71-89.

13 O.K. Kwon, H.S. Kong and P.K. Oh, Condition monitoring techniques for an internal combustion engine, *Tribology International*, Vol. 20, 1987, pp. 153-159.

14 H. Levinsohn, Limitations of atomic absorption spectrophotometry applied to spectrometric oil analysis, *Transactions ASLE*, Vol. 27, 1984, pp. 24-32.

15 J.R. Brown, C.S. Saba, W.E. Rhine and K.J. Eisentraut, Particle size independent spectrometric determination of wear metals in aircraft lubricating oils, *Analytical Chemistry*, Vol. 52, 1980, pp. 2365-2370.

16 E. Jantzen, V. Buck and S.H. Kaegler, Influence of particle size on wear assessment by spectrometric oil analysis, I: Atomic absorption spectrometry, *Wear*, Vol. 87, 1983, pp. 331-338.

17 M. Lukas, State of the art oil analysis instrumentation, Proc. Int. Tribology Conf., Melbourne, 2-4 December 1987, IEA Conf. Publ. No. 87/18, pp. 316-324.

18 W.W. Seifert and V.C. Westcott, A method for the study of wear particles in lubricating oil, *Wear*, Vol. 21, 1972, pp. 22-42.

19 V.C. Westcott and W.W. Seifert, Investigation of iron content of lubricating oil using Ferrograph and an Emission Spectrometer, *Wear*, Vol. 23, 1973, pp. 239-249.

20 C.H. Evans, E.R. Bowen, J. Bowen, W.P. Tew and V.C. Westcott, Synovial fluid analysis by ferrography, *J. Biochem. Biophys. Methods*, Vol. 2, 1980, pp. 11-18.

21 C.H. Evans, D.C. Mears and J.L. McKnight, A preliminary ferrographic survey of the wear particles in human synovial fluid, *Arthritis and Rheumatism*, Vol. 24, 1981, pp. 912-918.

22 T.B. Kirk, G.W. Stachowiak, A. Warton and G.J. Carroll, Ferrographic and cytological analysis of asymptomatic and arthritic human synovial fluid: a preliminary report, *Journal of Orthopaedic Rheumatology*, Vol. 2, 1989, pp. 91-103.

23 G.W. Stachowiak and J. Baranski, An apparatus for automatic assessment of machine condition, Proc. IEA-ASLE 2nd International Tribology Conference, Melbourne, 2-4 Dec. 1987, IEA Nat. Conf. Publ. 97/18, pp. 311-315.

24 D. Scott and V.C. Westcott, Predictive maintenance by ferrography, *Wear*, Vol. 44, 1977, pp. 173-182.

25 Quantitative Ferrography - Its Meaning and Application, Technical Information, TI 612-120, Foxboro, 1981.

26 W.R. Jones (Jr.) and S.H. Loewenthal, Analysis of wear debris from full-scale bearing fatigue tests using the ferrograph, *ASLE Transactions*, Vol. 24, No. 3, 1981, pp. 323-330.

27 D.P. Anderson, Wear Particle Atlas, Spectro Inc., USA, 1991.

28 T.M. Hunt, Handbook of Wear Debris Analysis and Particle Detection in Liquids, Elsevier Applied Science, 1993.

29 E.R. Bowen and V.C. Westcott, Wear Particle Atlas, Naval Air Engineering Centre, Lakehurst, N.J. July 1976.

30 D. Scott, P.J. McCullagh and G.W. Campbell, Condition monitoring of gas turbines, an exploratory investigation of ferrographic trend analysis, *Wear*, Vol. 49, 1978, pp. 373-389.

31 A.L. Price and B.J. Roylance, The rotary particle depositor - a response to problems experienced with wear particle deposition, Proc. Int. Conf. on Condition Monitoring, Swansea UK, April, 1984, pp. 596-607.

32 P.S. Baur, Ferrography: Machinery-wear analysis with a predictable future, *Power Magazine, The Magazine of Power Generation and Plant Energy Systems*, October, 1982, pp. 1-4.

33 R.E. Kauffman, Techniques to improve the wear metal detection capabilities of rotating disk electrode-emission spectrometers, *Lubrication Engineering*, Vol. 46, 1990, pp. 173-180.

34 C.S. Saba, Improving sample introduction for total wear metal determination by atomic emission spectroscopy, *Lubrication Engineering*, Vol. 48, 1992, pp. 227-232.

35 L.G. Hampson, Industrial uses of ferrography, *Chartered Mechanical Engineer*, May 1981, pp. 60-63.

36 M. Kurahashi and M. Ando, Contamination measuring method - Ferrography, *Trans. JSLE*, Vol. 29, 1984, pp. 919-923.

37 A.W. Batchelor and G.W. Stachowiak, Revealing the hidden world of wear and friction, *Mechanical Engineering Transactions, Journal of the Inst. of Engineers, Australia*, Vol. ME 20, No. 1, 1995, pp. 5-13.

38 P.R. Roberge, C.G. Selkirk and G.F. Fisher, Developing an expert system assistant for filter debris analysis, *Lubrication Engineering*, Vol. 50, 1994, pp. 678-683.

39 B.J. Roylance, I.A. Albidewi, M.S. Laghari, A.R. Luxmoore and F. Deravi, Computer-aided vision engineering (CAVE) - Quantification of wear particle morphology, *Lubrication Engineering*, Vol. 50, 1994, pp. 111-116.

40 S. Raadnui and B.J. Roylance, The classification of wear particle shape, *Lubrication Engineering*, Vol. 51, 1995, pp. 432-437.

41 J.C. Russ, Image Processing Handbook, CRC Press, Inc., 1995.

42 H.H. Hausner, Characterization of the powder particles shape, *Planseeberichte fur Pulvermetallurgie*, Vol. 14, 1966, pp. 75-84.

43 G.W. Stachowiak, Numerical characterization of wear particle morphology and angularity of particles and surfaces, *Tribology International*, Vol. 31, 1998, pp. 139-157.

44 J.C. Russ, Computer Assisted Microscopy, Plenum Press, New York, 1990.

45 B.J. Roylance, G. Wang and C.H. Bovington, The determination of particle morphological parameters to assist in the elucidation of the wear process, Proc. 18th Leeds-Lyon Symposium, 1991, pp. 75-79.

46 B.B. Mandelbrot, Fractals Form, Chance and Dimension, Publ. W.H. Freeman and Company, San Francisco, 1977, pp. 25-35.

47 T.B. Kirk, G.W. Stachowiak and A.W. Batchelor, Fractal parameters and computer image analysis applied to wear particles isolated by ferrography, *Wear*, Vol. 145, 1991, pp. 347-365.

48 T.B. Kirk, G.W. Stachowiak, Fractal computer image analysis applied to wear particles from arthritic and asymptomatic human joints, *Journal of Orthopaedic Rheumatology*, Vol. 4, 1991, pp. 13-30.

49 M.G. Hamblin and G.W. Stachowiak, Comparison of boundary fractal dimension from projected and sectioned particle images, Part II - Dimension changes, *Journal of Computer Assisted Microscopy*, Vol. 5, No. 4, 1993, pp. 301-308.

50 B.H. Kaye, A Random Walk Through Fractal Dimensions, 1989, VCH, Verlagsgesellschaft, Weinheim, Germany.

51 B.H. Kaye, Applied fractal geometry and the fine particle specialist, Part I: Rugged boundaries and rough surfaces, *Particles and Particle System Characterisation*, Vol. 10, 1993, pp. 99-110.

52 N.N. Clark, Three techniques for implementing digital fractal analysis of particle shape, *Powder Technology*, Vol. 46, 1986, pp. 45-52.

53 B.B. Mandelbrot, The Fractal Geometry of Nature, W.H. Freeman, New York, 1982.

54 G.W. Stachowiak, T.B. Kirk and G.B. Stachowiak, Ferrography and fractal analysis of contamination particles in unused lubricating oils, *Tribology International*, Vol. 24, 1991, pp. 329-334.

55 M.G. Hamblin and G.W. Stachowiak, Variation of particle shape parameters when viewed as a projection and a section, *Journal of Computer Assisted Microscopy*, Vol. 6, No. 2, 1994, pp. 51-59.

56 M.G. Hamblin and G.W. Stachowiak, Fractal dimensions from images, Image Analysis '94, editors: A. van Riessen and J.R. Hatch, WASEM, 1994, p. 4 and pp. 43-61.

57 M. Allen, B.J. Brown and N.I. Miles, Measurement of boundary fractal dimensions: review of current techniques, *Powder Technology*, Vol. 84, 1995, pp. 1-14.

58 B.H. Kaye, Fractal description of fine particle systems, Modern Methods in Fine Particle Characterization, editor: J.K. Bedow, CRC Press, Boca Raton, FL, 1983.

59 L.C. Chan and N.W. Page, Fractal measures of particle shape and their relation to interparticle friction in granular materials, Proc. of the First Australasian Congress on Applied Mechanics, 21-23 February 1996, Melbourne, Australia, 1996, pp. 475-480.

60 L.E. Vallejo, Fractal analysis of granual materials, *Geotechnique*, Vol. 45, 1995, pp. 159-163.

61 M.G. Hamblin and G.W. Stachowiak, Measurement of fractal surface profiles obtained from scanning electron and laser scanning microscope images and contact profile meter, *Journal of Computer Assisted Microscopy*, Vol. 6, No. 4, 1994, pp. 181-194.

62 M.G. Hamblin and G.W. Stachowiak, Application of the Richardson technique to the analysis of surface profiles and particle boundaries, *Tribology Letters*, Vol. 1, 1995, pp. 95-108.

63 P. Pfeifer and D. Avnir, Chemistry in non-integer dimensions between 2 and 3, Part 1: Fractal theory of heterogeneous surfaces, Part 2: Fractal surfaces of absorbants, *J. Chem. Phys.*, Vol. 79, 1983, pp. 3558-3571.

64 P. Podsiadlo and G.W. Stachowiak, Scale-invariant analysis of wear particle surface morphology. I: Theoretical background, computer implementation and technique testing, *Wear*, Vol. 242, 2000, pp. 160-179.

65 P. Podsiadlo and G.W. Stachowiak, Scale-invariant analysis of wear particle surface morphology. II: Fractal dimension, *Wear*, Vol. 242, 2000, pp. 180-188.

66 P. Podsiadlo and G.W. Stachowiak, Scale-invariant analysis of wear particle surface morphology. III: Pattern recognition, *Wear*, Vol. 242, 2000, pp. 189-201.

67 G.W. Stachowiak and P. Podsiadlo, Surface characterization of wear particles, *Wear*, Vol. 225-229 Part II, 1999, pp. 1171-1185.

68 J.C. Russ, Fractal Surfaces, Plenum Press, New York, 1994.

69 P. Podsiadlo and G.W. Stachowiak, The development of modified Hurst orientation transform for the characterization of surface topography of wear particles, *Tribology Letters*, Vol. 4, 1998, pp. 215-229.

70 G.W. Stachowiak and P. Podsiadlo, Characterization and classification of wear particles and surfaces, *Wear*, Vol. 249, 2001, pp. 194-200.

71 P. Podsiadlo and G.W. Stachowiak, Scale-invariant analysis of tribological surfaces, Thinning Films and Tribological Interfaces, editor: D. Dowson et al, Tribology Series 38, Elsevier, 2000, pp. 546-557.

72 M.G. Hamblin and G.W. Stachowiak, Comparison of boundary fractal dimension from projected and sectioned particle images, Part I - Technique evaluation, *Journal of Computer Assisted Microscopy*, Vol. 5, No. 4, 1993, pp. 291-300.

73 G.W. Stachowiak and M.G. Hamblin, Application of fractals to the description of shape of the particles found in tribological systems, *Mechanical Engineering Transactions, Journal of the Inst. of Engineers, Australia*, Vol. ME 20, No. 2, 1995, pp. 127-136.

74 L.F. Richardson, The problem of contiguity: an appendix of statistics of deadly quarrels, *General Systems, Yearbook*, Vol. 6, 1961, pp. 139-187.

75 T.B. Kirk and G.W. Stachowiak, The applications of fractals to the morphological description of particles generated in tribological systems, *Mechanical Engineering Transactions, Journal of the Inst. of Engineers, Australia*, Vol. 16, 1991, pp. 279-293.

76 B.H. Kaye, Specification of the ruggedness and/or texture of a fine particle profile by its fractal dimension, *Powder Technology*, Vol. 21, 1976, pp. 207-213.

77 T.B. Kirk and G.W. Stachowiak, Computer fractal image analysis of wear particles from synovial joints, Proc. The 1992 Joint Conference on Electron Microscopy and Cell Biology, ACEM-12&ANZ5CB-11, University of Western Australia, 10-14 February 1992, pp. 84-85.

78 P. Podsiadlo and G.W. Stachowiak, Evaluation of boundary fractal methods for the characterization of wear particles, *Wear*, Vol. 217, 1998, pp. 24-34.

79 S.K. Kennedy and W.H. Lin, FRACT - a FORTRAN subroutine to calculate the variables necessary to determine the fractal dimension of closed forms, *Computers and Geosciences*, Vol. 12, 1986, pp. 705-712.

80 M.Q. Zhang, Z.P. Lu and K. Friedrich, On the wear debris of polyetheretherketone - fractal dimensions in relation to wear mechanisms, *Tribology International*, Vol. 30, 1997, pp. 87-102.

81 G.W. Stachowiak, G.B. Stachowiak and P. Campbell, Application of numerical descriptors to the characterization of wear particles obtained from joint replacements, *Proceedings of the Institution of Mechanical Engineers - Part H - Engineering in Medicine*, Vol. 211, 1997, pp. 1-10.

82 G.W. Stachowiak and P. Podsiadlo, Analysis of wear particle boundaries found in sheep knee joints during in-vitro wear tests without muscle compensation, *Journal of Biomechanics*, Vol. 30 1997, pp. 415-419.

83 P. Podsiadlo, M. Kuster and G.W. Stachowiak, Numerical analysis of wear particles from non-arthritic and osteoarthritic human knee joints, *Wear*, Vol. 210, 1997, pp. 318-325.

84 H.K. Wickramasinghe, Scanned-probe microscopes, *Scientific American*, Vol. 261, 1989, pp. 98-105.

85 Z. Peng, T.B. Kirk, and Z.L. Xu, The development of three-dimensional imaging techniques of wear particle analysis, *Wear*, Vol. 203-204, 1997, pp. 418-424.

86 P.J. Caber, Interferometric profiler for rough surfaces, *Applied Optics*, Vol. 32, 1993, pp. 3438-3441.

87 P. Podsiadlo and G.W. Stachowiak, Characterization of surface topography of wear particles by SEM stereoscopy, *Wear*, Vol. 206, 1997, pp. 39-52.

88 P. Podsiadlo and G.W. Stachowiak, 3-D imaging of wear particles found in synovial joints, *Wear*, Vol. 230, 1999, pp. 184-193.

89 A. Boyde, Quantitative photogrammetric analysis and quantitative stereoscopic analysis of SEM images, *Journal of Microscopy*, Vol. 98A, 1973, pp. 452-471.

90 B.B. Mandelbrot, Self-affine fractals and fractal dimensions, *Physica Scripta*, Vol. 32, 1985, pp. 257-260.

91 G. Borgerfors, Distance transforms in arbitrary dimensions, *Comp. Vision, Graphics Image Proc.*, Vol. 27, 1984, pp. 321-345.

92 P. Grassberger and I. Procaccia, Characterisation of strange attractors, *Phys. Rev. Letters*, Vol. 50, 1983, pp. 346-349.

93 K. Judd, An improved estimator of dimension and comments on providing confidence intervals, *Phys. D*, Vol. 56, 1992, pp. 216-228.

94 P. Prusinkiewicz and A. Lindenmayer, The algorithmic beauty of plants, Springer-Verlag, New York, 1990.

95 M.F. Barsney and L.P. Hurd, Fractals Everywhere, Academic Press, San Diego, 1988.

96 Y. Fisher (editor) Fractal Image Compression. Theory and Application, Springer-Verlag, New York, 1995.

97 P. Podsiadlo and G.W. Stachowiak, Hybrid fractal-wavelet method for characterization of tribological surfaces - a preliminary study, *Tribology Letters*, Vol. 13 (4), 2002, pp. 241-250.

98 P. Podsiadlo, and G.W. Stachowiak, Fractal-wavelet based classification of tribological surfaces, *Wear*, Vol. 254, 2003, pp. 1189-1198.

99 G.W. Stachowiak and P. Podsiadlo, Classification of tribological surfaces, *Tribology International*, Vol. 37, 2004, pp. 211-217.

100 P. Podsiadlo and G.W. Stachowiak, Classification of tribological surfaces without surface parameters, Proceedings of the 6th International Tribology Conference AUSTRIB 2002, 'Frontiers in Tribology', (editor: G.W. Stachowiak), 2-5 December, Western Australia, 2002, pp. 109 117.

101 N.K. Myshkin, O.K. Kwon, A.Y. Grigoriev, H.-S. Ahn and H. Kong, Classification of wear debris using a neural network, *Wear*, Vol. 203-204, 1997, pp. 658-662.

102 Z. Peng and T.B. Kirk, Automatic wear particle classification using neural networks, *Tribology Letters*, Vol. 5, 1998, pp. 249-257.

103 X. Kun, A.R. Luxmoore and F. Deravi, Comparison of shape features for the classification of wear particles, *Engineering Applications of Artificial Intelligence*, Vol. 10, 1997, pp. 485-493.

104 B.J. Roylance, I.A. Albidewi, A.L. Price and A.R. Luxmoore, The development of a computer-aided systematic particle analysis procedure - CASPA, *Lubrication Engineering*, Vol. 48, 1992, pp. 940-946.

105 Z. Peng and T.B. Kirk, Wear particle classification in a fuzzy grey system, *Wear*, Vol. 225-229, 1999, pp. 1238-1247.

106 G.W. Stachowiak and P. Podsiadlo, The application of the modified partitioned iterated function system (PIFS) to the characterization and classification of wear particles and surfaces, Proc. of the International Tribology Conference in Nagasaki, 2000, Japanese Society of Tribologists, Vol. 1, 2001, pp. 251-256.

107 M.G. Hamblin and G.W. Stachowiak, A multi-scale measure of particle abrasivity, *Wear*, Vol. 185, 1995, pp. 225-233.

108 M.G. Hamblin and G.W. Stachowiak, A multi-scale measure of particle abrasivity and its relation to two body abrasive wear, *Wear*, Vol. 190, 1995, pp. 190-196.

109 M.G. Hamblin and G.W. Stachowiak, Description of abrasive particle shape and its relation to two-body abrasive wear, *Tribology Transactions*, Vol. 39, No. 4, 1996, pp. 803-810.

110 G.W. Stachowiak, Particle angularity and its relationship to abrasive and erosive wear, *Wear*, Vol. 241, 2000, pp. 214-220.

111 D.W. Luerkens, J.K. Beddow and A.F. Vetter, Invariant Fourier descriptors, *Powder Technology*, Vol. 31, 1982, pp. 209-215.

112 P.A. Swanson and A.F. Vetter, The measurement of abrasive particle shape and its effect on wear, *ASLE Transactions*, Vol. 28, 1984, pp. 225-230.

113 S. Bahadur and R. Badruddin, Erodent particle characterization and the effect of particle size and shape on erosion, *Wear*, Vol. 138, 1990, pp. 189-208.

114 V.F. Leavers, Use of two-dimensional Radon transform to generate a taxonomy of shape for the characterization of abrasive powder particles, IEEE *Transactions of Pattern Analysis and Machine Intelligence*, Vol. 22(12), 2000, pp. 1411-1423.

115 D.V. De Pellegrin and G.W. Stachowiak, New technique of grit-profile analysis by cone-fitting, *Wear*, Vol. 247, 2001, pp. 109-119.

116 D.V. De Pellegrin and G.W. Stachowiak, Abrasiveness of particles measured by cone-fit analysis (CFA), From Model Experiment to Industrial Problem, editor: D. Dowson et al, Tribology Series 39, Elsevier, 2001.

117 E.C. Aifantis, Strain gradient interpretation of size effects, *International Journal of Fracture*, Vol. 95, 1999, pp. 299-314.

118 D.V. De Pellegrin and G.W. Stachowiak, Assessing the role of particle shape and scale in abrasion using 'Sharpness Analysis', Part I - Technique development, *Wear*, Vol. 253, 2002, pp. 1016-1025.

119 G.W. Stachowiak and A.W. Batchelor, Engineering Tribology, Butterworth Heinemann, 2001.

120 D.V. De Pellegrin and G.W. Stachowiak, Assessing the role of particle shape and scale in abrasion using 'Sharpness Analysis', Part II - Technique evaluation, *Wear*, Vol. 253, 2002, pp. 1026-1034.

121 D.V. De Pellegrin and G.W. Stachowiak, Sharpness of abrasive particles and surfaces, Proceedings of the 6th International Tribology Conference AUSTRIB 2002, 'Frontiers in Tribology', (editor: G.W. Stachowiak), 2-5 December, Western Australia, 2002, pp. 203-210.

11 SIMULATION OF WEAR AND FRICTION EXPERIMENTS

and Presentation of Tribological Data

11.1 INTRODUCTION

A basic objective of tribological experimentation is the accurate simulation or reproduction of wear and friction phenomena as found in machinery and other subjects of investigation. The difficulty is that wear and friction are controlled by so many variables that virtually every dynamic contact presents a unique form of wear and friction. The task of laboratory experiments is to simulate the friction and wear with sufficient accuracy to allow reliable interpretation of the effects of, for example, new lubricants or materials. If the variability of wear and friction with basic parameters such as load and sliding speed is not given sufficient consideration then even elementary questions such as the effect of hardness on wear might be incorrectly answered with misleading data. Apart from the well known parameters such as load and speed, there are also a multitude of other factors affecting wear and friction. For example, material's grain size or different method of lubricant supply may have a significant influence. Given this uncertainty about the similarity between laboratory tests and the object of study, it is necessary to apply specifically designed tests that would allow some correspondence or similarity between the two (or more) dynamic contacts involved.

In most cases the complexity of wear and friction data defies immediate analysis. It is usually not possible to obtain a quantitative or semi-quantitative relationship between controlling parameters and dependent parameters by simple inspection or graphical analysis. Specialised techniques of data analysis based on statistical methods and the theory of dimensional groups can, however, provide much useful information that may lead to a definition of wear or friction in terms of combinations of controlling variables. In tribology, the task of conducting the experiment simulating the real situation is often encountered. A researcher is then faced with the problem of selecting the test operating conditions and the test samples so that the results obtained will be relevant to the real, e.g. industrial, situation.

In some cases, usually where the major controlling parameters have been identified, dimensional analysis can be used to optimize the design of a tribological experiment. It is also possible, using dimensional analysis, to obtain quantitative numerical expressions for wear and friction in terms of the controlling parameters. Tests of parameter significance can be applied to exclude unimportant parameters from further analysis. Numerical tests of similarity between experiment and real object can confirm whether the experimental data from a model test can be used to reliably predict the characteristics of the original friction or wear problem.

Given the complexity of much tribological data, great care is needed in the presentation of data. Graphical presentation is effective but can only be used to display the influence of one, perhaps two, controlling parameters on friction or wear. With this restriction, it is vital to adjust the graphical display until key features of the friction and wear data are adequately revealed.

This chapter attempts to provide an answer or a means of finding the answer to questions such as: How can the investigator be sure that the test experiment matches the real situation? How can meaningful answers be obtained with the minimum number of tests? How can the data be used to construct useful predictive formulae? How should the data be presented so that non-specialists as well as specialists in tribology can readily understand the salient features of the wear or friction problem?

11.2 BASIC PRINCIPLE BEHIND THE ACCURATE SIMULATION OF FRICTION AND WEAR

Wear and friction can be classified into a number of mechanisms, each of which has different controlling influences and resulting characteristics. This distinction is exemplified by the difference between abrasive wear and fretting wear. Even within the comparatively restricted range of sliding between smooth surfaces of equal or similar hardness, several wear mechanisms can operate. A fundamental characteristic of wear or friction is that there are some critical levels of controlling parameters where the mechanism of wear or friction can change, for example, from adhesive wear to delamination wear. These relatively sharp transitions form the basis of so called wear mechanism diagrams or 'wear maps' [1,2]. Wear maps are an attempt to provide a general view of wear behaviour by plotting wear rates and observed mechanisms as a function of basic parameters such as contact stress and sliding speed. An example of a wear map, illustrating the wear of high speed steel cutting tools in dry air, is shown in Figure 11.1.

It can be seen from Figure 11.1 that the rate of wear and its controlling mechanism change as basic parameters such as cutting speed and tool feed rate are varied. The underlying physical phenomena which control the wear process differ fundamentally as operating conditions are varied. For the high speed steel cutting tool wear, wear is determined by either mechanical strength or diffusion and solubility of tool material into the workpiece material. From a practical viewpoint, the optimisation of tool material should allow for this variation in controlling physical properties. There is little use in rising mechanical strength if wear is determined by material's solubility.

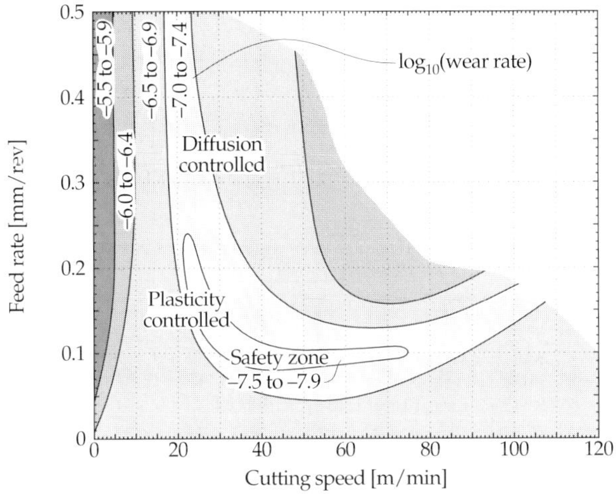

Figure 11.1 Wear map of high speed steel cutting tools in open air (adapted from [2]).

If the dynamic contact under investigation, e.g. a contact between a cam and tappet, results in a certain wear mechanism then the experimental model simulated, for example, on a pin-on-disc machine must reproduce the same type of wear mechanism. Although many of the wear maps, commonly produced, tend to illustrate wear as a function of only load or speed it needs to be remembered that many other parameters, for example, ambient temperature and humidity, are also involved in the control of wear mechanisms. A fundamental problem in designing valid friction and wear experiments is to ensure the adequate similarity between the mechanism of wear and friction found in the machinery under investigation and the wear and friction occurring in a test apparatus (see also Chapter 2).

11.3 LIMITATIONS OF LABORATORY EXPERIMENTS AND TEST EQUIPMENT

There are many problems associated with the interpretation of data obtained from standard test rigs working under a limited range of operating conditions. As many standard tests were designed about 50 years ago they provided only a limited information about lubricant's or material's behaviour. An example of these problems was discussed in Chapter 1. The problem described was associated with comparing an additive enriched oil with a plain oil when the ratio of rolling to sliding was not specified. In that example it was illustrated how an apparently invariant relationship such as the ranking of lubricants is in fact dependent on many factors, e.g. contact load. Another problem associated with laboratory testing is that the tests conducted are not sensitive enough to factors that strongly influence wear in the real equipment. This is illustrated in Figure 11.2 showing the wear rate results of marine engine cylinders versus 'O.K. load' obtained on a four-ball tester.

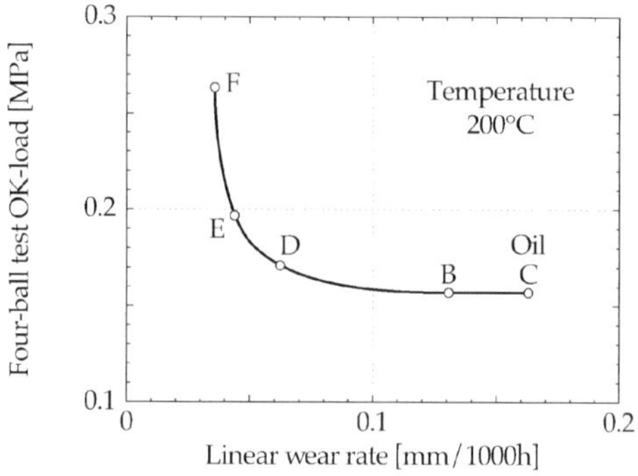

Figure 11.2 Wear rate results of marine engine cylinders versus four-ball test OK-load (adapted from [3]).

It can be seen from Figure 11.2 that below a certain critical load, the four-ball test OK-load data provides no discrimination between differing wear rates. This discrepancy can be explained by the fact that wear is subject to far more influences in the complex environment of a marine cylinder than under the simplified conditions of a four-ball tester. The practice of comparing a limiting contact stress for low friction sliding and a wear rate is also questionable since friction and wear do not always exhibit similar characteristics even though they share the same cause, i.e. dynamic contact. For example, corrosive wear is characterized by high wear and low friction. The use of a four-ball tester without careful consideration of similarity between wear processes occurring is an example of a fallible method of tribological testing. Where test equipment is chosen and used according to reasons of convenience and familiarity, some quite amazing results can be obtained. In one series of such tests, it was proven that conventionally accepted standard test machines were capable of indicating that milk and beer were more effective lubricants than fully formulated mineral oils [4]. Milk contains fat which can act as an effective lubricant but the ability of beer to provide lubrication appears to challenge existing concepts of friction and wear.

11.4 SYSTEMS ANALYSIS OF A DYNAMIC CONTACT AND EXPERIMENT DESIGN FOR SIMILARITY BETWEEN TEST AND THE REAL PROBLEM

The systems method of analysis, originally advocated as a solution to tribological problems and developed by Czichos [5], envisages a tribological phenomenon as controlled by a hierarchy of mechanisms. The concept of the mechanism hierarchy or 'system envelopes' is illustrated schematically in Figure 11.3.

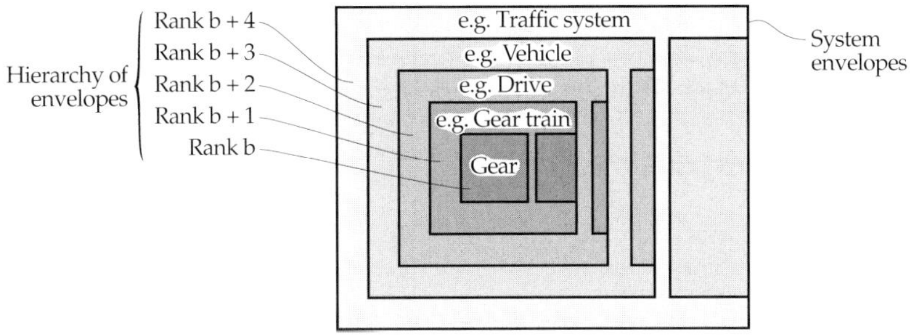

Figure 11.3 Systems theory of a hierarchy of mechanisms controlling wear and friction (adapted from [5]).

The system envelopes, when applied to a dynamic contact, provide a structured model of the relationship between fundamental aspects of friction and wear and the external behaviour of a machine which contains one or more dynamic contacts. A common example of the envelope model of friction and wear is provided by the wear of vehicle engines. At the outer envelope, the influences of external factors such as traffic conditions, climate and pollution or dust are significant. At an inner level, the design of the engine and the imposed loads are the controlling factors. At the innermost or core level, the fundamental aspects of dynamic contact such as load, sliding speed, temperature and lubricant determine friction and wear. It is obvious that while climatic and traffic conditions will affect the optimisation of a lubricant and engine design, the converse that lubricant and engine design influence climate and traffic conditions is not true.

In any given experiment, it is usually not possible to study all levels of a friction and wear process. For example, in the case of vehicle engines, practical experience shows that full-scale vehicle tests, i.e. fleet tests, usually sustain too much traffic interference to provide reliable information on, for example, a comparison of friction losses with different lubricants. If the driver of a test car accelerates through a traffic light changing from amber to red, the data on fuel economy as a function of engine friction may become corrupted. Fundamental aspects of friction and wear of vehicle engines are therefore more effectively studied on specialised apparatus in a controlled and isolated environment. Studies on external factors controlling friction and wear are best investigated by direct simulations of the particular problem. In these cases the original equipment, e.g. car, is usually employed as the test apparatus and gear change, accelerator and brakes are applied by a computer-controlled set of levers simulating an urban or a highway traffic.

Similarity between real test and experimental simulation is partly determined by the requirements of the real test. If a comparative study on vehicle engine wear in moderate traffic conditions represented by freeway driving and severe traffic conditions in a congested city is required, then some specific criterion of similarity between test traffic conditions and general traffic conditions has to be defined. The detailed mechanics contained in each 'envelope' should accurately simulate the actual phenomena involved. In other words, controlling parameters such as sliding speed, load and environmental conditions should all be carefully analyzed and set to specific levels. The same process can be applied in reverse to assess the validity of an existing experiment for a new test application.

In most tribological studies, however, the specific characteristics of friction and wear are of interest rather than indirect influences represented by the outer envelope shown in Figure 11.3. To ensure that the friction and wear characteristics simulated experimentally are fundamentally similar to those found in the operating machinery it is necessary to apply specific tests of similarity in wear and friction. Very often, the friction and wear test apparatus has very little resemblance to the operating machinery. This is particularly true in the case of four-ball tester, described in Chapter 3, used in simulation of vehicle engine tests. This machine is used to assess the performance of lubricants in engines and gears. However, operating conditions in dynamic contact in combustion engines and gears are different from those encountered in contact between bearing balls in a four-ball tester. Thus it is not always possible to assume similarity on the basis that dynamic contacts of similar design will share similarity in friction and wear.

11.5 EXPERIMENTAL CONFIRMATION OF SIMILARITY BETWEEN 'REAL' AND SIMULATED WEAR

The degree of similarity between the laboratory experiment and a real tribological problem occurring in machinery is not usually immediately apparent. In many cases, there is only a brief description of the original wear and friction problem occurring. Often, discussion of the problem is confined to verbal comments about excessive noise, heat from a component prior to failure or actual failure. However, much more detailed information about the prevailing friction and wear is required to assess the similarity between wear and friction in real machinery and simulated on a test rig. Current research efforts aim to overcome these practical difficulties and ensure reliable experimental data from any tribological investigation.

Comparison of Wear Rates and Distribution of Wear

The wear rate of the real dynamic contact and the wear rate measured during the laboratory experiments should be similar, i.e. within an order of magnitude or closer. Friction coefficients should also be similar, e.g. no more than 10% difference. If there is a larger difference, it is likely that the dynamic contacts simulated in the laboratory environment and the original machine are operating under different wear and friction mechanisms. Accelerated wear tests are inadvisable for this reason and often much patience is required with time-consuming wear tests. A classical example of slow wear tests are lifetime tests of orthopaedic endoprostheses (artificial joints) where the individual tests can easily last a year [6].

The only exception to the rule of close correspondence in friction and wear coefficients is when a study of some specific aspect of the lubrication mechanism is required. For example, E.P. (Extreme Pressure) and A.W. (Anti-Wear) characteristics of crankcase oils are often studied using the four-ball tester which involves much higher friction coefficients and wear rates than typically found in internal combustion engines. During these tests the limiting frictional seizure characteristics of the lubricants are being assessed. As the data from accelerated tests involving four-ball machines is still subject to controversy some compromise of data quality is inevitable.

The distribution of wear between two surfaces in a dynamic contact should also be similar for both laboratory experiments on test apparatus and original equipment [7]. For example, in one work during the experiments on the dry wear of railway wagon wheels by brake pads, a block-on-ring apparatus was used to model the wheel wear [7]. The ring was used to simulate the wheel, while the blocks were made of brake pad material. With the initial design of block-on-ring test rig, negligible wear of the ring occurred while the block wore rapidly. Analysis of the test rig design revealed that the block length in the direction of sliding was too small compared to the circumference of the ring. As a result, ring surface temperatures were lower than temperatures on the wagon wheels and this resulted in a reduced wear. Reduction of the ring diameter while maintaining the same block size allowed wear to occur on the ring. These modifications to the test rig are schematically illustrated in Figure 11.4.

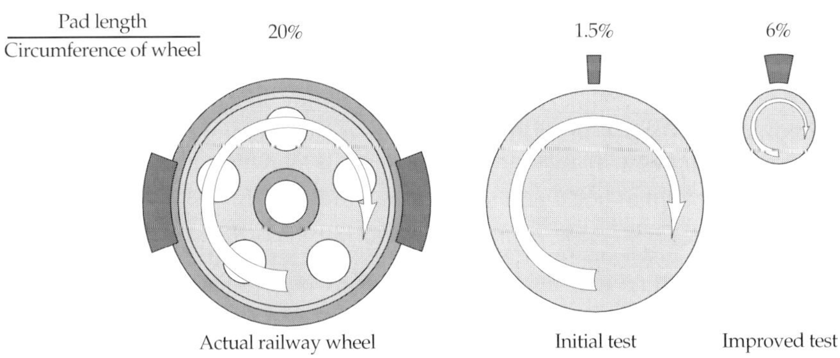

Figure 11.4 Schematic illustration of the modifications to test to ensure accurate modelling of wear (adapted from [7]).

Simulation of similarity between test apparatus and original equipment can be extended even further. In principle, any other characteristic of wear and friction should display close correspondence between tests conducted in the laboratory on test rigs and original equipment. Features of wear and friction, such as trends in initial values of friction and wear or the presence of stick-slip, should all be noted and tested for similarity.

Micrographic Examination and Comparison of Worn Surfaces

The morphology of worn surfaces is perhaps the most sensitive index allowing comparison to be made between different dynamic contacts. As discussed already in Chapter 8, microscopic examination can reveal a wealth of detailed information about the wear and friction mechanisms occurring. Such evidence allows for a very close comparison of the dynamic contacts on test rig and original equipment to be made. Micrographic examination of worn surfaces has also a major practical advantage of providing detailed information about the wear mechanism occurring in the original equipment, without the need for instrumented tests. If tribological similarity is evaluated based on wear and friction measurements, it is then necessary to make detailed measurements inside the original equipment. In most cases, this is impractical, expensive or undesirable since in the case of industrial equipment it could result in production losses. Usually, all that is available to the tribologist is a selection of failed components with evidence of a wear or friction problem. Therefore the evaluation of any tribological similarity is often forced to depend on the information that can be gleaned from the examination of failed components.

The micrographic appearance of the worn surfaces obtained after the laboratory experiments conducted on a test rig and the original equipment should exhibit a fundamental resemblance [5,8]. Since wear is a chaotic process an exact correspondence between any two worn surfaces is extremely unlikely. What is intended, is a functional similarity, i.e. if the wear scar on the surface obtained from the original equipment exhibits cracks and pits indicative of surface fatigue, then similar cracks and pits should be present on the wear scar on the test specimen. It is important that worn specimens from the original equipment that have been retrieved are not only those from the final destructive stages of failure. If only tests specimens from the final stages of failure are chosen, then the model test conditions will tend to represent only extreme operating conditions [7]. For example, a comparative display of worn surfaces is shown in Figure 11.5.

In Figure 11.5a a three-body abrasion surface of high chromium white cast iron from the ball cratering test is shown while in Figure 11.5b a worn surface of high chromium white cast iron from a slurry pump is presented. It can be seen from Figures 11.5a and 11.5b that removal of the matrix material and carbides exposure by the action of abrasive grits are evident in both worn surfaces. Sufficient degree of tribological similarity between test and original equipment can be concluded on the basis of this visual evidence.

Tribological similarity is not always achieved at the first attempt of the test rig design. A methodology of successive refinement of test operating conditions to achieve the required degree of similarity has been proposed [5]. Using wear scar micrographs from the original equipment as a benchmark, experimental conditions such as load, sliding speed or even apparatus design are adjusted until tribological similarity is obtained. For example, if the wear scar on the component surface from the original equipment reveals signs of adhesive wear with extensive surface roughening while the wear scar on the test specimen remains smooth, then an increase in contact stress or a decrease of velocity during laboratory tests to initiate adhesive wear may be appropriate. The process of obtaining the required degree of tribological similarity is illustrated schematically in a flow chart shown in Figure 11.6.

a)
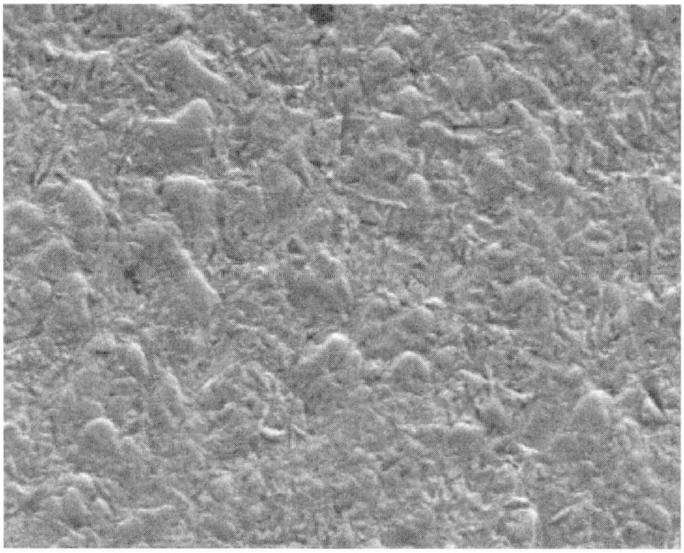

50μm

b)
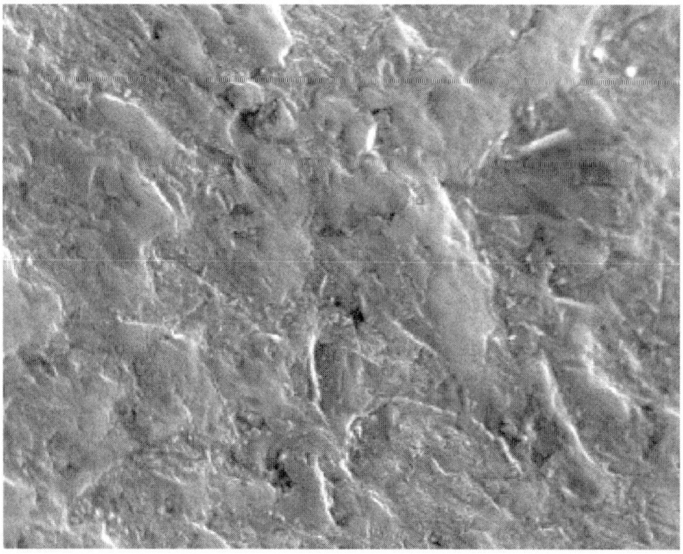

50μm

Figure 11.5 Example of comparison between the worn surfaces (SEM images) obtained from the test specimen (ball cratering test) and original equipment (a slurry pump used in mineral processing).

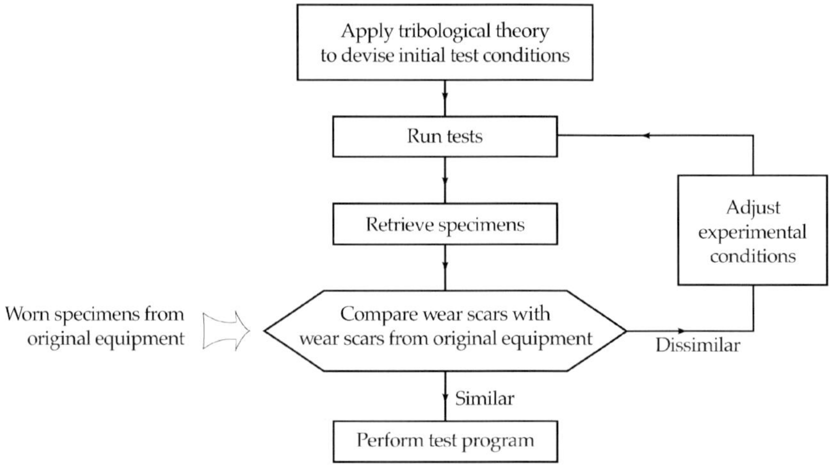

Figure 11.6 Schematic illustration of the iterative procedure for achieving the tribological similarity between model test and the problem.

Tests of micrographic similarity should also include cross sectional views of the micrographic structure. Some forms of wear are associated with severe subsurface deformation or thermal alteration of surface microstructure. Subsurface changes are not usually visible from micrographs of worn surfaces so that a sectional micrographic examination is required for true confirmation of similarity.

Examination of Wear Debris as a Test of Similarity

In some instances, the retrieval of worn specimens from the original equipment to monitor progress of wear may present difficulty or is undesirable. For example, it is very difficult to study the progress of wear in vivo in orthopaedic endoprostheses (artificial joints) or arthritic synovial joints. In both cases, the retrieval of wearing components is undesirable as there is some pain and risk involved together with the difficulty of relocating the components correctly. In such cases progress of wear can be monitored and studied through the analysis of wear particle morphology. As wear particles form during wear their morphology contains lots of information about the processes involved in their formation. Particle analysis is widely used in machine condition monitoring as it allows predictions to be made as to whether the wear occurring in some machine in operation will cause its imminent failure (see also Chapter 10). Wear debris from a real machine can be retrieved and compared with wear debris obtained from a laboratory test rig providing information on the level of similarity between wear mechanisms occurring in both instances. The same principles of micrographic examination developed for worn surfaces apply to wear particles analysis. A functional similarity between wear particles based on evidence from common processes of deformation and fracture can be assessed. This is illustrated in Figures 11.7 and 11.8 where the similarity between wear particles obtained from grade 3 osteoarthritic human joints and sheep joints subjected to accelerated wear tests is apparent [9,10].

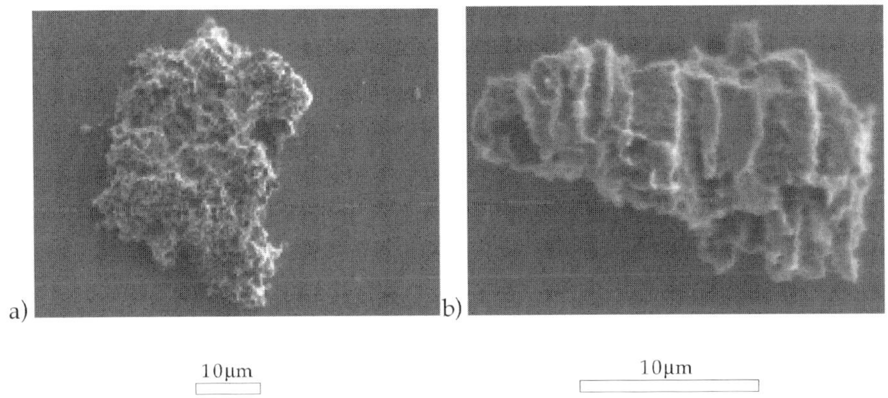

Figure 11.7 SEM images of chunky, cartilaginous wear particles found in; a) grade 1 osteoarthritic and b) grade 3 osteoarthritic human joints [9].

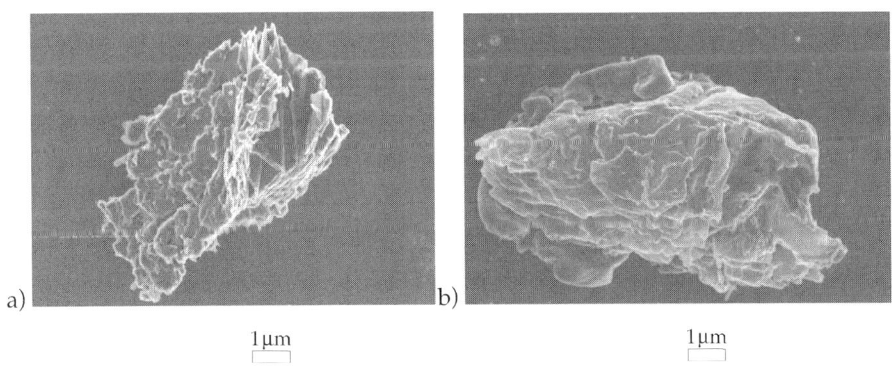

Figure 11.8 Examples of FESEM images of typical synovial fluid wear particles from sheep joints injected with lipid solvent and subjected to; a) 18,000 wear cycles and b) 36,000 wear cycles.

For the collection and wear debris analysis the ferrography technique, described in Chapter 10, is usually used.

Fractal Examination of Similarity Between Worn Surfaces

As discussed already in Chapter 10 the major difficulty associated with the visual comparison between worn surfaces or wear particles lays in the subjective nature of the process. In particular the evaluation of similarity between the worn surfaces depends on expert interpretation which may not always be available.

In characterization of worn surfaces fractal measures were applied to differentiate between surfaces. For example, a surface roughness profile measured by profilometer (e.g. Talysurf) of a worn rubber surface was taken and analyzed to find its fractal dimension [11]. The fractal dimension was determined from the size and number of rectangular outlines required to completely enclose the surface profile, as schematically illustrated in Figure 11.9 [12]. It was found that the method of rectangular outlines is suitable for a surface profile where the distance normal to the centreline average is usually magnified more than the distance along the surface profile.

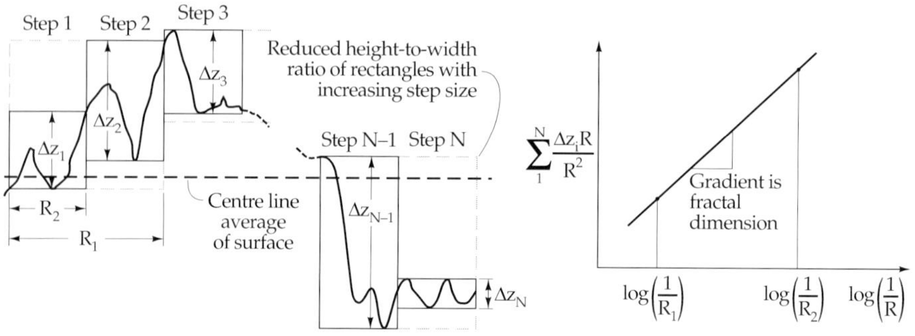

Figure 11.9 Schematic illustration of the method of fractal analysis applied to a surface profile.

The fractal dimension of the worn rubber surface was found to change as the wear mode altered from a wear caused by abrasion combined with chemical degradation of the rubber to mechanical removal of material by abrasive wear when the level of frictional energy dissipation was raised. It has been shown that it is possible to distinguish between different test conditions on the basis of variations in fractal dimensions [11]. In a similar work it was found that fractal dimensions of polished copper surfaces varied as a function of polishing [13]. Since polishing is in effect a mild form of wear process, it may be concluded that there is some relationship between surface fractal dimensions and severity of wear which can then be used to assess the similarity of wear and friction mechanisms operating on two worn surfaces.

Another method, called 'Richardson's technique', used to evaluate the fractal dimension from a surface profile involves the following steps. First, profile data in the vertical direction is normalized by its range, rendering all values between 0 and 1. Next, the absolute vertical difference between every point along the horizontal axis (i.e. step size = 1) is summed giving the maximum vertical variation [14]. The procedure is then repeated for increased step sizes, i.e. every second, third, etc., point, until some limit value is reached as illustrated schematically in Figure 11.10 [14].

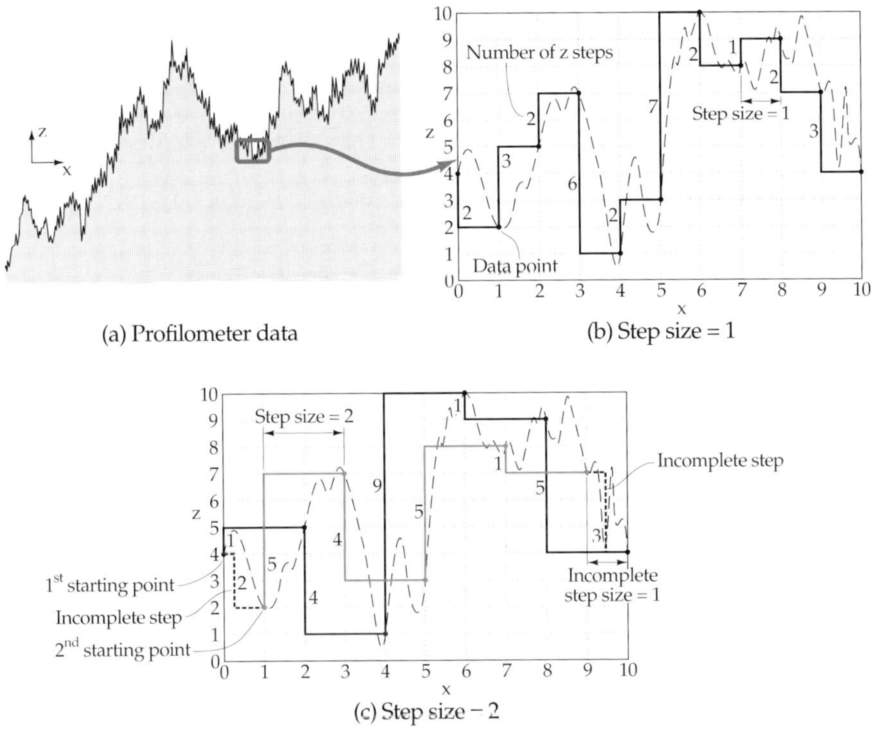

Figure 11.10 Schematic illustration of evaluating the vertical variation of a profile; a) a profile at step size = 1 and vertical variation of $\Sigma z_{ss=1} = 30$, b) a profile at step size = 2, first starting point: $\Sigma z = 20$, second starting point: $\Sigma z = 15 + (2 \times 1 + 3 \times 1)/2 = 17.5$ (where 1 is the incomplete step size). Average vertical variation, $\Sigma z_{ss=2} = (20 + 17.5)/2 = 18.75$ as there are two possible starting points (adapted from [14]).

This technique is similar to that involved in calculating the fractal dimension from a particle boundary, described in Chapter 10. All starting points for a particular step size are used to evaluate the vertical variation of the profile and then an average vertical sum is calculated [14]. The results obtained are then plotted versus step size on a log-log scale and from the slope '**m**' of the curve the fractal dimension '**D**' of a profile is calculated as **D = 1 - m** [14].

The problem is that fractal dimensions calculated from 2-D surface profiles exhibit some fundamental limitations especially when applied in the characterization of worn surfaces [15-17]. As these surfaces exhibit anisotropic characteristics fractal dimensions calculated from surface profiles fail to distinguish between them [18]. The methods described above, since they deal with single surface profiles, may be applicable to isotropic surfaces which exhibit the same surface characteristics in all directions. However, the majority of tribological surfaces are anisotropic exhibiting different surface patterns along different directions [19]. In general, characterization of surface topography based on surface profiles, provides only limited information about the 3-D characteristics of real surfaces no matter how refined the technique is. The basic problem is that a surface is a 3-D object and its characterization, based on 2-D profiles taken from certain regions of the surface, is not adequate.

In attempting to solve this problem special fractal methods allowing the calculation of the surface fractal dimension were developed and applied to characterization of 3-D surface topographies [16,19-21]. Amongst those methods the most well known are: the ε-blanket [22], box-counting [23], two-dimensional Hurst analysis [16], triangular prism surface area [24] and variation method [25], generalized fractal analysis based on the Ganti-Bhushan model [21] and the patchwork method [26]. However, the basic limitation of these methods is that they work well only with isotropic surfaces, i.e. with surfaces, which exhibit the same statistical characteristics in all directions [27]. This limitation was partially overcome with the development of a modified Hurst Orientation Transform (HOT) method. HOT method allows the calculation of the Hurst coefficients (H), which are directly related to surface fractal dimensions, i.e. $D = 3\text{-}H$, along all possible directions [28]. The coefficients calculated are then plotted as a function of orientation, revealing surface anisotropy [27,28] as illustrated in Figure 11.11.

It can be seen from Figure 11.11 that for a surface texture which is approximately isotropic (Figure 11.11a) the corresponding shape of the rose plot is approximately circular (Figure 11.11c). On the other hand, for a surface texture exhibiting anisotropy (Figure 11.11b) the shape of the rose plot can be approximated by an ellipse, (Figure 11.11d) [27]. This means that the approximately isotropic surfaces exhibit similar values of Hurst coefficients in all the directions, while for the anisotropic surfaces Hurst coefficient values vary in all the directions. Based on Hurst coefficient rose plots some comparison of similarity between tribological surfaces can be made.

Since the Hurst coefficients are related to fractal dimension, they were also used as a relative measure of surface roughness, i.e., the rougher the surface the lower the Hurst coefficient values are [27-30].

Classification of Tribological Surfaces

The difference between worn and unworn surfaces is usually established based on a number of parameters obtained from surface roughness measurements or visual inspection. New computer-based image analysis methods, currently being developed, aim to eliminate the need for experts and remove subjectivity from the visual surface inspections. Attempts have been made to classify wear particles and surfaces using neural networks [31], wavelet transforms [32], ring-shape filters [33], Gabor filters [34,35], etc. However, none of the techniques developed, so far, provides the desirable results [36]. The hybrid fractal-wavelet method (HFW), mentioned in Chapter 10, allows an unclassified tribological surface to be assigned into a specific group/class according to a predefined criterion such as, for example, wear mechanism, surface texture, etc.

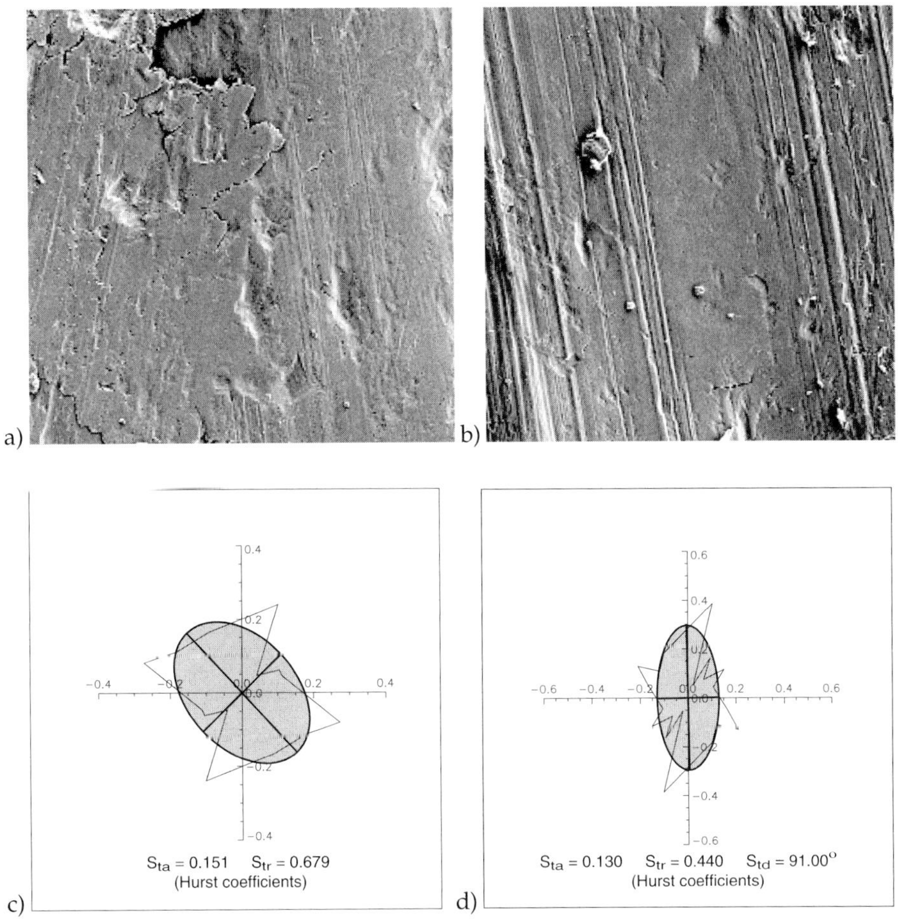

Figure 11.11 Examples of SEM images of tribological surfaces exhibiting different surface topography together with the corresponding Hurst coefficient rose plots.

As already described in Chapter 10, the 3-D surface topography data is first decomposed into ranges of different scales (roughness, waviness, etc.) and then a fractal method (a Partition Iterated Function System (PIFS)) is used to build a mathematical model for each range of scales [37]. Wavelets allow surfaces to be characterized at each individual scale while fractals allow to characterize surfaces in a scale-invariant manner. The combination of these two methods allows the 3-D surface topography to be described with unique accuracy and precision [36].

Based on the HFW method an attempt to build an automated classification system for tribological surfaces has been made [37-39].

A number of tribological surfaces were initially classified, based on visual observation, into different groups using the level of wear damage as a classification criterion [38]. Range images of these surfaces were first decomposed into individual frequency components over ranges of different scales using wavelets and then fractals were applied to construct a mathematical model (PIFS) for each frequency component. The PIFSs obtained from these surfaces formed a data base. The unclassified surface was first decomposed with the wavelets. Next, each PIFS from the data base was decoded using each different frequency component from the unclassified surface as an initial image. Only one iteration was allowed. The resulting decoded images were then compared with the unclassified surface image and a difference measure, i.e. Baddeley distance (BD), was calculated [36]. The lowest value of the difference measure indicates to which group, of already classified surfaces, the unclassified surface belongs. This process is schematically illustrated in Figures 11.12 and 11.13 [36].

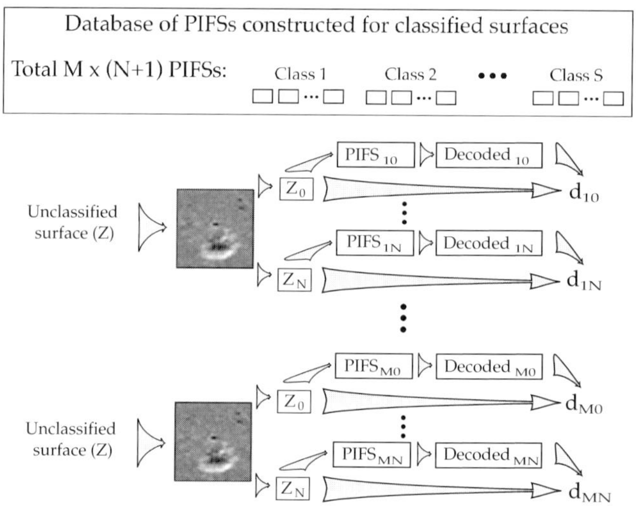

M - number of classified surfaces in the database

N - number of scales over which the surface is decomposed

$\{Z_0,\ldots,Z_N\}$ - different scale components of the unclassified surface (Z) obtained after the application of wavelets

$\{d_{10},\ldots,d_{MN}\}$ - difference measures calculated between the unclassified surface decomposed into different scale components and decoded images

Unclassified surface is assigned to a class of surfaces for which 'd' is minimum, i.e. $\min\{d_{10},\ldots,d_{MN}\}$

Figure 11.12 Schematic illustration of the classification process based on fractal-wavelet method (adapted from [36]).

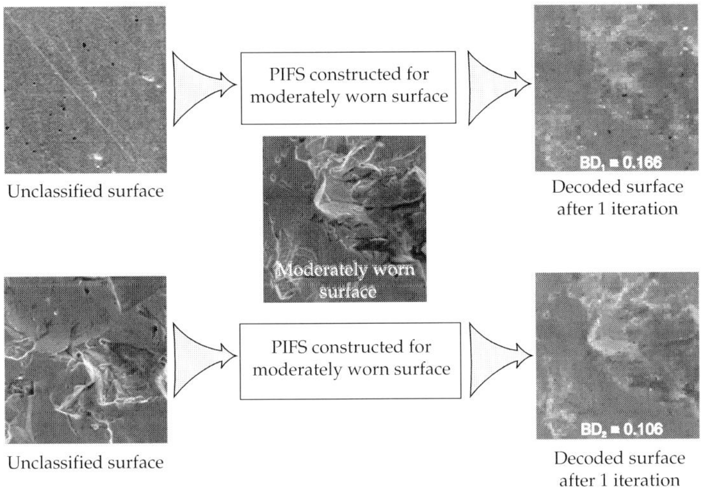

Figure 11.13 Schematic illustration of surface classification procedure. Unclassified surfaces are assigned to the class (group) for which BDs take the smallest value. The second unclassified surface has been assigned to the class of moderately worn surfaces ($BD_2 < BD_1$) (adapted from [36]).

11.6 APPLICATION OF DIMENSIONAL ANALYSIS TO MODELLING OF TRIBOLOGICAL PROCESSES

Tribological problems are often complex and their understanding and solution rely on experimental data obtained from laboratory tests. The purpose of these laboratory experiments is, based on information obtained, to gain the understanding or describe the behaviour of other, 'outside the laboratory', similar systems. Laboratory experiments can be treated as 'models' and are used to study a specific phenomenon under precisely controlled conditions. Based on the data from these experiments empirical equations describing the 'model' behaviour often result. From the information gained it is hoped that predictions can be made about the behaviour or characteristics of a similar 'outside laboratory' (real life) system. However, in order to do so the relationships between the laboratory and real systems need to be found first. The mathematical technique which may help to establish these relationships is dimensional analysis. This technique is particularly useful in situations where modelling and scaling-up problems need to be solved.

As described in numerous textbooks the technique works in the following manner. Any physical process can be described by a set of variables having specific dimensions, e.g. of pressure, force, velocity, etc. These dimensions are represented by a system of fundamental units, for example, a pressure unit [Nm^{-2}] is represented by [m], [kg] and [s].

Let the physical process be expressed as a function 'Φ' of variables '$\mathbf{v}_1, \mathbf{v}_2, \dots, \mathbf{v}_n$', i.e.:

$$\mathbf{Z} = \Phi(\mathbf{v}_1, \mathbf{v}_2, \dots, \mathbf{v}_n) \tag{11.1}$$

where:

$\mathbf{Z} \in \Pi_s$	is the value of function 'Φ';
$\mathbf{v}_1, \mathbf{v}_2, \dots, \mathbf{v}_n \in \Pi_s$	are the variables of function 'Φ';
$\mathbf{n}$	is the number of variables;
Π_s	is the dimensional space with 's' units [40].

The variables '$\mathbf{v}_1, \mathbf{v}_2, \dots, \mathbf{v}_n$' are expressed in terms of specific dimensions, e.g. of pressure, force, velocity, etc. From the variables '$\mathbf{v}_1, \mathbf{v}_2, \dots, \mathbf{v}_n$' the subset of dimensionally independent variables '$\mathbf{A}_1, \mathbf{A}_2, \dots, \mathbf{A}_m$', called 'dimensional base', is initially selected. Assuming that '$\mathbf{E}_1, \mathbf{E}_2, \dots, \mathbf{E}_u$' represent the set of units then the independent variables $\mathbf{A}_1, \mathbf{A}_2, \dots, \mathbf{A}_m$ can be expressed in the following form [41]:

$$\mathbf{A}_i = \alpha_i \prod_{q=1}^{u} \mathbf{E}_q^{e_{iq}} \quad \mathbf{i} = 1, 2, \dots, \mathbf{m} \tag{11.2}$$

where:

$\alpha_i \in \Pi_0$	are the values of the dimensionally independent variables;
$\mathbf{e}_{iq} \in \mathbb{R}$	are the exponents of the units '$\mathbf{E}_q$' for the independent variables '$\mathbf{A}_i$'.

For the set of variables '$\mathbf{A}_i$' to be dimensionally independent, i.e. to form a dimensional base, the condition is that the determinant of matrix made up from exponents '$\mathbf{e}_{iq}$' (11.2) must be different from zero, i.e.:

$$\det \begin{bmatrix} \mathbf{e}_{11} & \mathbf{e}_{12} \dots \mathbf{e}_{1u} \\ \mathbf{e}_{21} & \mathbf{e}_{22} \dots \mathbf{e}_{2u} \\ \vdots & \vdots \\ \mathbf{e}_{m1} & \mathbf{e}_{m2} \dots \mathbf{e}_{mu} \end{bmatrix} \neq 0 \tag{11.3}$$

As can be seen from equation (11.2) the number of dimensionally independent variables is equal to a number of fundamental units '$\mathbf{u}$' used in the process description, e.g. if [kg], [m] and [s] are used then the number is three. The remaining variables '$\mathbf{B}_1, \mathbf{B}_2, \dots, \mathbf{B}_r$' of the function '$\Phi$' are dimensionally dependent and the equation (11.1) can then be re-written in the following form:

$$\mathbf{Z} = \Phi(\mathbf{A}_1, \mathbf{A}_2, \dots, \mathbf{A}_m, \mathbf{B}_1, \mathbf{B}_2, \dots, \mathbf{B}_r) \tag{11.4}$$

where:

$\mathbf{A_i} \in \Pi_s$	are the dimensionally independent variables, $\mathbf{i = 1, 2, \dots, m}$;
$\mathbf{B_j} \in \Pi_s$	are the dimensionally dependent variables, $\mathbf{j = 1, 2, \dots, r}$;
$\mathbf{m}$	is the number of dimensionally independent variables;
$\mathbf{r}$	is the number of dimensionally dependent variables, i.e. $\mathbf{n - m = r}$.

A theorem, called 'π' theorem, developed by Buckingham [42] and proved in [40] allows the variables $\mathbf{B_1, B_2, \dots, B_r}$ and the value of the function 'Φ' to be expressed in terms of dimensionally independent variables $\mathbf{A_1, A_2, \dots, A_m}$ in the following manner [42]:

$$\mathbf{B_j = \phi_j \prod_{i=1}^{m} A_i^{a_{ji}} \quad j = 1, 2, \dots, r} \tag{11.5}$$

$$\mathbf{Z = F(\phi_1, \phi_2, \dots, \phi_r) \prod_{i=1}^{m} A_i^{a_i}} \tag{11.6}$$

where:

$\phi_j \in \Pi_0$	are the non-dimensional parameters (also called invariances of similarity);
Π_0	is a dimensional space with zero units;
$\mathbf{a_{ji}} \in \mathbb{R}$	are the dimensional base exponents of the dependent variables '$\mathbf{B_j}$';
$\mathbf{F(\phi_1, \phi_2, \dots, \phi_r)}$	is a numerical function of the non-dimensional parameters;
$\mathbf{a_i} \in \mathbb{R}$	are the dimensional base exponents of the independent variables '$\mathbf{A_i}$'.

Rearranging (11.5) yields:

$$\phi_j = \frac{\mathbf{B_j}}{\prod_{i=1}^{m} \mathbf{A_i^{a_{ji}}}} \quad \mathbf{j = 1, 2, \dots, r} \tag{11.7}$$

Substituting for 'ϕ_j' (11.7) into (11.6) yields a general expression for a process function expressed in terms of non-dimensional parameters:

$$\mathbf{Z = F\left(\frac{B_1}{\prod_{i=1}^{m} A_i^{a_{1i}}}, \frac{B_2}{\prod_{i=1}^{m} A_i^{a_{2i}}}, \dots, \frac{B_r}{\prod_{i=1}^{m} A_i^{a_{ri}}}\right) \prod_{i=1}^{m} A_i^{a_i}} \tag{11.8}$$

It can be seen that the application of the 'π' theorem resulted in the reduction in the number of '**v**' variables of the function 'Φ' (eq. 11.1) from '**n**' to '**r = n - m**'. This form of the function 'Φ' with '**r**' arguments is very useful as it can be determined experimentally by treating non-dimensional parameters 'ϕ_j', j = 1, 2, ... , r as inputs and

$$\frac{Z}{\prod_{i=1}^{m} A_i^{a_i}} \tag{11.9}$$

as an output. Rearranging equation (11.8) yields:

$$y = \frac{Z}{\prod_{i=1}^{m} A_i^{a_i}} = F\left(\frac{B_1}{\prod_{i=1}^{m} A_i^{a_{1i}}}, \frac{B_2}{\prod_{i=1}^{m} A_i^{a_{2i}}}, \ldots, \frac{B_r}{\prod_{i=1}^{m} A_i^{a_{ri}}}\right) \tag{11.10}$$

where:

y is the output value of the function '**F**'.

Summarizing, the process under consideration is now expressed in terms of non-dimensional parameters, being combinations of dimensionally dependent and dimensionally independent variables, rather than individual variables. If an arbitrary physical process is expressed by three variables and three fundamental units then only one non-dimensional group results, giving an accurate description of this process. However, if the number of variables is greater than the number of fundamental units then the number of possible dimensional base choices dramatically increases. The number of these choices is equal to the number of the '**m**' element combinations in an '**n**' element set, i.e.:

$$T = \binom{n}{m} \tag{11.11}$$

One can intuitively feel that only one of these dimensionally independent variables combinations, i.e. dimensional base, is optimal for a given process and would give its most accurate description. The question is: which dimensional base, of these many possible base choices, is an optimal one? How can we determine this base?

The problem of the optimal dimensional base selection has been addressed in [43-46]. The procedure boils down to finding the function '**F**' (eq. 11.6) in the form of polynomial (if linear approximation is assumed) with respect to the non-dimensional parameters '$\phi_1, \phi_2, \ldots, \phi_r$'. Initially, from the set of process variables '**v**' dimensionally independent parameters $A_1, A_2, \ldots, A_m$ are chosen and checked whether they form a dimensional base, i.e. the determinant of their dimension exponents is different from zero, $\mathbf{det\{e_{iq}\} \neq 0}$ (11.3).

Next a family of approximating functions is assumed, e.g. linearly parametrized functions. Based on the experimental data it is possible to define a function '$\mathbf{F^*}$' which would describe the process under consideration, i.e. it would approximate the function 'F' (11.6). In other words, for the arbitrarily chosen class of functions '$\mathbf{F^*}$', the numerical coefficients can be determined. The accuracy of a mathematical model of the process under consideration would depend on the selection of the appropriate approximating function. In practice there is a large number of function forms which can be used to approximate the experimental results. For example, linearly parametrized functions can be used as a class of approximating functions. However, it is best to use the class of functions which would provide close convergence with the experimental results [47].

Once the type of approximating function has been decided, e.g. polynomial, based on the experimental data, for each possible choice of dimensional base a numerical model together with the relative error of approximation, i.e. minimised criterion 'ε_p', is computed. The dimensional base for which the value of the minimised criterion 'ε_p' is the lowest is considered as the optimal for the assumed form of the approximating function '$\mathbf{F^*}$'. If the value of minimised criterion is not satisfactory the process is repeated for other forms of approximating function '$\mathbf{F^*}$'. The flow chart outlining this process is shown in Figure 11.14.

Example of the Application of Dimensional Analysis to Modelling of Erosive Wear

There is a number of tribological problems to which dimensional analysis has been applied [e.g. 48-53]. Erosion by solid abrasive particles is a good example since most of the controlling variables can be identified from current understanding of the physics of the problem, provided that the erosion occurs in the absence of corrosive media. If corrosive agents are present, often more rapid corrosive-erosion wear occurs [54] and, in a manner similar to boundary lubrication and scuffing, poorly understood mechano-chemical effects become significant. In one work, aiming to develop a model to predict the erosion of materials, erosive wear rate by solid particles has been expressed as a function of seven independent variables [55], i.e.:

$$\mathbf{w} = \Phi\ (\mathbf{H_t}, \gamma, \mathbf{v}, \mathbf{R}, \mathbf{H_p}, \mathbf{D}, \mathbf{Q}) \qquad (11.12)$$

where:

- $\mathbf{w}$ is the erosive wear rate [m^3kg^{-1}];
- $\mathbf{H_t}$ is the hardness of the target [$kgm^{-1}s^{-2}$];
- γ is the apparent particle impingement angle [dimensionless];
- $\mathbf{v}$ is the effective impact velocity of the particles [ms^{-1}];

R is the roundness (particle shape) [dimensionless];

$\mathbf{H_p}$ is the effective hardness of the particle [$kgm^{-1}s^{-2}$];

D is the erosion resistance per unit volume of the target [$kgm^{-1}s^{-2}$];

Q is the volume of the target affected [m^3].

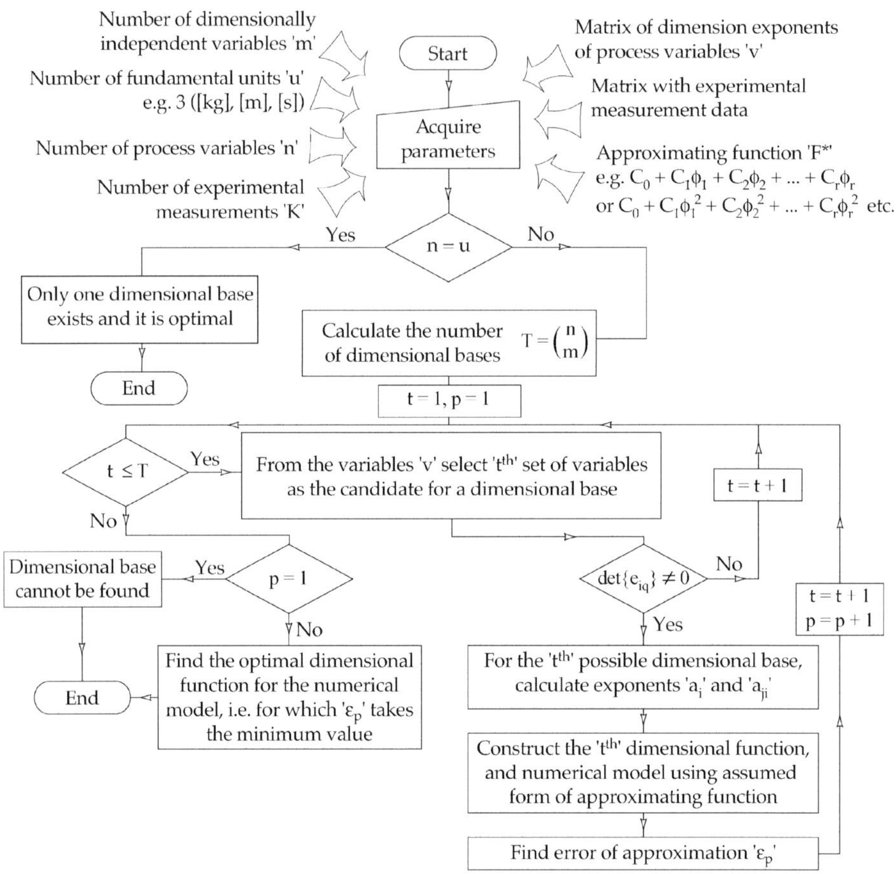

Figure 11.14 Flow chart of determining a dimensional base and a numerical model of a tribological process, where '**t**' is the number of possible dimensional base choices, i.e. **t = 1, 2, ... ,** T, '**T**' is the total number of possible dimensional base choices, '**p**' is the number of actual dimensional bases, i.e. **p = 1, 2, ... , P** and '**P**' is the total number of actual dimensional base choices, i.e. variable combinations which form the dimensional base, where **P ≤ T**.

In equation (11.12) most of the controlling variables involved in this mechanism of wear have been identified. Amongst the physical variables parameters such as roundness of the particles and the modulus of toughness of the target material have also been included [55]. Particle roundness is an early version of particle shape descriptor defined by the following formula [55,56]:

$$\mathbf{R} = \frac{\Sigma \mathbf{r_i}}{\mathbf{I_r N}} \tag{11.13}$$

where:

$\mathbf{r_i}$	is the radius of curvature of the individual corners [m];
$\mathbf{I_r}$	is the radius of the maximum inscribed circle [m];
N	is the number of corners measured [dimensionless].

The erosion resistance '**D**' per unit volume of the target is defined arbitrarily as a relationship between hardness '$\mathbf{H_t}$' of the target, the apparent impingement angle 'γ' (in radians) and the target modulus of toughness '$\mathbf{M_t}$' [55], i.e.:

$$\mathbf{D} = (\mathbf{H_t} - \mathbf{M_t}) \cos\gamma + \mathbf{M_t} \tag{11.14}$$

or as:

$$\mathbf{D} = \mathbf{H_t} - \gamma\,(\mathbf{H_t} - \mathbf{M_t})/1.5708 \tag{11.15}$$

where:

$\mathbf{H_t}$	is the hardness of the target [Pa];
γ	is the apparent impingement angle [radians];
$\mathbf{M_t}$	is the target modulus of toughness [Pa].

The modulus of toughness is defined by Murphy's criterion [55], i.e.:

$$\mathbf{M_t} = \varepsilon_u(\sigma_y + \sigma_u)/2 \tag{11.16}$$

where:

ε_u	is the strain at failure [dimensionless];
σ_y	is the yield strength at 2% permanent strain [Pa];
σ_u	is the ultimate tensile strength [Pa].

The above formulae were developed on an intuitive basis and served equally well in describing the target resistance to erosion. It can be seen from equations (11.14) and (11.15) that at zero impingement angle all the erosion resistance is contributed to by the target hardness while at normal impingement the toughness of the target material alone is responsible for resisting erosion [55].

It can be seen from equation (11.12) that there are seven variables and three dimensions involved in defining the process of erosive wear, i.e. mass [kg], length [m] and time [s]. According to equation (11.11) the number of possible dimensional base choices, i.e. dimensionally independent sets of variable combinations, in this case is not higher than:

$$t \leq \binom{7}{3} = 35 \qquad (11.17)$$

Out of these 35 possible dimensional base choices the condition $\mathbf{det\{e_{iq}\} \neq 0}$ is satisfied only in three cases, i.e. only three sets of dimensionally independent variables exist for this particular example described by function (11.12). Possible dimensional bases and dimensional functions are listed in Table 11.1 [57].

Table 11.1 Possible dimensional functions and dimensional bases for erosive wear models.

Dimensional base	Dimensional function
H_t, v, Q	$w = F\left(\gamma, R, \frac{H_p}{H_t}, \frac{D}{H_t}\right)\frac{v^2}{H_t}$
v, H_p, Q	$w = F\left(\gamma, R, \frac{H_t}{H_p}, \frac{D}{H_p}\right)\frac{v^2}{H_p}$
v, D, Q	$w = F\left(\gamma, R, \frac{H_t}{D}, \frac{H_p}{D}\right)\frac{v^2}{D}$

Out of the functions shown in Table 11.1 only one is optimal for either brittle or ductile erosive wear model. To determine which one, for each of these functions and assumed form of approximating polynomial, using experimental data from [55], a numerical model together with relative error of approximation have been computed. Two sets of data were used in calculations, i.e. ductile and brittle target erosion data. The first set of data consisted of erosion tests results involving metallic targets such as aluminium alloy, stainless steel and beryllium copper alloy while the second group consisted of erosion tests results involving plate glass targets [55]. Examples of approximating polynomials, used in calculations, together with relative errors of approximation are shown in Table 11.2.

It can be seen from Table 11.2 that the relative errors of approximation can vary significantly depending on dimensional base choice and the form of approximating polynomial. The minimum values of relative errors of approximation found, based on the experimental data from [55], are: for ductile targets 16.83% and for brittle targets 0.06%.

Table 11.2 The relative errors of approximation for possible dimensional bases and different forms of approximating polynomials for ductile and brittle targets (adapted from [57]).

Approximating polynomial	**Relative errors of approximation 'ε_p'**					
	Ductile targets			**Brittle targets**		
	Dimensional base			**Dimensional base**		
	H_t,v,Q	**v,H_p,Q**	**v,D,Q**	**H_t,v,Q**	**v,H_p,Q**	**v,D,Q**
$C_0 + C_1\phi_1 + C_2\phi_2 + C_3\phi_3 + C_4\phi_4$	22.27	43.55	276.12	17.02	20.12	28.17
$C_0 + C_1\phi_1\phi_2 + C_2\phi_1\phi_3 + C_3\phi_1\phi_4 + C_4\phi_2\phi_3 + C_5\phi_2\phi_4 + C_6\phi_3\phi_4$	19.70	38.44	163.55	2.09	11.70	74.32
$C_0 + C_1\phi_1 + C_2\phi_2 + C_3\phi_3 + C_4\phi_4 + C_5\phi_1\phi_2 + C_6\phi_1\phi_3 + C_7\phi_1\phi_4 + C_8\phi_2\phi_3 + C_9\phi_2\phi_4 + C_{10}\phi_3\phi_4$	18.54	25.22	88.85	0.30	0.06	3142.39
$C_0 + C_1\phi_1 + C_2\phi_2 + C_3\phi_3 + C_4\phi_4 + C_5\phi_1^2 + C_6\phi_2^2 + C_7\phi_3^2 + C_8\phi_4^2$	22.12	31.69	204.85	5.92	15.93	129.86
$C_0 + C_1\phi_1 + C_2\phi_2 + C_3\phi_3 + C_4\phi_4 + C_5\phi_1^2 + C_6\phi_2^2 + C_7\phi_3^2 + C_8\phi_4^2 + C_9\phi_1\phi_2 + C_{10}\phi_1\phi_3 + C_{11}\phi_1\phi_4 + C_{12}\phi_2\phi_3 + C_{13}\phi_2\phi_4 + C_{14}\phi_3\phi_4$	16.83	20.83	30.61	51.81	33.60	733.99

where:

C are the polynomial coefficients;

ϕ are the non-dimensional parameters for appropriate dimensional base choice (Table 11.1).

The optimal dimensional functions for the models of erosive wear by solid particles are in the following form:

$$w = F\left(\gamma, R, \frac{H_p}{H_t}, \frac{D}{H_t}\right)\frac{v^2}{H_t} \quad \text{ductile metallic targets} \quad (11.18)$$

$$w = F\left(\gamma, R, \frac{H_t}{H_p}, \frac{D}{H_p}\right)\frac{v^2}{H_p} \quad \text{brittle targets (plate glass)} \quad (11.19)$$

The resulting empirical descriptions of erosive wear models for ductile and brittle targets which gave the minimal values of relative errors of approximation are in the following forms [57]:

$$w = \left[-0.028 - 0.0012\gamma + 2.35R - 0.0047\frac{H_p}{H_t} - 0.97\frac{D}{H_t} - 0.66\times10^{-4}\gamma^2 - 1.56R^2 + 0.39\times10^{-3}\left(\frac{H_p}{H_t}\right)^2 + 0.34\left(\frac{D}{H_t}\right)^2 - 0.0012\gamma R - 0.1E\text{-}3\gamma\frac{H_p}{H_t} + 0.0032\gamma\frac{D}{H_t} + 0.018R\frac{H_p}{H_t} - 0.15R\frac{D}{H_t} - 0.0095\frac{H_p}{H_t}\frac{D}{H_t}\right]\times\frac{v^2}{H_t} \quad (11.20)$$

$$w = \left[-434.49 - 352.75\frac{H_t}{H_p} + 3.72\gamma + 1161.68R + 242.42\frac{D}{H_p} + 6.89\frac{H_t}{H_p}\gamma - 322.21\frac{H_t}{H_p}R + 510.06\frac{H_t}{H_p}\frac{D}{H_p} - 11.16\gamma R + 0.06\gamma\frac{D}{H_p} + 793.17R\frac{D}{H_p}\right]\times\frac{v^2}{H_p} \quad (11.21)$$

Plots of experimental and predicted (from equations (11.20) and (11.21)) values of erosive wear rates for ductile and brittle metallic targets are shown in Figure 11.15.

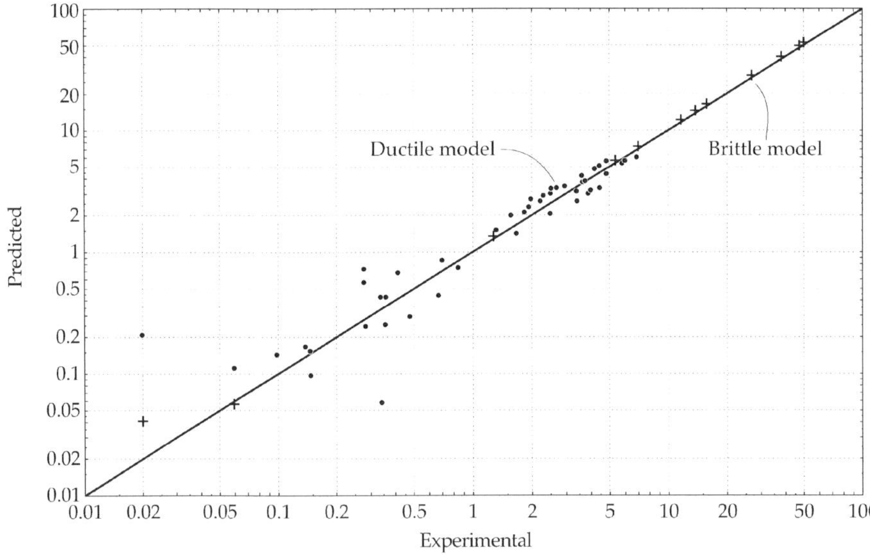

Figure 11.15 Comparison between experimental and predicted erosive wear rates for ductile and brittle metallic targets (adapted from [57]).

As it can be seen from Figure 11.15 good agreement has been achieved between the model and experimental data. Finding an optimal dimensional function yields another benefit as these functions in their parametric form [41] can be quite helpful in tribological experiment design. They provide conditions which need to be satisfied to obtain the similarity between the processes taking place 'outside the laboratory' on a real object and during the laboratory simulation on a test rig [57].

11.7 NON-LINEAR MODELLING OF FRICTION AND WEAR

The limitation with the regressional and dimensional analysis models is that they often provide only linear approximations of the processes under consideration, i.e. the models developed are only accurate for a local neighbourhood of data points. In order to develop more accurate models allowing more precise predictions in wider range of data a different approach, i.e. non-linear modelling, is needed.

An important example of the validity of non-linear modelling to wear is frictional instability at high levels of frictional energy dissipation. For any lubricated metallic sliding system (but not necessarily for other systems, e.g. dry polymers), the coefficient of friction rises precipitately with the temperature of the sliding system. Simultaneously, the temperature of the sliding system rises with the level of friction. This can be expressed in mathematical terms in the following manner:

$$\delta\mu/\delta T > 0 \; ; \; \delta^2\mu/\delta T^2 > 0 \qquad (11.22)$$

(the same applies for some or all of the higher derivatives)

$$\delta T / \delta\mu = \mathbf{f}\,(\mathbf{U}, \mathbf{W}, \mathbf{I}) > 0 \qquad (11.23)$$

where:

- μ is the coefficient of friction [dimensionless];
- T is the temperature [K];
- U is the sliding speed [m/s];
- W is the load [N];
- I is the thermal impedance [K/W], i.e.: 'I' is a function of convection, conduction, radiation and possibly evaporation if there is a volatilizing lubricant.

Without attempting to analyze these equations in further detail, it is apparent that most lubricated metallic sliding systems have an inherent tendency towards instability as the level of frictional energy dissipation rises. Detailed modelling of these phenomena would involve the non-linear analysis of systems with positive feed-back.

The distribution of wear on a surface is also found to be controlled by non-linearity. A common assumption is of a uniform distribution of wear, which allows a plane surface to remain plane after wear. A model of corrugation in rolling contact between a metal wheel and rail has been developed based on non-linear contact mechanics [58]. It was found that corrugations occur when an initial disturbance in rolling motion leads to a self-propagating oscillation in wear. A non-linear model of corrosion deposit formation inside boiler tubes has also been developed [59]. Transfer functions, which are an application of Control Theory, were used to develop a model which predicted the remaining life of the boiler tube before it was clogged by corrosion deposits. Other non-linear analyses of wear and friction have also been proposed [60,61].

Attempts were also made to predict wear [e.g. 59-61]. In non-linear approximations, numerical methods, e.g. radial basis function (RBF) networks [62], multilayer feed-forward neural networks (MFNN) [63,64], wavelets [65] and fractals [66], can be used. These methods can approximate any arbitrary non-linear function to any desired accuracy. The RBF networks are used to approximate the process data as a linear combination of one variable function, called a radial basis function. Some typical choices of this function are the thin-plate spline function and the Gaussian function. The performance of the RBF network is very slow if the input signal is of dimension more than four. In the MFNN method, the process data is approximated by a multilayer network. Each layer of this network contains computing nodes and the strength between nodes of adjacent layers is defined by weighted parameters. These parameters are updated to minimize the difference between the process data and the output of the network using a backpropagation algorithm. The updating of parameters can take a long time before the difference is less than a given number [64]. When the wavelet based approximation is used, an inner product of the process data and a wavelet mother function is first calculated and then it is multiplied by the wavelet function. This process is repeated for a number of shifted and scaled versions of the mother wavelet function (called the wavelet basis) yielding an approximation of the process data. The accuracy of this approximation depends on the type of mother wavelet function used and the grid on which the wavelet basis is defined [65]. In fractal-based approximation, the process data is expressed as a collection of affine transformations. Each transformation converts a smaller part of the data into a larger part of the same data. This conversion is based on the assumption that the process data contains transformed copies of parts of themselves, i.e. they are 'piecewise self-transformable'.

Aside from the detailed mathematics of non-linear analysis it is important to note that any numerical model is only as accurate as is allowed by its assumptions and initial postulates. A wearing contact is not only a mechanical system but can also be modelled as a miniature chemical reactor where reactions are accelerated by heat, forced mixing and mechanical activation. A comprehensive non-linear model should include chemical parameters as well as purely physical parameters. This is still in the area of intense research and hopefully models allowing an accurate prediction of wear and friction behaviour will become available in the future.

11.8 PRESENTATION OF WEAR AND FRICTION DATA

The method of presenting wear and friction data has a strong effect on how we perceive friction and wear phenomena. A common example is the listing of friction and wear coefficients in scientific and engineering handbooks. This gives the impression that friction and wear coefficients are constants in the same manner as the Faraday constant or the permittivity of a vacuum. However, the fact, known by almost any investigator of wear and friction, is that friction and wear coefficients are based on an idealized linear model of friction and wear. In reality friction and wear vary in a complex manner, displaying a dependence on time and a host of other parameters, some known and some unknown. The critical skill in effective presentation of friction and wear data is to clearly show the most important controlling factors while minimizing the complexity entailed.

Statistical Representation of Wear Data

In practical applications a statistical analysis is used for the presentation of the experimental data obtained. This method allows the investigator to discriminate between random data variation and systematic influences. The mode and scale of the random variation can also be quantified. Great deal of information about the tribological process occurring can be learned from the careful statistical analysis of tribological data.

For example, data from the dry and wet sliding tests between polyethylene pins and smooth metallic or ceramic counterfaces revealed that wear rates exhibit a log-normal distribution [67,68]. A log-normal distribution is a distribution where the logarithm of the variable displays a Gaussian distribution. A log-normal distribution indicates that wear is controlled by several factors acting by multiplication rather than addition. A log-normal distribution also causes the arithmetic average wear rate to be greater than the statistically most frequent wear rate (the median).

When presenting a wear rate data it would be useful not only to give the values of the average logarithm of wear rate but also the standard deviation of this logarithm. Kurtosis is a useful statistical parameter. Although in some cases the log-normal distribution may show no significant kurtosis it still should be considered for any new series of tests. It is also useful to compare the nature of the wear rate variation data with random data simulations as wear rate variation is not always perfectly random [67,69]. Defining the controlling statistical distribution of wear data is particularly useful in the development of a numerical model of wear, which can be used by engineers.

- *Presentation of Data for the Demonstration of True Wear and Friction Coefficients*

Before calculating any values of wear rates, it has to be confirmed experimentally that the wear progresses linearly with sliding distance. In most cases a simple plot of wear depth or wear volume versus sliding distance should reveal the progress of the wear. Wear never progresses in a perfectly linear manner, instead random variation, cyclic variation and step-like progression may occur. As indicated in the early chapters, for almost all wear and friction tests, the initial values of wear (and friction) will differ significantly from their long-term values. Plots of wear rates and coefficient of friction versus sliding distance should therefore be used to obtain long-term values of wear and friction coefficients [70]. Visual inspection should be used to confirm that the wear or friction has indeed converged to a long-term steady value before a data value is recorded [70].

The gradient of the graph of wear depth, which is directly proportional to the wear coefficient, will approximate to the line of statistical regression. For friction data, the direct value of the friction force is proportional to the friction coefficient. The commonly used parameter sliding distance may not always be appropriate in representation of wear data. For example, in 3-body abrasion the volume of abrasive used in the sliding contact may be more appropriate, while in impact wear the number of impact cycles should be used. Where no convenient geometrical parameter, e.g. sliding distance, volume of abrasive, etc., is available, wear rate can be graphed as a function of time [70]. Although most emphasis is given to wear, the same statistical and graphical analysis can be applied to friction coefficients [70].

· *Presentation of Multiple Influences on Friction and Wear*

A convenient means of presenting the effect of two parameters on friction and wear are graphical maps, where the wear or friction coefficient is assigned to the z-axis while the x and y axes are used for e.g. load and sliding speed. Another possible combination of variables used for this purpose could be load and temperature or load and pH (acidity) for corrosive wear. 3-dimensional (3-D) presentation of wear and friction data has been popularized by Czichos [71] and co-workers.

In one work the 3-D graphing of wear and friction data has been combined with micrographic studies of wear scars to show how variation in friction and wear correlates with change in friction and wear mechanism [1,72]. In this style of presentation the map of wear (or friction) data is overlaid with a map of wear scar classifications. Both the wear coefficients and wear scar classifications are then smoothed and matched to produce a diagram, which explicitly relates change in wear rate (or friction coefficient) to change in wear mechanism. To facilitate graphical presentation, the z-axis is often substituted by contours of constant wear rate and boundary lines are added to demarcate the different wear mechanisms in action. This type of diagram, known as a wear map, is very useful for summarising otherwise scattered data for a potential use by the engineers. These graphs allow the easily evaluation of the materials wear and friction behaviour under certain operating conditions. The wear maps have been found to be very useful for evaluation of, for example, wear characteristics of cutting tools, engineering ceramics, surface coatings, and others [73].

While the wear maps are very useful as a general guide, it is easy to overlook that each map only refers to one combination of sliding materials under a narrow range of conditions. Thus the use of the maps outside their intended scope can be misleading.

· *Comparative Presentation of Data With a Model of Friction and Wear*

Due to the empirical nature of most of the wear studies, it is not a common practice to directly compare theoretical values of wear or friction with the measurements. In some cases a statistical model, based on, for example, the Weibull distribution is used and a set of upper and lower limits for wear are compiled [74]. The use of maximum and minimum limits is very useful for identifying whether there is some agreement between data and theory. In most cases, however, there is so much variability in experimental wear rates that any agreement with a single theoretical value would be difficult to recognize.

· *Impartiality in Presentation of Tribological Data*

Presentation of tribological data can be used to not only enhance friction and wear data but also to bias the perception of the data. A simple example is the use of logarithmic graphing scales to obscure the variability of the data. Friction and wear coefficients can be reconfigured in parameters such as 'relative performance index' to create a false impression of the data. Also logarithmic scales are not the linear scales and even small variations in data presented can represent order of magnitude changes.

11.9 INDUSTRIAL TESTING IN TRIBOLOGY

Testing in an industrial context is usually performed for different purposes than scientific research where the objective is to confirm or develop a model of friction, wear or related phenomena (e.g. acoustic emission). Industrial testing has the purpose of confirming the performance of a component or system. Typical example might be to confirm whether a lubricant can withstand 5,000 hours (approximately 6 months) of service in one specific item of equipment. By contrast a scientific investigation might be directed to mechanisms of oil degradation as a function of mechanical stress and ambient conditions.

Industrial testing methodology is primarily empirical, especially in tribology where reliable calculation methods are lacking of many important issues. For example, when a literal confirmation of performance is generally required, a fleet of cars might be organized to test the durability of a crankcase lubricant or a new wear-resistant brake lining. Where the costs of full-scale testing become prohibitive or testing is too slow, such as determining the lifetime of a component meant to last for 5 years, a model test is accepted. A standardized apparatus such as the Falex tribometer might be selected for the tests or else an in-house device would be constructed. The advantages of a standardized test are the ease of preparation (it is merely necessary to buy the device and follow the manufacturer's instructions) and ability to compare results with other data from the same test procedure. An in-house device and test procedure is usually only arranged when the standardized tests are perceived to be inadequate. Perceptions of inadequacy may arise when either the standardized tests address only different issues from that of immediate concern or when the consequences of failure to accurately predict machine performance are extremely severe. An example of the former case might be a problem where sliding is combined with impulsive loads from a hammer. An example of the latter may be the wear induced failure of an electrical relay contact for a safety control system. In the latter case, an elaborate full-scale test would be the only reasonable option. Another important difference between industrial testing and scientific investigation is that industrial testing generally relates to commercial products, which may be incompletely specified or else the specification is confidential. A common example are oil additives, which are generally sold as a 'package' without a publicly available specification. The uncertainty about test materials further heightens the need for literal testing as opposed to the application of scientific models since the necessary parameter values are unavailable.

Industrial Testing of Erosion in Pipes

An important characteristic of erosion is the variation in erosive wear rate with position of eroded material. This variation is largely controlled by erosive particle trajectories, which can be numerically predicted in some instances but, in general, they are usually hard to predict. Industrial testing is usually directed at finding the critical locations for erosive wear for a specified erosive particle speed, concentration of erosive particles, type of erosive particle and type of gaseous or liquid media. A test section of pipe, either full size or scale model is subjected to pumped flow of a standardized erosive medium (e.g. an acidic slurry). The thickness of the pipe is then monitored by e.g., eddy current or ultrasound detectors. The performance of protective coatings inside the pipe can also be evaluated by this test.

Industrial Engine and Gear Testing

Engines and gearboxes contain many parts with complex phenomena occurring that cannot be adequately investigated using a simple tribometer. For example, an interaction between the lubricating oil and the combustion cycles of an internal combustion engine is difficult to replicate with a tribometer. An engine or gearbox test, although expensive, will reveal much more data than can be obtained from a tribometer. Direct testing of gearboxes and engines is the only means of finding important practical information such as the relationship between fuel or power consumption, the level of oil in the engine sump or gearbox and the overall performance of the whole assembly.

For most test-work, the engine or gearbox is mounted in a purpose built test-bed with instrumentation to record temperatures, forces, flow rates of lubricants and fuels, etc. A standardized loading cycle is applied to determine critical parameters such as the maximum power output and maximum torque. In many cases, the test procedures do not replicate actual conditions but instead are designed to quickly reveal salient characteristics of the engine or gear. The operating conditions of the engine or gearbox will also differ from actual usage, e.g. a specialized cooling system may be fitted to ensure constant operating temperature where the actual machine displays a cyclic temperature variation. However, because of the cost involved, these tests are used only when sufficient information cannot be obtained from simplified bench tests. In many cases, the decisive motivation to perform full-scale gear tests is a need to demonstrate that the company's gearboxes are superior to those of its competitors.

Tribological Testing of Seals Performance

Seals are widely considered to be the most direct application of data from model friction and wear experiments since the sliding interface of a seal is comparable in size to the seal itself. However, seals exhibit some unique phenomena, which cannot be easily replicated on a standard tribometer. The most critical problem is blistering where 'blisters' form on the surface of a seal after prolonged use. Blisters are commonly found on carbon seals, the brittleness of the carbon leads to crack formation from the blisters [75,76]. Tests involving a model seal are used to investigate blister formation, where variables such as pressure differential, sliding speed, seal geometry, seal material and lubricant can be investigated under conditions closely approximating to actual service conditions.

Tribological Testing in Metalworking

Metalworking involves many frictional conditions, which cannot be accurately replicated by a tribometer. Special tests have therefore been developed to simulate metalworking conditions. For example, an important test is the tapping test, where screw-thread cutting taps are systematically driven into drilled holes while a test cutting-fluid is applied to the tapped workpiece. Empirical measures of lubricant performance, such as tapping torque and number of tapped holes till a maximum torque is exceeded, are collected. This data is then used to evaluate the utility of extreme pressure additives and other lubricant additives to ensure long working life of taps and accurate screw-thread cutting.

Other tests may involve measurement of power consumption, surface finish and accuracy of a machined work-piece as a function of cutting fluid composition or wear-resistant coating on the cutting tool. Similarly, the surface finish of pressed sheet steel is used to evaluate the suitability of lubricants for press-work. More information on tribological testing in metalworking can be found in [e.g. 77-79].

To some extent, tribological testing in a metalworking context differs from pure tribology. Tribology is a study of the factors controlling wear and friction, whereas in metalworking the parameters of interest are surface finish, precision, tool-life and power-consumption. Questions of primary importance in metalworking are e.g.: the number of times the same die can be used to form headlamp lenses of sufficient optical quality, the duration that a cutting tool can be used until a built-up edge causes unacceptable roughness of manufactured components, the maximum cut depth and cutting speed that can be allowed without exceeding dimensional tolerances on a component, etc. However, these issues are largely outside the scope of this book.

Testing of Wear in Orthopaedic Implants

After Charnley's unfortunate experience with polytetrafluroethylene (PTFE) implants where rapid wear necessitated surgical revision for many patients, manufacturers are well aware of the need to predict wear in orthopaedic implants [80,81]. One method of measuring this wear is to examine retrieved implants from patients and this procedure has demonstrated the durability of UHMWPE as a bearing material inside the human synovial joint. The problem with joint retrieval data is that it only refers to designs of implants that are perhaps 10 to 20 years old. A manufacturer needs to know much earlier the performance characteristics of new designs of implant. It is also unethical to expect patients to tolerate an unsatisfactory design of an implant for the sake of gathering some experimental data. Animal models, e.g. a cow or a dog, are often used for initial testing but there is a significant variation in loads and joint flexure speeds between humans and other animals.

Slow data collection from implants that are designed to be very wear-resistant is compensated by setting up a multi-stage simulator. In this apparatus, 10 or more orthopaedic implants (artificial joints) are continuously flexed at loads and speeds closely resembling the human gait. A hydraulic servo-controlled loading system is often fitted to enable the complex loading cycle of the human gait to be accurately simulated. There are several designs of joints simulator [e.g. 82-85]. Some degree of standardization in apparatus is provided by the International Standards Organization [86]. Most implant simulators can test a variety of different implant designs and may incorporate features such as misaligned articulation of the implant to measure the effect of inaccurate placement of the implant by a surgeon [83]. A model lubricant, e.g. saline solution or bovine serum, is also reticulated to each flexing implant in order to model synovial fluid. The choice of model lubricant and the test atmosphere is significant since they control the long-term wear characteristics of implant materials. The model of synovial fluid, e.g. saline solution or bovine serum, may be sufficiently remote from the real synovial fluid to cause significant experimental artefact. For instance, it was recently observed that friction levels in a Durham hip simulator were significantly increased by the addition of proteins to the test lubricant [84,87]. Elevated friction levels would cause over-heating in the implant simulator unless extra forced cooling is applied. Biomedical wear phenomena are very sensitive to temperature, so any uncontrolled temperature variation would spoil the experimental data.

The complexity of service conditions inside a synovial joint, whether in the natural state or fitted with an implant, remains a challenge for designers of tribometers. Most implants are designed and optimized using data from tribometers that can only offer a loose approximation to synovial joint conditions. There remains the disturbing possibility that a new implant design, which performs well during tests, may generate problems for future patients.

11.10 SUMMARY

The sensitivity of friction and wear to operating conditions, material and lubricant necessitate a detailed confirmation of similarity between experiment and real problem. If dissimilar wear or friction modes occur in the laboratory experiment and a real problem, entirely misleading results may be obtained. Friction and wear are subject to various influences ranging from fundamental phenomena to human factors. The meaning of similarity between laboratory experiment and real problem is at least partly determined by the objective of the investigation. In general, no single experiment can provide data relating to all influences or factors controlling wear and friction. Although the method of confirming tribological similarity will depend on the type of investigation, for most tribological studies at a fundamental level a common method exists. Evidence of wear and friction processes contained in wear scars and wear particles can be used to check that similar wear and friction mechanisms are found in both laboratory experiment and real problem. A method of successive adjustments can then be applied to ensure experimental similarity, if it does not occur on first trial. Many existing test methodologies do not follow carefully enough similarity principles described in this chapter and the data obtained is a subject to controversy.

Dimensional analysis, based on the experimental data, can provide the numerical models of the tribological process under consideration together with the appropriate dimensional function and invariances of similarity. Based on the parametric form of the process dimensional function, conditions necessary for designing experiment and the physical model to study the tribological process can be derived. The dimensional analysis, when properly applied, can be a useful tool in tribological process identification, tribological experiment design and processing the tribological data. For accurate models giving more precise predictions over a wide range of data values, non-linear approximations appear to be more appropriate.

Advanced graphical techniques such as 3-dimensional plots and contour maps can effectively be used to show the controlling effects of parameter combinations. A significant caveat with such graphs is the idealizations required, where other controlling factors are easily overlooked. It is possible then for conclusions to be based on secondary features of the data while primary features of the data remain unobserved.

In many cases of practical concern, there is not sufficient time to establish an accurate degree of similarity between a model test apparatus and the test object. It is then necessary to apply more empirical methods where lack of knowledge about the controlling phenomena is compensated for with careful replication of the original system. An example of this is the test engine where the service characteristics of crankcase lubricants are investigated. This field of testing is known as 'industrial tribology' and is practised in many industries ranging from lubricants to orthopaedic implants. The critical advantages of industrial tribology are speed and flexibility to adapt to any real problem. The major problem of industrial tribology is the difficulty in identifying the controlling phenomena to obtain a full understanding of the problem.

REFERENCES

1 S.C. Lim and M.F. Ashby, Wear-mechanism maps, *Acta Metallurgica*, Vol. 35, 1987, pp. 1-24.

2 S.C. Lim, Y.B. Liu, S.H. Lee and K.H.W. Seah, Mapping the wear of some cutting-tool materials, *Wear*, Vol. 162-164, 1993, pp. 971-974.

3 S. Mitsutake, F. Takahashi and K. Inaba, A ring-liner test rig for evaluation of cylinder oil load-carrying properties, Proc. JSLE Int. Tribology Conf., Tokyo, 8-10 July 1985, Elsevier, 1985, pp. 295-300.

4 E. Van Damme, Surface engineering, gear wear simulations, Proc. 2nd Australian International Tribology Conference, Melbourne, 2-4 Dec. 1987, IEA Nat. Conf. Publ. 87/18, pp. 391-396.

5 H. Czichos, Tribology: a systems approach to the science and technology of friction, lubrication and wear, Elsevier Sequoia, Amsterdam, 1978.

6 D. Dowson, B. Jobbins and A. Syed-Harraf, An evaluation of the penetration of ceramic femoral heads into polyethylene acetabular cups, *Wear*, Vol. 162-164, 1993, pp. 880-889.

7 S. Hogmark and S. Jacobson, Hints and guidelines for tribotesting and evaluation, *Lubrication Engineering*, Vol. 48, 1992, pp. 401-409.

8 G. Heinke, Verschleiss - eine systemeigenschaft auswirkungen auf die verschleissprufung, *Journal of Materials Technology/Zeitschrift fur Werkstofftechnik*, Vol. 5, 1975, pp. 164-169.

9 P. Podsiadlo, M. Kuster and G.W. Stachowiak, Numerical analysis of wear particles from non-arthritic and osteoarthritic human knee joints, *Wear*, Vol. 210, 1997, pp. 318-325.

10 G.C. Ballantine and G.W. Stachowiak, The effects of lipid depletion on osteoarthritic wear, *Wear*, Vol. 253, 2002, pp. 385-393.

11 P.R. Stupak, J.H. Kang and J.A. Donovan, Fractal characteristics of rubber wear surfaces as a function of load and velocity, *Wear*, Vol. 141, 1990, pp. 73-84.

12 B. Dubuc, J.F. Quiniou, C. Roques-Carmes, C. Tricot and S.W. Zucker, Evaluating the fractal dimension of profiles, *Physical Review A (General-Physics)*, Vol. 39(3; 1), 1989, pp. 1500-1512.

13 B.H. Kaye, The description of 2 dimensional rugged boundaries in fine particle science by means of fractal dimensions, *Powder Technology*, Vol. 46, 1986, pp. 245-254.

14 M.G. Hamblin and G.W. Stachowiak, Application of the Richardson technique to the analysis of surface profiles and particle boundaries, *Tribology Letters*, Vol. 1, 1995, pp. 95-108.

15 M.G. Hamblin and G.W. Stachowiak, Measurement of fractal surface profiles obtained from scanning electron and laser scanning microscope images and contact profile meter, *Journal of Computer Assisted Microscopy*, Vol. 6, No. 4, 1994, pp. 181-194.

16 J.C. Russ, Fractal Surfaces, Plenum Press, New York, 1994.

17 W.P. Dong, P.J. Sullivan and K.J. Stout, Comprehensive study of parameters for characterising three-dimensional surface topography, II: Statistical properties of parameter variation, *Wear*, Vol. 167, 1993, pp. 9-21.

18 C. Tricot, P. Ferland and G. Baran, Fractal analysis of worn surfaces, *Wear*, Vol. 172, 1994, pp. 127-133.

19 G.W. Stachowiak and A.W. Batchelor, Engineering Tribology, Butterworth Heinemann, 2001.

20 J. Lopez, G. Hansali, H. Zahouani, J.C. Le Bosse and T. Mathia, 3D fractal-based characterisation for engineered surface topography, *International Journal of Machine Tools and Manufacture*, Vol. 35, 1995, pp. 211-217.

21 S. Ganti and B. Bhushan, Generalized fractal analysis and its applications to engineering surfaces, *Wear*, Vol. 180, 1995, pp. 17-34.

22 S. Peleg, J. Naor, R. Harley and D. Avnir, Multiresolution texture analysis and classification, *IEEE Transactions on Pattern Analysis Machine Intelligence*, Vol. 4, 1984, pp. 518-523.

23 J.J. Gangepain and C. Roques-Carmes, Fractal approach to two dimensional and three dimensional surface roughness, *Wear*, Vol. 109, 1986, pp. 119-126.

24 K.C. Clarke, Computation of the fractal dimension of topographic surfaces using the triangular prism surface area method, *Computers and Geosciences*, Vol. 12, 1986, pp. 713-722.

25 B. Dubuc, S.W. Zucker, C. Tricot, J-F. Quiniou and D. Wehbi, Evaluating the fractal dimension of surfaces, *Proc. Roy. Soc. London*, Series A425, 1989, pp. 113-127.

26 C.A. Brown, P.D. Charles, W.A. Johnsen and S. Chesters, Fractal analysis of topographic data by the patchwork method, *Wear*, Vol. 161, 1993, pp. 61-67.

27 G.W. Stachowiak and P. Podsiadlo, Surface characterization of wear particles, *Wear*, Vol. 225-229 Part II, 1999, pp. 1171-1185.

28 P. Podsiadlo and G.W. Stachowiak, The development of modified Hurst orientation transform for the characterization of surface topography of wear particles, *Tribology Letters*, Vol. 4, 1998, pp. 215-229.

29 J.C. Russ, Characterizing and modelling fractal surfaces, *Journal of Computer-Assisted Microscopy*, Vol. 4, No. 1, 1992, pp. 73-126.

30 J.C. Russ, Surface characterization: fractal dimensions, Hurst coefficients, and frequency transforms, *Journal of Computer Assisted Microscopy*, Vol. 2, No. 3, 1990, pp. 161 183.

31 N.K. Myshkin, O.K. Kwon, A.Ya. Grigoriev, H.-S. Ahn and H. Kong, Classification of wear debris using a neural network, *Wear*, Vol. 203-204, 1997, pp. 658-662.

32 Y.M. Niu, Y.S. Wong, G.S. Hong and T.I. Liu, Multi-category classification of tool conditions using wavelet packets and ART2 network, *Journal of Manufacturing Science and Engineering, Transactions of the ASME*, Vol. 120, 1998, pp. 807-816.

33 D.M. Tsai and C.F. Tseng, Surface roughness classification for castings, *Pattern Recognition*, Vol. 32, 1999, pp. 389-405.

34 D.M. Tsai and S.K. Wu, Automated surface inspection using Gabor filters, *Advanced Manufacturing Technology*, Vol. 16, 2000, pp. 474-482.

35 K. Wiltschi, A. Pinz and T. Lindeberg, An automatic assessment scheme for steel quality inspection, *Machine Vision and Applications*, Vol. 12, 2000, pp. 113-128.

36 G.W. Stachowiak and P. Podsiadlo, Classification of tribological surfaces, *Tribology International*, 2004, Vol. 37, pp. 211-217.

37 P. Podsiadlo and G.W. Stachowiak, Hybrid fractal-wavelet method for characterization of tribological surfaces - a preliminary study, *Tribology Letters*, Vol. 13 (4), 2002, pp. 241-250.

38 P. Podsiadlo and G.W. Stachowiak, Fractal-wavelet based classification of tribological surfaces, *Wear*, Vol. 254, 2003, pp. 1189-1198.

39 P. Podsiadlo and G.W. Stachowiak, Classification of tribological surfaces without surface parameters, Proceedings of the 6th International Tribology Conference AUSTRIB 2002, 'Frontiers in Tribology', 2-5 December, Western Australia, 2002, pp. 109-117.

40 S. Drobot, On the foundation of dimensional analysis, *Studia Mathematica*, Vol. 14, 1953, pp. 84-99.

41 B. Lysik, Uber eine algebraische Form der Dimensionanalysis, *Wissenschaftliche Zeitschrift der Hochschule fur Bauwesen Leipzig*, Heft 3, 1972, pp. 219-223.

42 E. Buckingham, On physically similar systems, illustrations of the use of dimensional equations, *Phys. Rev.*, Vol. 4, 1914, pp. 345-376.

43 M. Kokar, Determining the mathematical model with the aid of dimensional analysis, (Ustalanie postaci modelu matematycznego z wykorzystaniem analizy wymiarowej) *Chemical Engineering*, (Inzynieria Chemiczna) Vol. 5, 1975, pp. 103-119 (in Polish)

44 W. Kasprzak and B. Lysik, On the computer oriented procedure for dimensional analysis, *Bulletin de L'Academie Polonaise des Sciences, Serie des Sciences Techniques*, Vol. 22, 1974, pp. 767-773.

45 M. Kokar, The Optimisation of the Dimensional Base Selection in the Case of Identification of Dimensional Process, Ph.D. Thesis (in Polish), Wroclaw Technical University, 1973.

46 W. Kasprzak, B. Lysik and M. Szata, Problems of the selection of a dimensional base in experiment design and planning, *Bulletin de L'Academie Polonaise des Sciences, Serie des Sciences Techniques*, Vol. 25, 1997, pp. 631-639.

47 W. Kasprzak and B. Lysik, Construction of a mathematical model for process description in continuous media, *Bulletin de L'Academie Polonaise des Sciences, Serie des Sciences Techniques*, Vol. 24, 1991, pp. 23-31.

48 H. Moes, Optimum similarity analysis with applications to elastohydrodynamic lubrication, *Wear*, Vol. 159, 1992, pp. 57-66.

49 N.Z. Wang and C.Y. Chang, An application of Newton's method to the lubrication of air-lubricated bearings, *Tribology Transactions*, Vol. 42 (2), 1999, pp. 419-424.

50 H.S.S. Hsiao, Optimum dimensional analysis with applications to elastohydrodynamic lubrication, *Journal of Tribology, Transactions of the ASME*, Vol. 123 (4), 2001, pp. 822-827.

51 J.C. Gibbings, Dimensional analysis of wear by particle impact in fluid flows, *Journal of Mechanical Engineering Science*, Vol. 13, No. 4, 1971, pp. 234-236.

52 M.C. Shaw, Dimensional analysis for wear systems, *Wear*, Vol. 43, 1977, pp. 263-266.

53 D.D. Drucker, Elementary results of dimensional analysis for friction and wear in steady state sliding, *Wear*, Vol. 40, 1976, pp. 129-133.

54 I. Clark, K. White and A.J. Bursle, The use of corrosion resistant steel in sugar milling, *Proc. Aus. Soc. Sugar Cane Technologists*, 1980, pp. 253-259.

55 W.J. Head and M.E. Harr, The development of a model to predict the erosion of materials by natural contaminants, *Wear*, Vol. 15, 1970, pp. 1-46.

56 H. Wadell, Sphericity and roundness of rock particles, *J. Geol.*, Vol. 41, 1933, pp. 310-331.

57 G.W. Stachowiak and A.W. Batchelor - Dimensional analysis in modelling tribological data, Proc. of the International Tribology Conference in Melbourne, The Institution of Engineers, Australia, National Conference Publication No. 87/18, December 1987, pp. 255-259.

58 J.B. Nielsen, Evolution of rail corrugation predicted with a non-linear wear model, *Journal of Sound and Vibration*, Vol. 227, No. 5, 1999, pp. 915-933.

59 P. Neuenschwander, D.E. Maurer and L. Rychlicki, Long-term wear monitoring and wear prediction by means of wear models, *Control Engineering Practice*, Vol. 3(7), 1995, pp. 1011-1015.

60 C.A. de Saracibar and M. Chiumenti, On the numerical modelling of frictional wear phenomena, *Computer Methods in Applied Mechanics and Engineering*, Vol. 177(3-4), 1999, pp. 401-426.

61 N. Viswanath and D.G. Bellow, Development of an equation for the wear of polymers, *Wear*, Vol. 181, Part 2, 1995, pp. 42-49.

62 J. Park and I.W. Sandberg, Universal approximation using radial-basis-function networks, *Neural Computation*, Vol. 3, 1991, pp. 246-257.

63 K. Hornik, M. Stinchcombe and H. White, Multilayer feedforward networks are universal approximators, *Neural Networks*, Vol. 2, 1989, pp. 359-366.

64 R.P. Lipmann, An introduction to computing with neural nets, *IEEE ASSP Magazine*, Vol. 4, 1987, pp. 4-22.

65 S. Mallat, A Wavelet Tour of Signal Processing, Academic Press, 2nd Edition, San Diego, 1999.

66 M.F. Barsney and L.P. Hurd, Fractals Everywhere, Academic Press, San Diego, 1988.

67 N.C. Wallbridge and D. Dowson, Distribution of wear rate data and a statistical approach to sliding wear theory, *Wear*, Vol. 119, 1987, pp. 295-312.

68 E. Rabinowicz, Wear coefficients - metals, ASME Wear Control Handbook, editors: M.B. Peterson and W.O. Winer, American Society of Mechanical Engineers, New York, 1980, pp. 475-506.

69 N.C. Wallbridge, A.J. Winn and S.S. Hassan, The variability of wear rate data in simultaneous experiments, Interface Dynamics, Proceedings of the 14th Leeds-Lyon Symposium on Tribology, Elsevier, Amsterdam, 1988, pp. 245-252.

70 H. Czichos, Presentation of friction and wear data, ASM Handbook - Friction, Lubrication, and Wear Technology, ASM International, Vol. 18, 1992, pp. 489-492.

71 H. Czichos, Design of wear and friction experiments, ASM Handbook - Friction, Lubrication, and Wear Technology, ASM International, Vol. 18, 1992, pp. 480-488.

72 S.C. Lim and M.F. Ashby, Wear-rate transitions and their relationship to wear mechanisms, *Acta Metallurgica*, Vol. 35, 1987, pp. 1343-1348.

73 S.C. Lim, Recent developments in wear-mechanism maps, New Directions in Tribology, editor: I.M. Hutchings, Plenary and Invited Papers from the First World Tribology Congress, London, MEP Publications Ltd., 1997, pp. 309-320.

74 C. Lhymn, Statistical analysis of wear rate data, *Trans ASME, Journal of Tribology*, Vol. 109, 1987, pp. 594-597.

75 P.J. Guichelaar, P.J. Wilde and M.W. Williams, The detection and characterization of blisters on carbon-graphite mechanical seal faces, *Tribology Transactions*, Vol. 43(3), 2000, pp. 395-402.

76 P.J. Guichelaar, P.J. Wilde and M.W. Williams, The effect of oil type on blister formation in carbon-graphite mechanical seals, *Lubrication Engineering*, Vol. 57(8), 2001, pp. 21-27.

77 M.H. Jones and D. Scott (editors), Industrial tribology: the Practical Aspects of Friction, Lubrication and Wear, Elsevier Scientific Pub. Co., 1983.

78 J.A. Schey, Tribology in Metalworking: Friction, Lubrication and Wear, Metals Park, Ohio, American Society for Metals, 1983.

79 J.D. Summers Smith, An Introductory Guide to Tribology in Industry, Mechanical Engineering Publications, 1994.

80 J.C. Charnley, Arthroplasty of the hip: A new operation, *The Lancet*, Vol. 1, 1961, pp. 1129-1132.

81 J.C. Charnley, Tissue reactions to polytetrafluoroethylene (letter), *The Lancet*, Vol. 2, 1963, p. 1379.

82 A.A.J. Goldsmith and D. Dowson, A multi-station hip joint simulator study for the performance of 22-millimetre zirconia-UHMWPE total replacement hip joints, *Proc. Instn. Mech. Engineers, Journal of Engineering in Medicine*, 1999, pp. 77-90.

83 M. Pellicciari, G. Barbanti and A.O. Andrisano, Design and development of a multi-axis hip joint simulator, Proc. 10th International Conference on Biomedical Engineering, 6-9 December 2000, Singapore, National University of Singapore, editor: J.C.H. Goh, pp. 159-160.

84 S. Scholes, A. Unsworth and A.A.J. Goldsmith, A frictional study of total hip joint replacements, *Phys. Med. Biol.*, Vol. 45, 2000, pp. 3721-3735.

85 V. Saikko, T. Ahlroos and O. Calonius, A three-axis knee wear simulator with ball-on-flat contact, *Wear*, Vol. 249, 2001, pp. 310-315.

86 Technical reports ISO TR 9325, Implants for surgery, partial and total hip joint prostheses, recommendations for simulators for evaluation of hip joint prostheses.

87 S. Hargreaves, It's hip to be smooth, *The Lancet*, Vol. 356, No. 9244, 2000, p. 1826.

INDEX